English Agriculture
in 1850-51

JAMES CAIRD

CAMBRIDGE
UNIVERSITY PRESS

CAMBRIDGE UNIVERSITY PRESS

Cambridge, New York, Melbourne, Madrid, Cape Town, Singapore,
São Paolo, Delhi, Dubai, Tokyo, Mexico City

Published in the United States of America by Cambridge University Press, New York

www.cambridge.org
Information on this title: www.cambridge.org/9781108024730

© in this compilation Cambridge University Press 2010

This edition first published 1852
This digitally printed version 2010

ISBN 978-1-108-02473-0 Paperback

CAMBRIDGE LIBRARY COLLECTION

Books of enduring scholarly value

History

The books reissued in this series include accounts of historical events and movements by eye-witnesses and contemporaries, as well as landmark studies that assembled significant source materials or developed new historiographical methods. The series includes work in social, political and military history on a wide range of periods and regions, giving modern scholars ready access to influential publications of the past.

English Agriculture in 1850-51

Sir James Caird (1816–92) was a Scottish agriculturalist and M.P., who wrote widely on agricultural matters, not only in Britain but in Ireland, Canada, America and India. British agricultural incomes had been falling owing to low grain prices since 1846, and Caird was commissioned by *The Times* to undertake a survey of English agriculture. His county-by-country reports were published in 1852 as *English Agriculture in 1850–51*. The work was also published in America, and in German, French and Swedish versions. Changing patterns of trade meant that British agriculture had to adapt to compete with cheap imports, and tenant farmers needed greater security. Caird campaigned in Parliament for regular and official agricultural statistics to be collected, so that the agricultural economy could be made more efficient, though it was nine years before this happened. Caird was knighted in 1882, and served on many official committees.

Cambridge University Press has long been a pioneer in the reissuing of out-of-print titles from its own backlist, producing digital reprints of books that are still sought after by scholars and students but could not be reprinted economically using traditional technology. The Cambridge Library Collection extends this activity to a wider range of books which are still of importance to researchers and professionals, either for the source material they contain, or as landmarks in the history of their academic discipline.

Drawing from the world-renowned collections in the Cambridge University Library, and guided by the advice of experts in each subject area, Cambridge University Press is using state-of-the-art scanning machines in its own Printing House to capture the content of each book selected for inclusion. The files are processed to give a consistently clear, crisp image, and the books finished to the high quality standard for which the Press is recognised around the world. The latest print-on-demand technology ensures that the books will remain available indefinitely, and that orders for single or multiple copies can quickly be supplied.

The Cambridge Library Collection will bring back to life books of enduring scholarly value (including out-of-copyright works originally issued by other publishers) across a wide range of disciplines in the humanities and social sciences and in science and technology.

OUTLINE MAP OF ENGLAND,

Shewing the distinction between the Corn and Grazing counties; and the line of division between high and low Wages.

All to the East of the black line, running from North to South, may be regarded as the chief Corn Districts of England; the average rental per acre of the cultivated land of which is 30 per cent. less than that of the counties to the West of the same line, which are the principal Grazing, Green Crop, and Dairy districts.

The dotted line, running from East to West, shows the line of Wages; the average of the counties to the North of that line being 37 per cent. higher than those to the South of it.

See pages 480. 512. 514. and 516.

ENGLISH

AGRICULTURE

IN

1850–51.

BY JAMES CAIRD, ESQ.

THE "TIMES" COMMISSIONER.

———————

" Books will not teach farming, but if they describe the practices of the best farmers they will
make men think, and show where to learn it."—PH. PUSEY.

———————

LONDON:

LONGMAN, BROWN, GREEN, AND LONGMANS.

1852.

LONDON:
SPOTTISWOODES and SHAW,
New-street-Square.

THIS VOLUME

IS DEDICATED,

WITH RESPECT AND ESTEEM,

TO

JOHN WALTER, ESQ., M.P.

BY WHOSE PUBLIC SPIRIT

THE INQUIRY WHICH IT EMBODIES

WAS UNDERTAKEN.

PREFACE.

In the beginning of 1850, the low prices of agricultural produce and the serious complaints of farmers and landlords, indicated the necessity of some inquiry into the actual state of agriculture in the principal counties of England. In order to ascertain the extent and true cause of the distress, this inquiry was originated by "THE TIMES." Having been invited to undertake a task so extensive and at that time so difficult on account of the excited state of the agricultural mind — I ventured to consult the late Sir Robert Peel, whether such an inquiry, conducted in the fair and temperate spirit which was desired, might not be beneficial to those connected with land, and to the tenant farmers in particular, in whose prosperity I knew that both his feelings and his interests were deeply engaged. With the concurrence of his literary Executors, I am enabled to lay before the Public the following letter from that lamented statesman : —

> "Drayton Manor, Fazeley,
> "January 6. 1850.

"DEAR SIR,

"I am inclined to advise the acceptance of the offer conveyed in the inclosed.

"There is so little intercommunion between agricul-

A 4

turists in different parts of the country, and such a
general unwillingness on the part of ordinary farmers
to travel beyond the bounds of their own parish — that
much good might, I think, be done by presenting to them
in an attractive form, the observations of practical men
on the different systems of farming, and the different
usages which prevail in various parts of the country.

"You will find immense tracts of good land in
certain counties (Lancashire and Cheshire for example),
with good roads, good markets, and a moist climate,
that remain pretty nearly in a state of nature, undrained, badly fenced, and wretchedly farmed.

"Nothing has hitherto been effectual in awakening
the proprietors to a sense of their own interest. I
cannot help thinking that a dispassionate and temperate
contrast between the productiveness of their properties
and that of others in less favoured positions, and the
conclusive proof that might be exhibited that protection
had in their cases not stimulated improvement, but
had probably been the parent of neglect, might reconcile them to the withdrawal of it, and induce them
to look out for more certain aid in 'good farming under
liberal covenants.'

"The main consideration is the character and qualifications of those with whom you would be associated.

"I presume the character and the interests of the
paper are so deeply concerned, as to insure every attention to this important matter.

"Faithfully yours,

"ROBERT PEEL.

"JAMES CAIRD, ESQ."

In the first part of the inquiry I was associated with Mr. J. C. M'Donald, of the Inner Temple, whose literary abilities contributed much to the success of the Letters. The latter part I conducted alone; and the whole having been re-written by me, I am now solely responsible for the opinions they contain.

With the view of rendering these Letters permanently useful, not merely as exhibiting the state of Agriculture throughout England, of which, since Arthur Young's Tours, upwards of eighty years ago, they afford the only general account, I was careful to note good examples of farming in the several counties, and have described them in minute detail, for the information of farmers in the same and other counties. Many eminent practical men have already acknowledged the benefit they have received, by combining with their own the practice, in some particular department, of good farmers in other counties, thus brought under their notice. I have also sometimes noticed objectionable practices in order to reprobate them. A copious index has been added, which renders the work a book of reference for the best systems of agriculture at present practised in the various counties in England. The arrangements between landlords and their tenants have also been fully discussed; and the condition of the labourer has obtained a due share of attention.

To the liberality of " THE TIMES " I feel deeply grateful, for the ample means placed at my disposal for conducting this inquiry, and for the perfect freedom with which I was permitted to express my opinions, irrespective of their political bearing.

As the object was to obtain facts, and the field so extensive, it was thought that the clearest and most methodical description of English agriculture would be got by a separate examination of each county. The Southern counties were first examined, — then the Eastern and Midland counties, — next the Western and Northern, — and last Derbyshire and Northampton, and some of the corn-growing counties near the metropolis which had not been previously visited.

All the matter contained in these Letters was obtained by personal inquiry and inspection, principally by walking or riding carefully over individual farms, in different districts of each county, accompanied by the farmers, — by traversing estates with the landlord or his agent, — and by seeking access to the best and most trustworthy sources of local information.

Two points call for special remark here, which are the cause of much national loss to Agriculture, — the general absence of Leases throughout England, and the immense mass of fertilizing matter which runs to waste from all the large towns of the Kingdom. To the first of these I have endeavoured to draw attention in the following Letters, and the second has likewise been noticed. The general adoption of the one, and the preservation and application of the other as manure, would cause a great addition to the annual produce of England.

The concluding Letters bring into one view the general results of the investigation.

Baldoon, Wigtown,
January, 1852.

CONTENTS.

LETTER VIII.

DORSETSHIRE.

LETTER IX.

DORSETSHIRE — *continued.*

LETTER X.

WILTSHIRE.

LETTER XI.

WILTSHIRE — *continued.*

LETTER XII.

HAMPSHIRE.

LETTER XIII.

NORTH HANTS. — BERKSHIRE.

LETTER XVI.

BERKSHIRE.

LETTER XV.

SURREY.

LETTER XVI.

SUSSEX.

LETTER XVII.

ESSEX.

LETTER XVIII.

ESSEX — SUFFOLK.

LETTER XXII.

LINCOLNSHIRE.

LETTER XXIII.

LINCOLNSHIRE — *continued.*

LETTER XXIV.

NOTTINGHAMSHIRE.

LETTER XXV.

NOTTINGHAMSHIRE — *continued.*

LETTER XXVI.

LEICESTERSHIRE.

LETTER XXVII.

WARWICKSHIRE.

LETTER XXVIII.

STAFFORDSHIRE.

LETTER XXIX.

STAFFORDSHIRE — *continued.*

LETTER XXXIII.

LANCASHIRE — *continued.*

LETTER XXXIV.

YORKSHIRE.

LETTER XXXV.

YORKSHIRE — *continued.*

LETTER XXXVI.

YORKSHIRE — *continued.*

LETTER XXXVII.

YORKSHIRE — *continued.*

LETTER XXXVIII.

DURHAM.

LETTER XXXIX.

DURHAM — *continued.*

LETTER XLVI.

NORTHAMPTONSHIRE.

LETTER XLVII.

NORTHAMPTONSHIRE — *continued.*

LETTER XLVIII.

BEDFORDSHIRE.

LETTER LII.

CONCLUSIONS.

LETTER LIII.

THE LANDLORD.

LETTER LIV.

THE FARMER.

LETTER LV.

THE LABOURER.

LETTER LVI.

CONCLUSION.

ENGLISH AGRICULTURE

IN

1850—51.

LETTER I.

BUCKINGHAMSHIRE.

VALE OF AYLESBURY — PROPORTION OF PASTURAGE TO TILLAGE — MODE
OF DRAINAGE — FARM BUILDINGS, THEIR DEFICIENCY — TENURE — RENT
— SIZE OF FARMS — VARIOUS KINDS OF STOCK AND MODE OF MANAGE-
MENT — BUTTER DAIRIES — HAY — LOSS BY FATTENING CATTLE —
IMPERFECT MANAGEMENT OF MANURE — LABOURERS AND THEIR WAGES
— TILLAGE — RENTS, RATES, AND DRAINAGE OF ARABLE LANDS — COURSE
OF CROPS AND THEIR MANAGEMENT — STOCK — FIELD ROADS.

AYLESBURY, Jan. 21. 1850.

THE soil, in the Vale of Aylesbury, is a strong clay loam,
varying in depth from two feet to a few inches, of rich earth,
generally incumbent on stiff clay. The vale is celebrated for
the excellence of its pastures, for which it is better adapted
by nature than for tillage. In three parishes which we visited,
the proportion of tillage to pasture was very small, there being
in the first only 8 acres in 2000 under the plough, in the second
90 acres in 900, and in the third no tillage whatever. It is a
country of rich pastures, laid out in large fields, devoted to the
feeding of stock and to dairy farming, and therefore affording
only limited employment to a scanty population. There is still

B

great room for improvement by drainage, though the grass lands, especially those of prime quality, have generally been drained by wedge or wood drains. These, though not so durable as tiles, are found very beneficial; and as the tenant has hitherto been left in this district to make nearly all permanent improvements at his own cost, he looks to cheapness as much as to permanence. In some cases the landlord provides the tiles, the tenant agreeing to put them into the ground; but there has been little intercourse between landlord and tenant, and the latter is left pretty much to himself in matters of improvement. The tillage lands (where drained at all) are usually drained with tiles.

The farm buildings are very inadequate, and generally constructed of rough timber, covered with thatch. They have evidently grown up by successive additions, as the necessities of the occupier dictated, without much regard to shape or situation, proximity to each other, or economy of space; and are often intermingled in confusion with stacks of hay and beans. One farm of 600 acres, all under grass, for which the tenant paid 1500*l.* a year of rent and tithe, was extremely deficient in this respect; and though the tenant had expended 200*l.* in making his house habitable on taking possession, his landlord refused to make any outlay in providing accommodation for his stock. On the whole, the tenantry of this grazing district have had little co-operation or enlightened sympathy from their landlords.

The tenures are principally from year to year, and such tenures are preferred by the farmers to leases. One farmer told us that his father and he had occupied the same farm for upwards of seventy years, and though during that time they had an indulgent landlord, large improvements in drainage, a rent not increased for thirty years, and a considerable capital invested in the soil, they had made nothing more than a respectable living by the business.

The rent of land ranges from 10*s.* an acre, for the lowest quality of undrained clay lands under tillage, to 50*s.*

for prime old grazing lands. The amount of poor rates, though not much complained of, appeared high considering the small population. Tithes were in some instances paid by the landlord, though generally by the tenant. The farms vary in extent from 300 to 600 acres. Many of the farmers are wealthy; but others, we were told, have never had any capital of their own, having purchased their stock with borrowed money, for which, being without good security to offer, they pay exorbitant interest.

The grass farms in the Vale of Aylesbury are stocked one third with ewes, and two thirds with dairy and fattening cattle. The quality of the soil on each farm determines the precise apportionment of stock; the best land being chiefly devoted to fattening cattle and sheep, the secondary and the worst to dairy purposes. On all the grass farms here the practice with regard to sheep is to purchase ewes early in autumn, which drop their lambs in January, and, after the lambs are disposed of in the London market, are fattened and sold during the summer. The stock is thus changed every year. The ewes are fed on the pastures summer and winter, getting, occasionally, corn in troughs in severe winter weather, and about the time of lambing. They receive no turnips during the winter; but the practice of farmers differs in regard to the summer food of the sheep, being regulated by the extent of their tillage land. Where fallow is made, the land is previously sown with winter vetches, on which the sheep are folded in the early part of summer, thus securing an excellent bite for the sheep, and at the same time giving a dressing of manure to the land. The lambs are sent to the London market as soon as they are ready. The prices which one of the most intelligent farmers whom we met considers remunerative, are — 32s. for lambs; 2s. advance in the difference between the selling and buying price of the ewes; and 4s. each ewe for wool. Below these rates he thinks the farmer will not be paid. It must be remarked that very many of the ewes have twin lambs. The present rates for ewes and lambs are about 25 per cent. under the above.   Wool maintains its price.   The ewes kept are principally of the Southdown breed.

Dairy farming is the most important branch of rural industry in this neighbourhood.   The farmers do not breed the stock, but buy young cows, and sell them as soon as they begin to fail as milkers.   The entire produce is converted into butter, which is sent up to the London market during the season, an agent in London being commissioned to sell it.   The cows are fed on the fine pastures of this district during the summer, and tied up in sheds for five or six months during the winter, where they are regularly supplied with hay.   No green food, wurzel, carrots, or turnips are grown on the farms, or given to the stock, but some good feeders supply them with a portion of oil-cake in addition to the hay.   The hay is of the finest quality.   In some farms the cows go loose during the winter in open yards, with sheds to retire under.   They are in all cases attended to by men, who feed, clean, and milk them.   One farmer employs twelve men to tend a herd of 100 cows during the winter season.

The milk, when carried to the dairy, is poured into large oblong shallow wooden vessels, lined with lead.   Twelve hours afterwards the cream is skimmed off; in twelve hours more it is again skimmed ; and the same process is repeated a third time. In the warm weather of summer this suffices ; but during winter a fourth, and sometimes a fifth, skimming is necessary before the careful dairymaid is satisfied that she has succeeded in extracting the whole of the cream.   The milk is then drawn off into a pipe, by which it is conducted to a tank out of doors, close to the feeding-troughs of the pigs.   The cream is churned by horse-power.   We did not meet with an instance in which the temperature of the dairy was regulated by artificial means.

The price of butter in this district for the last ten years was read to us from the book of one of the farmers whom we visited. From 1839 to 1847, in the month of January in each year, there appeared scarcely any variation, beginning at 15*d*. to 16*d*., and ending with 16*d*. per pound.   In 1847 it fell as low as 13*d*.; in 1848 it rose to 17*d*.; and now it has fallen to 14*d*. per pound. Dairy farming is said by all parties to be the only department of their business which leaves them a profit at present.

A farmer holding 300 acres of grass land mows about 100 acres annually.   Part of this receives a top-dressing of dung during the previous winter.   The produce varies from ten cwt. to two tons per acre, and the cost of making and stacking the crop is about 15*s*. an acre.   The good farmers consume the whole of their hay on their own farms.   But a very small portion is sent from this quarter to the London market, and that said to be hay of inferior quality, produced on the poorest land, and parted with by the neediest farmers.   There was a good deal said with reference to the quality of London hay, most of our informants stating to us that no really good hay ever was sent up to London, and that it was almost impossible to distinguish the difference betwixt good and inferior hay, without being informed of the quality of the land on which it had been grown.   This had reference to natural upland or meadow hay, not the hay of artificial grasses or clover.

Where cattle are fattened, they are purchased in autumn, receive hay, and in some cases oil-cake, in yards during the winter, and are fed fat on the best grazing land during summer.   It will thus be seen that the farmer of grass lands in this quarter changes his stock of sheep and fattening cattle every year, and his dairy cows when they cease to yield a profitable return.   The fall in the price of butcher meat has therefore very seriously affected him for the season, as it has in some cases nearly extinguished his usual profits on fattening cattle, and greatly reduced them on sheep.   The produce of three acres of good grass land, summer and winter, is reckoned necessary for the keep of one cow.   A milch cow consumes much more than the produce of one acre of hay during the winter

The management of the tillage land on these farms forms quite a subordinate branch of their system. The crops usually grown are wheat and beans alternately, one field being set aside for the purpose of tillage, and kept constantly under the plough. The very small proportion under crop enables the farmer to manure it heavily, and accordingly the crops he produces are good, five quarters of wheat and as many of beans being the common yield. One farmer told us that he had profited by a lesson he got from witnessing the effects of deep tillage on one of the labourers' allotments in his neighbourhood. In consequence of this, he instructed his ploughman to go eight inches deep, instead of five, which is the usual depth turned up here in preparing for beans ; and though, to the dismay of the ploughman, one or two inches of fresh clay were turned up to the surface, the bean crop, notwithstanding a dry summer, proved excellent, while most of those on the surrounding farms were a failure.

The construction of the farm buildings is everywhere defective in arrangements for accumulating or saving manure. To this most important point no attention is paid, the solid manure lying about the yards, and the liquid draining itself off to the watering pond or nearest open ditch. The use of bones or guano seems scarcely known, and their value as a manure for the grass lands appears not to have been discovered. One farmer said it might pay a man with a lease to use such purchased manures, but not otherwise. There can be no doubt that great benefit would arise from the application of bones and guano to the lands intended for hay, the produce of which might by such means be greatly increased. If to this were added the consumption of cheap feeding stuffs by the ewes, they could be kept from roaming over the whole of the pastures in the months of spring, by which the growth of the grass is often so much retarded as not to afford a full bite to the dairy stock before the beginning of June.

The number of labourers employed on these grazing farms

exceeded what we should have anticipated, from ten to fourteen people being engaged on farms of from 300 to 400 acres, with not more than fifty acres of tillage.   In all the parishes in which we were, there were no labourers out of employment, if we except a few under-drainers, whose work has been for the present stopped by frost.   The rate of wages is from 9s. to 10s. a week, with breakfast and ale on Sundays to the men employed with the dairy cattle.   Wages have not fallen more than 1s. a week from the average, though they are 2s. to 3s. a week lower than they were during the high prices of 1847.

The country extending from Aylesbury to Wendover and the Chiltern Hills being chiefly in tillage, we had there an opportunity of examining the methods of husbandry pursued on the arable lands of Buckinghamshire.   The men who have capital to embark, carefully shun the wet and inferior clay farms, which thus fall into the hands of a poorer and less intelligent class of farmers. Such farms, too, are small, and being very expensive and unprofitable, their occupiers make a scanty and thankless living by them.   But in the district we now refer to, the character of the soil is favourable to tillage, there being combined a considerable variety of soils, and the land altogether seems of fair quality for the purposes of cultivation.   It is obviously much in want of drainage, and the farm buildings, which are of rough wood and thatch, are all of the most primitive description, and very inadequate to the requirements of modern farming.   These matters are not much thought of by the landlords of this district, who, if they get their rents, leave nearly everything else to be arranged by the tenant as it best suits himself, and chiefly at his own expense.   The proprietors and tenants seem, nevertheless, to be on good terms with each other.

The land is generally held from year to year, the farmer being sometimes bound to certain modes of cultivation, and sometimes left to his own discretion on this point.   The average rent may be stated at 30s. an acre.   The poor rates, though heavy in

Wendover parish, are light in others adjoining, which is caused
by the labourers being sent in from the close parishes, in which,
the whole property being in the hands of one or two landlords,
no additional cottages are allowed to be erected; and so the
work people are driven to reside in the town, the ratepayers
of which are heavily and unfairly mulcted for the labour of
another parish, which thus goes almost free. Farms which
come into the market are readily taken again at former rents.

For drainage, tiles or pipes are seldom used: the material
most in use for filling the drains is "rag" or lumps of hard chalk,
about the size of paving stones, which are carted three or four
miles for this purpose. When the expense of cartage is taken
into account, this substance must be nearly as costly as pipes, and
as the chalk by degrees melts or crumbles away, it cannot be
nearly so lasting as pipes. Where the operation is performed in
the best manner here practised, a drain is opened between each
"land," the distance apart being from twenty-four to thirty feet,
and the depth of the drain about two feet. The blocks of "rag"
are then laid to the depth of fully a foot, a little straw is strewed
over, and as much as possible of the noxious substratum which
had been thrown out in digging the drain is packed carefully
over the straw, the surface soil being then replaced in its former
position. The more general plan however is, to put in a few
drains where they appear to be required, cutting to a depth of
fifteen or eighteen inches, and filling up to the bottom of the
plough furrow with wood or hedge trimmings. These are said to
last for many years. Wet spots here and there throughout a field
are so treated, but it is not thought necessary to go regularly
over the whole field in the same way. Nothing is believed to
be more injurious than any admixture of the substratum with
the surface staple of the soil. Of course subsoil ploughing or
trenching are carefully abstained from, and the benefit of deep
disintegration as accessory to drainage is accordingly lost.

The course of cropping followed does not seem to be very
definite. Some landlords do not interfere with their tenants,

but allow them to pursue whatever system they find most advantageous. Others prescribe a certain course, which is termed "three crops and a fallow." It begins with bare fallow, then wheat, next beans, pease and clover, and last wheat or barley. This may be considered the standard, from which there are few deviations. The fallow is found the best and surest preparation for wheat, for which it is usually dunged. After the wheat is removed the land is ploughed and planted with winter beans. The beans are put in in rows with a dibble, at the rate of three to three and a half bushels an acre, for which the workman is paid 1s. 6d. a bushel, or 4s. 6d. to 5s. 6d. an acre. In spring and summer the beans are hand-hoed with a broad hoe twice or three times, the price paid for each hoeing varying from 3s. 6d. to 4s. 6d. an acre, according to the clean or foul state of the land. After the beans are removed the land is sown with wheat, which, if it escape the ravages of the slug, generally proves a good crop. To destroy the slugs quicklime is used by the best farmers, and is scattered thinly over the surface at the rate of from one to two quarters an acre. Where it is not thought advisable to sow wheat after beans, the land lies untouched during the winter, and is ploughed and sown with barley in spring. Part of the division allotted to beans (an acre or two on a farm of 200 acres) is sown with turnips, which are evidently not considered of much value on this kind of soil. No mangold or other root crop is cultivated, and scarcely a rood of potatoes. On one farm this crop is proscribed by conditions of tenure, the tenant being allowed to grow only half an acre on his farm—a privilege which he does not make use of. No artificial manure, bones, or guano are used, and scarcely any purchased feeding stuffs. The farms are laid out in fine open fields, varying from eight or ten up to thirty acres in extent. They are inclosed with good thorn fences, and suffer little injury from excess of wood.

The crop of wheat of the present year (crop 1849) is very deficient in yield, turning out little more than sixteen or twenty bushels an acre on land of excellent quality. Barley is also a

short crop, but beans a very full one.   The low price and the
deficient yield are the cause of the present complaints : twenty-
eight bushels of wheat, thirty-two of barley, and the same of
beans, are reckoned fair average crops.

In working the land it is found necessary to use two different
descriptions of plough; one an old-fashioned wooden plough for
winter, and the other a more modern iron-wheel plough for summer.   The wheel plough comes into use " with the cuckoo," the
ground being so soft in winter that the wheels will not then
work.   The depth of furrow turned up is from four and a half
to five inches ; the latter depth not being exceeded for fear of
bringing up the dreaded subsoil.   The surface did not appear of
a peculiarly stiff nature ; in fact, rather the contrary, having in
many cases a large admixture of flints.   Yet in winter there are
seldom fewer than four horses in a plough, and three roods are
reckoned a fair day's work.   In summer three horses are used,
and an acre is turned over in a day.   The number of labourers
employed varies a little.   On one farm, with 120 acres under
tillage, there are eight men and a boy, two ploughs, and seven
work horses, in regular employment throughout the year, and
these may be reckoned as nearly the proportions for the arable
land round Wendover.

The quantity of stock kept on these arable farms is quite inconsiderable.   Three or four cows, and their produce, with a few
scores of sheep in summer, comprise the whole for a farm of 150
acres.   The farm buildings enclose a large court, into which the
straw as it is thrashed is thrown out of the barn, and the cattle,
aided by ten or twelve excellent pigs, eat and tread it into
manure.   The watering pond usually forms the lowest part of
the yard, and of course receives all the drainings of the dung.
The crops are thrashed out with the flail.

Many of the farms are intersected by public roads, and are
thereby well supplied with means of access to the different fields.
But where they are not on the line of road the farmers suffer
great inconvenience from the want of proper farm roads.   The

consequence is that in harvest they are obliged to stack
their crops in the corner of the field where they are grown,
waiting for dry frosty winter weather to carry them home.   On
one farm which we examined the roads had become impassable,
and the carters had therefore been obliged to turn into an adjoining wheat field, along the headland of which the heavy waggons had done much injury to the young wheat plant.   If a
different system of husbandry, involving a greater extent of
root crops, were adopted, the want of good roads of access would
be found still more injurious than at present.

LETTER II.

BUCKINGHAMSHIRE — *continued.*

CHALK DISTRICTS. — EXTENT OF FARMS — INSUFFICIENT BUILDINGS — RATE OF RENT — POOR AND OTHER RATES — MANAGEMENT OF ARABLE LAND — SEVERE COURSE OF CROPS — MORE LAND PLOUGHED ON ACCOUNT OF LOW PRICES — MANAGEMENT OF STOCK AND MANURE — IMPLEMENTS — AVERAGE PRODUCE OF CROPS AND STOCK — WAGES — " PAYING PRICES" — NEATNESS OF FARM HOUSES.

HIGH WYCOMBE, Jan. 24. 1850.

THE southern and eastern portions of the county of Buckingham lie principally on the chalk, and the mode of husbandry differs considerably from that we have already described. The farms vary in extent from 100 to 200 and 300 acres, some of the tenants holding on lease, and some from year to year. The landlords do not interest themselves much in the permanent improvement of their farms, although as drainage on these soils is not much required, the necessity for outlay on their part is limited to the improvement of farm buildings. These, however, are neither substantial nor convenient, being generally old wood and thatch buildings, very unsuitable for the requirements of modern husbandry.

The rent varies from 15*s.* to 30*s.* an acre. The productive qualities of the soil are spoken of very slightingly, and considerable dissatisfaction is felt by the farmers on account of the landlords having made little or no reduction of rent in consideration of the present low prices. Poor and other rates are unusually high, in one case as much as 8*s.* 6*d.* an acre, caused in this case partly by great mismanagement in the affairs of the union a year or two ago.

There appeared to be a very uniform system of management
adopted on the chalk districts of Bucks.   The fields are not encumbered with too numerous hedgerows.   Stock farming is
adopted only as a means of forcing corn crops, from the latter
of which the farmer has hitherto looked for his remuneration.
By raising green crops and feeding them on the land with sheep,
he is enabled to draw from the soil crops of wheat, barley, and
oats, which without this enriching preparation, it would not produce.   His chief attention is therefore directed to the culture of
green crops for consumption on the land, as the foundation of his
after success.   The rotation followed is termed " a five-field
course," commencing with (1) turnips, followed by (2) barley,
which is sown out with (3) " seeds" (the first crop of which is mown
and the second eaten on the ground) ; after the seeds the land is
dunged, then ploughed, and sown with (4) wheat, and this again
is followed by (5) oats, which form the last crop of the course.
There are thus three corn crops and two green crops every five
years, and it is obvious that on thin chalk land this system cannot be successfully continued without a liberal expenditure in
artificial food or manure.   The land should, therefore, be twice
dunged in the course—first for the turnip, and second for the wheat
crop ; and, in addition to this, the best farmers give a large quantity of corn or cake to the sheep while feeding on the turnips.   If,
besides this, ample doses of artificial manures are used for the
turnip crop, the farmer finds his land improving, under what
might otherwise be thought a severe system of cropping.   The
fields had in many places, however, rather an exhausted appearance, very few of them being sown to grass, and those seldom
showing any signs of verdure or fertility.   This was more
apparent from the contrast presented by the rich green colour
of a field here and there throughout the district on which the
effect of more generous management at once displayed itself.
We were sorry to learn that, on account of the low prices
of corn, many of the occupiers had discontinued all expenditure, both in artificial manures and feeding stuffs ; that

their system formerly was to purchase the sheep stock they required by the sale of their corn crop ; but, so much more of that was now necessary to be sold for the payment of rent and labour, that a smaller balance was left for investment, and a scantier flock of course made to suffice.   Corn being the only thing the farmer of these soils can at present raise from them to convert into cash, we were informed that he is induced to plough up a larger proportion of land than formerly, thus extending his corn crops and diminishing the green or manure-making crops. It is said that necessity is driving many of the farmers to this, and that next year the chalk lands will have a larger average in corn than was ever known before.   It appears an anomalous state of matters that the farmers should sow more corn with diminishing prices ; but if the fact be so, it is a ruinous system on these thin lands, and shows more plainly than the loudest complaints the necessities to which many have been reduced by the sudden transition of prices.   They say—" We are quite conscious that nothing but increased expenditure in artificial food and manures can enable us to maintain a larger sheep stock, and increase the yield of our corn crops ; but meantime our landlords show us little sympathy, our capital is gone, and we have not the means of making the necessary outlay."

In all this district, some thirty miles distant from London, no vegetable produce or roots, early or late potatoes, are grown for sale by any one who considers himself entitled to rank as a *good* farmer.

The sheep stock kept are chiefly Southdowns, changed every year, though some tenants are turning their attention to breeding the stock they require, having found that the Hampshire breeder and the London butcher hitherto divided the whole profits of the animal, fattened at the expense of the Bucks farmer.   Very few cattle are kept, and these seem to be fed, in the large yard which occupies the centre of the homestead, chiefly on straw, which is thus, with the help of the horses and a few pigs, converted into a species of dung. The super-fluous liquid is suffered to run off into the nearest ditch or stream, no provision being made for saving it in any of the farm buildings which came under our notice. The implements in use are waggons and two-horse carts, cumbrous wheel ploughs, and the flail for thrashing.

The average produce of crops on the better class of chalk lands was stated to us at twenty to twenty-four bushels of wheat, thirty-two to forty of barley, and thirty-six to forty-eight of oats per acre. The number of sheep kept on a 200 acre farm, during winter and part of summer, is about 300.

Turnips are generally sown broadcast. One field of Swedes which we walked over, and which was one of the best we saw, would not exceed ten tons an acre. The bulbs were small and far apart, and, by the superiority of a portion running through the centre of the field, it was plain that if the whole field had been equally well treated, the crop might have been doubled. Labourers' wages have fallen 1s. since last year, and are now 8s. a week.

The complaints of low prices among the Buckinghamshire farmers are very general, and they put little faith in any remedy except higher prices. On the grass lands a reduction of rent might meet the present depreciation in the value of their produce, but on the tillage lands they think rent affords too small a margin for any beneficial relief. " Paying prices" they defined as meaning, with reference to wheat, from 56s. to 64s. a quarter. How these were to be got was the difficulty ;— for very few of the farmers entertained any hope of a return to protective duties; indeed, one of them, strongly opposed to free trade principles, said that the labouring classes were now so well educated, and read so many tracts and newspapers, that they would rise in a body to prevent it.

The advantages of railway communication are not much appreciated in Buckinghamshire, from the fact that it formerly enjoyed a comparative monopoly in the supply of the metropolis with various articles of produce which are now sent from much more distant localities by the aid of the railways.

The neatness and general appearance of the farm-houses are very creditable to the taste of the farmers. A good situation appears to have been chosen in nearly every case, and that is turned to the best account by laying out the gardens and ground in front with ornamental plants and walks.

LETTER III.

OXFORDSHIRE.

SOUTH-EASTERN DISTRICT. — GREAT VARIETY OF SOILS — EXTENT OF FARMS — LOW ESTIMATION OF COLD CLAY SOILS — RENT — RATES — DOUBTFUL BENEFIT OF DIVIDING SURPLUS UNEMPLOYED LABOUR AMONG RATE-PAYERS — WAGES — COURSE OF CROPS — FARMERS' REASONS FOR EMPLOYING FIVE HORSES IN A PLOUGH — INJURIOUS EFFECT OF THIS ON THE SUBSOIL — EXCELLENT MANAGEMENT OF TURNIP FARMS DETAILED — ADVANTAGE OF CUTTING TURNIPS TO THE SHEEP — EARLY LAMB BREEDING AND MANAGEMENT. — SHEEP FEEDING—QUANTITY FED — MANAGEMENT OF CATTLE DEFECTIVE — HORSES— PIGS — FARM BUILDINGS.

OXFORD, Jan. 26.

PASSING from High Wycombe over an elevated range of chalk country, we descended into the rich vales of the south-eastern portion of Oxfordshire. The fields are here of large size, and not surrounded injuriously with numerous hedges, or with much hedgerow timber. The great variety of soil in this county, lying as it does on substrata of chalk, greensand, Oxford clay, upper and lower oolite, and the lias formations, each succeeding the other from the south-eastern to the north-western boundaries of the county, exercises much influence on the character of farming. The same farm, and sometimes the same field, often requires a different treatment as to drainage and cultivation, so rapidly does the character of the surface vary.

The south-eastern district seems very fertile. It is said by the farmers to be easily wrought, quick in vegetation, and readily converted to any system of management. The farms are from 200 and 300 to 600 acres in extent, the larger proportion under tillage. They are generally held from year to year, leases being the exception. Where drainage is required, the landlord gives the tiles, the tenant the labour. Hitherto,

C

however, the tenant has been left to make nearly all permanent improvements at his own cost. The cold clay lands are greatly in need of drainage, and as they have gone quite out of request among the farmers, owing to the expense of working them, and the low value of their produce, their owners must either exert themselves now for their improvement, or find them thrown on their hands in despair. One farmer, who held a portion of clay in conjunction with stock land, estimated the difference in value at 25s. an acre in favour of the latter.

The rent of sound green crop land over this district varies from 30s. to 2l. an acre, tithe free. In many parishes there is no tithe charge, and where it exists it is not heavy. The rates also are moderate, but the amount of poor rate is not always a safe indication of the weight of that burden on the farmer. In many parishes the farmers by agreement divide the surplus labour of the parish among them, to prevent the rates being swelled by the expense of supporting the unemployed. In so far they are benefited by getting something for their money. But it may be doubted whether such an arrangement is compatible with that economical subdivision of labour which ought to prevail on a well regulated farm, or whether a greater loss is not sustained by the example of unwilling labourers operating on the regular strength of a farm, than all the benefit received from their assistance. One farmer of 500 acres told us that he gave employment to twenty-six men and seven boys, which were seven men more than he required to do the work of his farm. Another had so many hands thrown upon him by this arrangement that he resorted to spade husbandry as the most profitable mode in which he could employ them. The addition of seven men on a farm of 500 acres is equivalent to an increase of nearly 6s. an acre of rent, and that is a very heavy charge to be laid exclusively on the tenant.

Wages have fallen very little in this part of the county, 9s. a week being the general rate. A few have reduced to 8s., but even with that, we were told by an experienced bailiff, the labourer can purchase more food than when he was receiving 11*s.* and 12*s.* a week during the time of high prices.

On the better class of clay farms in the southern part of Oxfordshire the mode of management adopted is precisely the same as has been already described to be the practice on similar land in Buckinghamshire, viz., " three crops and a fallow."

It is proper to state the reasons which the farmers give for using five horses in a plough during the winter season. First, they allege the stiffness of the clay and the consequent heavy draught, second, that they find their horses stand the work much longer when not too hard pressed ; third, that part of the team consists of young horses, which are thus exercised, and assist in the labour without injury to themselves; and as to their being yoked in line a-head of each other, instead of two a-breast, it is so arranged to prevent the injury which would otherwise be done to the soft surface-soil by the feet of the " land" horse. In wet undrained land, the injury done in this way would, no doubt, be very considerable; and even when this heavy land is drained, the trampling of horses is hurtful in wet weather. But if we suppose a person, who was entirely ignorant of the operation of ploughing and its effects, looking at these five large horses as they follow each other in a straight line in the bottom of the newly turned furrow, and carefully watching the close succession in which their twenty heavy iron-shod feet beat into the waxy subsoil, he would conclude that the operation intended was to render that subsoil impervious, and that the turning over the furrow was merely a subsidiary process. And when one considers that the bottom of every furrow in the field is subjected to the same repeated pressure, he sees at once a reason for this soil being easily wet in winter, and suffering readily from drought in summer. If the soil is really of such a character that five horses are necessary to plough it, and if, to save the surface, it is requisite to sacrifice the subsoil, it becomes a question whether the spade and manual labour would not be found at once cheaper and infinitely more effectual.

On the poorest description of clay, oats are taken after a bare fallow instead of wheat ; and in favourable seasons, when the land has been well manured, the yield amounts to sixty-four bushels per acre.

The management of the fine turnip farms of this county forms a marked contrast to anything we have yet met with in our tour. The nature of the soil which we have already described, admits of a very perfect cultivation, and the level open character of the fields, and the large extent of each enclosure, are very favourable to the exertions of their highly intelligent occupiers. The system pursued is the four-course, worked with great industry and skill in the following manner : —

1. *Wheat, drilled.* — As soon as the crop is reaped the land is ploughed, and one division sown with rye, another with vetches, and another with hop trefoil, all of which are eaten off by sheep in succession the following spring. As each portion is cleared, it is ploughed and prepared for —

2. *Turnips.* — These are eaten on the ground by sheep, and the land ploughed and prepared for barley. Part of this division is sown with peas early in spring, which are drilled in rows twenty inches apart, and, when hoed the second time, white turnip seed is sown between the rows of peas, and covered by the hoe. As soon as the peas are reaped the white turnips are hoed, and being by this time well forward they prove a fair crop, and are eaten on the ground by sheep ; after which the whole division is sown with

3. *Barley.* — The half of this is laid out with clover, the other half, as soon as the barley is removed, is planted with winter beans, thus forming the fourth crop of the course, viz. : —

4. *Clover and Beans.* — The clover is once mown and the second crop eaten off with sheep, after which it is ploughed, and, along with the part in beans, is sown with wheat, which begins the course again.

By diligently following out this course the land is never suffered to lie idle. As soon as one crop is removed another takes its place, and, even before the pease crop is reaped, that which is to succeed it has been sown. The nature of the soil is admirably adapted for this constant succession, its dry friable texture favouring the extirpation of weeds, which yield at once to the skilfully applied labour of the farmer.   The rye, vetches, and late turnips are grown during the winter months, and derive much of their sustenance from the damp atmosphere at that season ; and, being all consumed by sheep on the ground, more is returned to the soil than was taken from it, especially if the sheep are at the same time fed with corn.   The droppings of the sheep and their treading of the land give it that richness and solidity which, on these warm soils, are eminently favourable to successful grain crops.   The crops, both white and green, are sown in rows, and carefully and frequently stirred and hoed, manual labour being lavishly expended to insure perfect cultivation.

The advantage of employing labour is proved by the more rapid progress of the sheep when fed on cut turnips, placed for them in troughs, as compared with the old practice of suffering the sheep to gnaw the turnip on the dirty ground.   A very skilful farmer told us that he had ascertained by trial that the same sheep would make equal progress fed on turnips so cut and prepared, *without* the addition of corn, as they would *with* corn when the turnips were not so prepared.

On the large farms, machinery is employed for thrashing the wheat crop, but barley is thrashed by the flail, both to give employment to the labourers, and because the machines in use cut the grain too short, and thus injure it for the maltster.   The average crops of wheat are twenty-eight to thirty-two bushels ; and of barley, forty to forty-eight bushels an acre.

Next to the cultivation of the land, the chief attention of the farmer is devoted to the management of the sheep stock, the most remunerative part of which is the breeding of early lambs for the London and Oxford markets.   In the beginning of January, the ewes, which are of the South Down breed, drop their lambs, having been previously placed in a dry well-littered yard, surrounded with warm sheds, cheaply constructed with hurdles, and roofed with loose straw.   Here the ewes are supplied with cut turnips in boxes, clover hay, and ground beans. The lambs learn to eat the beans, of which they are allowed to take what they like, and do consume sometimes as much as a pint a-day. With this high feeding they are soon fit for the market, being ready to be disposed of at Easter and the month following, and then bringing, on an average, 30s. each. The lambs thus early removed, the ewes are soon made fat, and are of course much easier kept on the pastures than if they were suckling their lambs. Some farmers have a second flock of ewes, which drop their lambs a month or six weeks later. They are wintered in a straw-yard, getting no food but bean or pease-straw until they lamb, when a few cut turnips and clover hay are added. This flock is pastured, and fatten their lambs during the summer, without receiving any corn. The store-sheep are purchased when required, all that are bred being sold fat as lambs.

The manner in which the store-sheep are fed depends altogether on the taste and means of the farmer. If he wishes to have his farm in the best condition, he supplies the whole flock daily with beans, in addition to their other food — rye, vetches, clover, or turnips, as it happens. If this is too expensive a system to pursue throughout, he reserves the beans for the *feeding* pen, into which the best sheep are draughted weekly to supply the place of those which are sent off fat, weekly, to London. In this pen the whole flock is finished with corn, each sheep being in it a month or so, and receiving a pint to a quart of beans daily. But if the farmer's lease is coming to a close no corn whatever is supplied, as in this county there is no compensation to the tenant for unexhausted improvements. And, of course, where the farmer has not even the security of a lease, this system of good farming cannot be entered on with safety at all.

The quantity of sheep fed on a farm varies with the amount of artificial food supplied to the flock. On one farm which we visited, as many as 3000 sheep and lambs had been sent fat to

London in the course of a season.   A farm of 500 acres passing off 1000 sheep in a year, or two sheep to the acre all round, is considered very fair management.   The value of sheep has been much depreciated, and severe loss was sustained by the large holder who bought dear and was obliged to sell cheap.   He will not lose as much now, as his last purchase was in the same proportion less with his sales.   The value of wool has increased —what brought only 18s. two years ago is now (1850) fetching 26s.

The management of cattle is not attended with any thing like the same skill displayed in the feeding of sheep.   The usual plan is to have a few running loose in the farm-yard, where they live on straw, few or no turnips, and sometimes a little hay. Nor does the same careful economy guide the operations of those who feed cattle; on one farm, otherwise conducted with very great skill and practical knowledge, we found a lot of large cattle being stall-fed on bean and barley meal and hay, without any turnips or other green food.   Each animal is supplied with 18 lbs. of meal daily, mixed with hay-chaff, and costs the feeder 10s. a-week.   This obviously cannot pay, especially with a low rate of prices.   Indeed, there would be something wrong if it did, for it can scarcely be right to expend as much on the food of a fattening ox as would well suffice a labourer and his family. The farm horses are in some cases kept in stables; in others they are put from their stables, loose, into the yard every night. Considerable numbers of pigs are kept on most farms.   They roam about the straw yard, picking up what they can get, and are fed on meal besides.

The farm buildings on the larger farms comprise two or three extensive barns, stables for the horses, cow-house, and a large straw-yard in the centre with shelter sheds.   The farmers do not complain of want of accommodation, as their system of sheep-feeding is chiefly conducted out of doors, and does not, of course, demand a great extent of farm buildings.

LETTER IV.

OXFORDSHIRE — *continued.*

WESTERN DISTRICT.—NATURE OF LAND—BLENHEIM—DIFFERENT MANAGE-
MENT BY THE SAME TENANT UNDER DIFFERENT LANDLORDS — FARMERS
CONSCIOUS OF BAD FARMING, THEIR EXCUSE — AVERAGE CROPS AND
STOCK — WAGES — NEATNESS OF FARM HOUSES — WANT OF PRACTICAL
KNOWLEDGE BY LANDLORDS AND AGENTS— FARMERS' OPINIONS OF THEIR
PROSPECTS — INJURIOUS EFFECT OF LAW OF DISTRAINT IN ENABLING
LANDLORD TO ENCOURAGE UNFAIR COMPETITION — DUTY ON FOREIGN
CATTLE — ALLOTMENTS TO LABOURERS.

STOW-ON-THE-WOLD, Jan. 29.

FROM Oxford by Woodstock and Chipping Norton towards
the border of Gloucestershire, we pass over the lower oolite
formation, the soil upon which is generally thin and light good
stock and barley land, but subject to blight from drought and
other causes. The farm buildings, in this country of walling
stone, are substantially constructed, though very imperfect in
extent and arrangement. Interspersed with the hedgerows are
occasional lines of stone wall, which become more frequent as
we proceed westward.

The first estate of great magnitude through which we pass
is that of the Duke of Marlborough at Blenheim. From various
causes very many of the farms on this great estate are being
surrendered to the Duke, who now holds under his immediate
management somewhat more than 5000 acres of his own land.
When his Grace succeeded to the property about ten years ago,
the rents were very low, the land being generally underlet.
From low rents, with probably indolent farming, the change
appears to have been too sudden, and in consequence of an
addition of a third being placed on the rental without the con-
comitant outlays to which landlords must generally submit for

the accommodation of larger stock and the manufacture of
heavier crops, many of the farmers left the estate, and those who
remained, in very numerous cases, permitted their land to fall
into a bad state of cultivation. Hard pushed by the times,
with higher rents and lower prices, the little capital that remained has been rapidly diminishing, the Duke declines to make
any abatement, and as soon as a farm is so completely reduced
as to be untenable, it reverts to the landlord. The country exhibits a poverty-stricken and neglected look, and there is no confidence of a friendly or even feudal character between landlord
and tenant. This is much to be regretted, as the farms are
many of them very desirable in point of extent and quality,
varying from 600 to 700 and 900 acres, and well adapted for
green crop and stock farming.

The greater part of this district is a turnip and barley soil,
and the course of cropping generally adopted is the four-field or
Norfolk system. It is carried out with more or less energy,
according to the security which the tenant feels that he shall reap
the profit of his own exertions. For instance, a tenant who holds
extensively under different landlords told us that from one of
these he had no lease, and paid a very high rent; he therefore
spent nothing in purchased food or manures, as he might be
obliged to go at any time, if his landlord took a fancy to the
farm, and he must in that case leave his improvements behind
him. His high rent also prevented expenditure, for he might
wish to leave the farm as soon as he could get a better, and he
was therefore on his guard against outlay which he could not
take with him. On another estate the same man has a farm,
at a moderate rent, with a lease, and this year he expended on it
100*l.* in purchased manures, and 200*l.* in purchased food. On
the first farm he is losing money; on the second he thinks, with
prices a little better, he might do well enough.

Great part of the land which forms the subject of this letter
is let on yearly tenure. Farming is not, on the whole, carried on
with any degree of spirit; and of this the farmers are themselves

quite conscious. " We are not farming," one of them said to
us, " We know that we are not farming; we are only taking
out of the land what we can get from it at the least cost, as
we don't know how long we may remain in possession, and
have no security for what we might be disposed to invest in
improved cultivation." Purchased manures and food, especially
the latter, are highly approved, and must be of peculiar service
on this thin, dry soil, but they are very scantily used. Lime
has not been tried, as it is supposed that the natural limestone
in the soil supplies all the calcareous matter necessary. We
have, however, seen important benefit derived from the application of burnt lime on soils even more calcareous than this.
Very little draining is said to be requisite, as the soil is thin
and the subsoil porous ; though, from the soapy nature of the
land in wet weather, in ploughing the turnip lands for barley it
is necessary to yoke the horses in line, in order to avoid the
injury which the treading of the " near" horse, when yoked
abreast, would otherwise do to the soft land.

The farmers have a strong dislike to deep ploughing, as
a matter of principle. Light ploughing, they say, is easier
to the horses, keeps the manure near the surface, where
it is at once within reach of the crop, and does not injure
the active soil by any admixture with the barren, hungry
subsoil. We did not hear of an instance where deep tillage
had been tried, followed up by ample manuring and cultivation, and had failed ; and we may therefore venture to say
that the farmers have no experience to warrant them in pre-
ferring their own practice to another which they have never
tried. We cannot help thinking that a deeper stirring of the
soil would materially lessen the injurious effects of drought
in summer, to which the land is said to be subject.

The average produce of wheat for several miles round Woodstock was stated to us at twenty bushels, and barley forty bushels
an acre. Turnips were a light poor crop. The number of sheep
kept on a farm is little more than one to an acre, or about 600

sheep on a 600-acre farm.   The breed is a cross with the long-woolled sheep, as the lambs are kept for stock, and a good fleece is therefore a profitable consideration.   Good year-old sheep will be sent off fat in a few weeks, being previously shorn.   They are worth 30s. each, and the fleece 7s. or 8s. more.   Cattle form here, as in other parts of the county, quite a secondary consideration.   A few are kept in the straw-yard to trample down the straw, but they get little green food, and very seldom any corn or artificial food.   The straw is wastefully consumed, and there are few proper buildings to admit of a different practice.   As much as possible of the labour of the farm is done by taskwork, at which threshers earn 11s. a-week; carters and ploughmen receive 10s. to 11s. a-week, but, if present prices continue, the farmers say they must lower their rate of wages.

As to the house accommodation and comforts of the farmers in Oxfordshire, they appear to be at least equal, if not superior, to what we saw in Buckinghamshire.   The same neatness and order characterise them internally, and the fronts of the houses are, in the majority of cases, so arranged as to command an agreeable prospect.   They are generally so placed as to be in proximity to, and in full view of, the farm-yard, so that everything that goes on there may be under the immediate eye of the owner.   The farm-house has attached to it a good-sized garden, in which vegetables for the consumption of the family are grown.   It is also not unusual to see a piece of orchard-ground close at hand, the fruit of which is disposed of in the London markets.

As a general rule, the landlords of this county interest themselves very little in agriculture.   Few of them are practically acquainted with, or engaged in, farming.   And what is equally unfortunate, as regards the improvement of the soil, and the welfare of the different classes engaged in its cultivation, they have not yet seen the necessity of making amends for their own defective knowledge by the appointment of agents better qualified than themselves. In the majority of cases the agents or stewards are lawyers, who, without practical knowledge of the business of farming, and in the endeavour to secure the landlord's apparent interests, bind down the tenant with conditions most injurious to him from their stringency, and with no corresponding benefit to the landlord.

The opinions prevalent among the farmers in regard to their own prospects, and the means by which they hope to get over their present difficulties, are very various. On one point they were all agreed, viz. that free trade had done every body good except themselves; for, if the landlord's rent remained the same, he was a gainer by cheap food, and if the labourer's wages were the same he was also a gainer, both at the expense and to the exclusive loss of the farmer. As the readiest means of retrievement, many spoke of measures by which the free use of capital as a safe investment by the farmer might be encouraged. The most judicious with whom we conversed readily admitted that much might be done by improved cultivation, and that there was great room for such improvement. " But," said one of them to us, and he was a strong Protectionist, "if a farm is to be let, and one man with 3000*l*. and another with 300*l*. bid for it, the right of *distress* possessed by the landlord makes him safe to pit the latter against the former, and the consequence is either that the first man takes the farm at a higher rent than it is worth, and thereby injures himself, or that the second man gets it at a rent which he has recklessly offered; he struggles on with inadequate stock for a few years, taking all he can out of the land; his rent falls into arrear, everything is seized to pay it, all other creditors (who have most probably advanced much of the means for carrying on the farm) are cheated, and the poor man himself is a beggar. Now, suppose I go into the market with my wheat; one man may offer me 14*l*. a load, which I reluctantly refuse, because I know he has no capital, and am content to take 10*l*. from a man who has capital; why should my landlord have the privilege of forcing me to pay

40s. an acre for his land, because a man who has scarcely any thing to lose offers that sum, when, if no such privilege existed, he would be obliged to satisfy himself with 30s. ?   If we are to have free trade, let us also have no unfair privileges." Such were the views we heard repeatedly expressed on this question in Oxfordshire.

We did not find that any very serious hope was entertained, even by the strongest Protectionists, that Parliament would restore the duties on corn, but a prohibitory duty on cattle and foreign provisions they looked to as a reasonable claim, and put in this light: — That a fair case had been made out for the bulk of the people to insure them untaxed bread; but butcher's meat, being more of the nature of a luxury, was chiefly consumed by the classes above the labourers, and who could afford to pay for it; that the reduced price both of corn and cattle at the same time was a heavier blow than the farmers could bear unaided; that an encouragement of cattle and sheep-feeding was the surest means of improving the land, and that the facilities under the new Corn Bill of importing cheap kinds of corn for feeding stock would tend rapidly to increase the home supply.   For these reasons many of the farmers in Oxfordshire seek a restrictive duty on foreign cattle and provisions for a few years, so that they may have time to adjust their affairs to the new order of things.

Allotments for labourers are let at rents varying from 2l. to 3l., and as much as 4l. per acre.   The farmers all complained of them as injurious to the steady industry of the labourer, and a heavy tax on themselves.   The labourer s half-acre allotment, they said, was dug and tilled in the morning and evening — before and after the day's work.   It was, therefore, in part, an exhaustion of that physical energy which a full day's work required, and by so much a positive loss to the farmer.   In almost every case, too, the allotments were let at extravagant rents, generally at least double the average of the surrounding land; in fact, they were in many cases given on bad land, which to a farmer was nearly worthless.   As the labourer must pay his rent before he reaps his crop, he is frequently obliged to borrow it from his master in advance of his wages, and this leads to jealousy and bad feeling between master and servant. A piece of garden-ground in the neighbourhood of their cottages would be much more beneficial to the labourer, as he could then, without fatigue, raise such potherbs as were requisite for his table, and most farmers would willingly give a portion of their green-crop land in which to plant his potatoes at a much more moderate rate than the rent of the allotment ground.

LETTER V.

GLOUCESTERSHIRE.

COTSWOLDS. — CLIMATE AND NATURE OF SOIL — RENT — CUSTOM OF COUNTRY — RATES — WAGES — COURSE OF CROPS—SYSTEM OF BURNING THE LAND — REPEATED AT REGULAR INTERVALS FOR A LONG PERIOD WITH ADVANTAGE — BENEFIT OF ARTIFICIAL MANURES — COTSWOLD SHEEP AND CATTLE—AGRICULTURAL COLLEGE AT CIRENCESTER—COURSE OF EDUCATION — THE COLLEGE FARM.

GLOUCESTER, Jan. 31.

FROM Stow-on-the-Wold westward towards the Severn, and south-west towards Cirencester, extends an elevated tract of undulating country possessing a distinctive character as an agricultural district, and known as the Cotswold Hills. The greater part of this district has a considerable elevation above the sea, in some places as much as 600 or 700 feet, which delays the harvest about a fortnight beyond the period in the surrounding low grounds. The appearance of the country is a series of level plains, falling at intervals into gentle valleys, through which the natural drainage of the adjoining lands is carried off. With a cool climate the Cotswolds have a light soil, not very productive naturally, but capable of easy cultivation, and, under a generous system of farming, likely to remunerate the skill and capital invested in it. At no very remote period the greater part of this district was devoted to the pasturing of sheep, a peculiar and very superior breed of which takes its name from the locality. The grass-lands have now been nearly all brought under the plough, the richer pastures in some of the valleys being the only portions left untilled. The fields are large, and are inclosed either by hedgerows or dry stone walls. In the valleys they are smaller, and the hedgerows encumbered

with wood; but as the land is used for pasturage or meadow,
the inconvenience is less felt than on arable land, especially as
the shelter of the hedgerows is found in this elevated district
beneficial to stock.

The rent of land on the Cotswolds ranges between 10s. and
25s. an acre, the average being from 16s. to 18s., tithe free.
Many of the most wealthy of the farmers hold their lands on
lease, but the tenures are chiefly from year to year.   Particular
rules are prescribed by the landlords as to the system of cultivation, but they are neither adhered to nor strictly enforced.
When a tenant enters to a farm the proprietor is understood to
put his buildings and fences in good repair, but every thing else
must be done by the tenant.   Drainage, except where a
bed of clay intervenes, is not needed.   There is no custom
of compensation to outgoing tenants for any thing except " acts
of husbandry," which include the expense of ploughings by the
farmer for the benefit of his successor, and of carting and supplying the manure.   The value of the manure itself is regarded
as the property of the land, so that the outgoing tenant, having
no interest in accumulating it during the last year of his lease,
leaves the farm sometimes badly provided in this respect.   The
incoming tenant having very little to pay for "acts of husbandry," men of small capital may venture to take farms for
which their means would be otherwise inadequate.

Rates of all kinds are moderate, and in few instances exceed
2s. in the pound, and much of the land is exempt from tithe.
All the labourers of the district, female as well as male, find employment on the farms, the best workers making by piece-work
from 10s. to 11s. a week, though the common rate for day
labourers does not exceed 6s. and 7s., and women only 6d. a day.
It is somewhat remarkable that labour should be so low priced
in a district where there is no surplus population, where the rates
are unusually light, and the rent moderate.   There is said to be
no want of capital among the Cotswold tenants, and farms, except
those of a stiff unkindly nature, are much in request.

The fields on the Cotswold farms present an unbroken surface, with no impediment of any kind to husbandry. The farms, consequently, are large, and the operations conducted on an extensive scale. The rotations followed vary with the quality of the soil, the best parts being managed under the alternate system of corn and cattle crops ; the inferior in a five, six, or seven years' course, as appears most advisable to the farmer. The five years' course is simply an extension of the four course, by permitting the grass to continue two years before being ploughed. The six years' embraces the same crops as the preceding, with the addition of a crop of oats after the wheat crop. The seven years' appeared to be — 1. turnips ; 2. barley ; 3. clover ; 4. wheat ; 5. oats ; 6. and 7. saintfoin. This, we were informed, is the prevalent mode of cropping followed on the Cotswolds, and as it has some peculiar features, we shall shortly describe it. In the seven years the land gives three corn crops, and four green or cattle crops.

The great feature of the management is the burning of the land. Beginning with the turnip crop, the preparatory process is to pare and burn the surface, in which is the tough sward of the two years' saintfoin. This process is commenced early in the spring, and is done by men with a breast plough. With this they pare off the turf, and then collect it into heaps and burn it. 16s. an acre is usually the cost of the paring and burning, which is done by taskwork. This process, besides providing an immense store of ashes for manure, likewise prepares the soil for being worked down and completely pulverised for the reception of the turnip seed. When that is accomplished, the land is covered with the burnt ashes, and the seed is sown either broadcast or in ridges. The ashes secure a fair crop, which is eaten on the ground by sheep, after which the land is sown with barley and clover seed.

The barley being removed, the clover is either mown or pastured, — usually the former, — and the land is then ploughed up for wheat. The wheat is followed, after proper tilth, by

D

a crop of oats, with which is sown the seed of the saintfoin. This is pastured for two consecutive years by the sheep stock, and completes the course, though a field of saintfoin occasionally remains unbroken for eight or ten years.

Before the introduction of guano and bones, this system depended altogether on the burning of the soil at the commencement of each course. The ashes thereby afforded secured the turnip crop, which being consumed on the ground, enriched it for the succeeding crops. A stranger to the character of the soil would not easily believe that such a course could be long continued without the aid of other manures, and might be apt to think that in process of time not only the organic matter but the thin soil itself would gradually be burnt away altogether. But we must not too hastily conclude that such is the effect. The best farmers on the Wold are the men who burn most extensively. On a 700 acre farm we were assured by its occupier that he every year burnt from 60 to 100 acres of land in preparation for turnips, and seldom failed to have a fair crop. One field over which we walked had been broken up from its natural state exactly fifty years ago ; it was then pared and burnt, and so started the first crop of turnips, which supported the other crops of the course. The same process had since, within the knowledge of our informant, been seven times repeated. No manure of any other kind had ever been applied, except such as arose from the consumption of its own produce on the ground, and the crops in each succeeding rotation had shown no sign of decreasing, the last having been an excellent crop of wheat. The soil which lies on the lower oolite formation is very thin, but as it is not more so than when first broken up, its depth must have been maintained by the ploughman, perhaps imperceptibly, bringing up some fresh subsoil after each burning. The value of the ashes as manure are undoubtedly enhanced by the effect of fire on the natural limestone of the soil.

The practical reader will see of how much benefit to the Cotswold farmer was the introduction of portable manures and cheap feeding stuffs. These he uses in too limited a degree — eight bushels of bones and two cwt. of guano per acre, being a common application to the turnip crop in addition to ashes. When dung is used, four bushels of bones and one cwt. of guano are deemed a sufficient additional supply. In the use of these manures the farmers of the Cotswolds are not decreasing their expenditure, for we learnt from a corn merchant who supplies them with guano and other similar substances that their orders this season had not fallen off from what they were last year. The increased produce which a liberal application of manure is sure to leave from the same extent of soil is cheaply purchased by the first cost, as the rent, the labour, and the other farm expenses remain nearly the same whatever the acreable produce may be. The present average crops of wheat are twenty bushels, and of barley thirty-two bushels an acre. The acreable produce of both might be considerably increased, and not only so, but a liberal expenditure in manure and feeding stuffs would enable the farmer to take these crops at shorter intervals, and, consequently, in increased breadths, from his farm every year.

The Cotswold breed of sheep is the principal description of live stock. A breeding flock is kept on each farm, the produce of which is sold as " teggs," or year old sheep, and generally bring, with their wool, from 37s. to 40s. It is reckoned a very poor farm indeed where one sheep to the acre all round cannot be kept. Very few cattle are fed; but we met with a new branch of cattle management here — viz., the rearing of heifer calves. These are bought from the dairy farmers of Bucks, they are reared on the Cotswolds till three years old, and then sold to the Wiltshire dairy farmers for milch cows. The regular cattle stock of a 700 acre farm, which we visited, consisted of twenty-four calves bought every year, kept on for the next two years, and sold out in the third to the dairyman at prices varying from 11l. to 15l. The management in the case here mentioned was the most methodical we met with, and this leads us to infer that cattle do not form a very important item in the profits of Cots-

wold farming. The stock were all tied up in open sheds, thus exposed to the wind, without the power of changing their position.
Their provender was a mixture of hay and straw cut into chaff by
a machine driven by horse power. Machines are also used for
thrashing wheat, but barley is thrashed with the flail. Twenty
men are regularly employed on the farm above mentioned, besides women for light work. Seven of these men are *breast
ploughers*, whose business during spring and the early part of
summer is the paring and burning of the land in preparation for
turnips; during the rest of the year they are employed in draining and other necessary operations.

While in the Cotswold district, we visited the Agricultural
College of Cirencester, which is the only institution of its kind
in England. It is a very handsome structure, in the Gothic
style of architecture, situated about a mile and a half from
Cirencester, and adjoining the park and woods of Earl Bathurst.
The principal front, 190 feet long, has a south aspect, and commands an extensive view over North Wiltshire. The buildings
include a large dining-hall, library, museum, lecture theatre,
laboratories, class-rooms, a chapel, and dormitories for about
200 students. The course of education extends over six sessions, of which there are two in each year. The first and second
sessions are chiefly devoted to instruction in practical agriculture,
which is given on the farm, and familiarises the student with
the manual operations of husbandry, the use of the best agricultural implements, and the most approved systems of management in the different departments of the farm. A laboratory,
conducted on Liebig's system, is appropriated to chymical
manipulation and analysis. Botany, geology, and zoology are
each made the subjects of practical instruction. Levelling,
surveying, and the measurement of land are also attended to ;
and to the advantages of actual practice are superadded the
lectures of the professors on every branch of science connected
with, or calculated to throw light upon, the cultivation of the
soil.

The college was originally founded in order to furnish a sound education in scientific agriculture for the sons of tenant farmers; but that class do not appear to have availed themselves of the advantages thus held out to them, nor was it altogether adapted for them; and the sixty students at present entered on the books (1850) are all the sons of solicitors, clergymen, officers, or landed proprietors. Most of them intend to engage themselves in the cultivation of the soil, either as owners or occupiers; and among them are a class of students who may yet prove very valuable to the community, as an educated and competent body of land agents and stewards, conversant with the details of agriculture.

The college farm extends to 700 acres, nearly all of which are under the plough. Three different rotations are adopted, both to suit variations in the soil, and to exhibit different practices in operation for the instruction of the students. The four and five courses are carried out on all the lighter lands of the farm, which comprise by much the larger portion of it. A three-field course is followed on the heavy land, viz. turnips, beans, wheat. The turnip crop is an early kind, sown early, for the purpose of being consumed on the ground by sheep before the month of November, as after the rains of that month the ground becomes too soft for sheep feeding. Beans are planted after the turnips, and wheat after the beans. We had not an opportunity of examining the farm minutely, but observed that the fields were large, the hedges narrow, and no land wasted at their roots; that the ploughs were drawn by two horses abreast; that the horses were in high spirit and condition, and turned over with ease a furrow three inches deeper than we saw five horses in line doing, in a different part of the Cotswolds, on soil of a similar character. We remarked that when the furrow was turned over (in preparation for wheat after carrots), the foot of the " near" horse left no injurious effect on the *deep* dry soil, and that depth had been gained by the use of the subsoil plough. Good roads, intersecting the farm, admit the use of the handy one-horse cart; the corn is sown in rows by the drill, and is

hoed cheaply and effectively by Garret's horse hoe ; drains are
made where required, and useless fences grubbed out and converted into useful land. The turnip crop we thought rather a
light one ; the manure used for it was twenty carts of dung and
3 cwt. of salt. We should certainly have preferred 3 cwt. of
guano, and would have added ten bushels of bones besides, in the
belief that the superiority of the crop would have amply compensated the additional cost of the manure.

The arrangement of the farm buildings is commodious, and
comprises sheds for all the implements of husbandry ; workshops
for the repair of these implements, and for the shoeing of the
farm-horses ; stores for portable manures ; a steam engine for
thrashing the crop, grinding the feeding stuffs, bruising coprolites, guano, or bones, and driving the chaff-cutter ; and, finally,
the waste steam is turned into vats to cook a mess of chaff and
meal for the live stock. The horses have each a loose box, the
cattle are partly fed in boxes and partly in stalls, a tramway
being laid from the turnip-house along the feeding-house to facilitate the feeder in bringing in the food, and afterwards carrying
out the dung. Sheep are fed in covered pens ; they stand on
sparred boards, and require no litter. But it may be remarked
with regard to this mode of feeding, that they are easily disturbed by the approach of any one ; and slipping about the
boards, they want that quiet docility which marks the fattening
animal. Lord Bathurst has his sheep tied up by the neck, like
stalled cattle, and in this position they soon become perfectly
quiet, and improve rapidly in condition. Sheds and yards for
pigs are provided, of which a very large stock is bred and fed on
the college farm. The system pursued on the farm is to breed
and fatten every animal which it supports, and a slaughter-house
is provided, in which the last process in the conversion of
vegetable into animal food is completed. The offal is thus kept
on the farm, and any portion of the meat which is not required by
the college establishment is sent to market. Besides the cattle
and pigs kept in the buildings, a large flock of sheep are fed on the

turnip-fields.   The farm is held on a lease for forty-seven years, at a rent varying from 20*s.* to 28*s.* an acre.   There is no tithe, and the rates are moderate.   A machine is placed at the entrance of the farm buildings, on which all the farm produce is weighed as it is brought in for consumption ; and the progress of experimental cattle is ascertained at any period by putting them on the scales.   A record is kept of weights, and, if the system is followed out as it might be, very valuable results may be expected.   A well digested system of farm accounts is also kept, and so arranged that it may with facility be applied to the ordinary receipts and expenditure of any other farm.

The Royal Agricultural College has hitherto been looked upon with suspicion by the farmers of the Cotswold Hills — a circumstance probably due in some degree to prejudice, but largely, also, to a persuasion that the college farming does not pay.   It is not difficult to understand how farming, for the double purpose of instruction and scientific experiment, leads to expense which, in the ordinary practice of individuals for remuneration, may be avoided.   The whole educational system has lately been reorganised, and placed upon a basis more extensive, and therefore likely to be more beneficial, than it was at the period of our visit.

The College, though patronised by Royalty, is not a Government institution, but originated in the enterprise of a body of agricultural improvers, who, in the face of many discouragements and much loss of capital, continue on public grounds to support it.   The assistance of Government, in extending similar institutions in various parts of the country, could not be given to any other educational purpose of greater public importance.

LETTER VI.

GLOUCESTERSHIRE — *continued.*

Vale of Gloucester — numerous hedgerows very injurious to the farmers — difference of character of landlords of little value if unaccompanied by the power of making improvements — rent — defective accommodation and consequent mismanagement of cows in winter — starving system, as regards both stock and land — improvements suggested — wages — course of crops — difference between vale and cotswold farmers in intelligence and enterprise — security for tenant's capital.

Berkeley, Feb. 1850.

The Vale of Gloucester extends in a south-westerly direction from Tewkesbury, on the borders of Worcestershire, to Thornbury, in the western division of Gloucestershire. It is a district of rich pastures, with a few fields of ploughed land here and there dotted through it. Even at this season, when the trees are leafless, it presents a densely wooded aspect, caused by the numerous hedgerows subdividing the small fields of pasture land, and which, in the luxuriant foliage of summer, must overshadow the surface, and draw from the soil much of that nutriment which the fields would otherwise yield to the farmer's stock. To give an idea of the extent in which the farmer's fields are here encumbered by hedgerow timber, it may be mentioned that 3500*l.* worth of timber was sold from the superfluous fences surrounding and subdividing Earl Ducie's model farm at Whitfield, only 260 acres in extent. Besides the evil of hedgerow timber, a condition is introduced into the agreements between landlord and tenant, by which the latter is bound not to prune his hedges oftener than once in seven years. The hedges in this sheltered vale are thus permitted to grow to a great size, becoming a harbour for game, which, from the small extent of the

land under tillage, are the more destructive to the crops of
the farmer. For this reason many landlords will not permit
the fences to be touched, thus adding to the evil, as with
such small and inconvenient fields the farmer, however anxious
to improve, is shut up to the old systems of management. Besides occupying space, these wide banks, covered with juicy
vegetation, abstract from the land much of its natural strength,
and of the manure which the farmer has added to it.   On all
sides the farmer suffers : he pays rent for space occupied by
his landlord's trees ; he provides harbour for his landlord's game,
which, in return, feed upon his crops ; if he attempts to plough
out inferior pasture, his crop becomes an additional feeding-
ground for the game ; whilst the small fields and crooked fences
prevent all efforts at economy of labour, and compel him either
to restrict his cultivation or to execute it negligently and un-
profitably.

The reader will not be surprised to learn that farming is here
in a very stationary position, and that even under landlords
differing in character the tenants preserve a remarkable same-
ness of no progress.   There are two noblemen in the Vale, the
one the founder of the model farm of Whitfield, a distinguished
short-horn breeder, and an advocate for improved farming ; the
other a great preserver of game, and said to interest himself
little in agricultural progress.   The tenantry of both are much
upon a level as regards improvement, both adopting the same
system which has prevailed unchanged for many years.   Both
landlords are owners of settled estates, and though the first
encourages improvement by precept and example, he cannot
afford that participation in its cost without which a tenant from
year to year is not justified in incurring large outlay.   The
tenants of the second are thus scarcely in a worse position than
those of the first, and their agricultural practice and skill are
nearly equal.

The rent of land in the Vale of Gloucester is about 30s. an
acre, and rates and tithe 10s. more.   The best quality lets as

high as 50*s.* inclusive of all rates. The farms are generally held
from year to year.

The Vale is celebrated for its dairy farming, which might
be described with perfect accuracy by an extract from the
County Report to the Board of Agriculture made nearly
forty years ago. Taken as a whole, it has undergone no
change in its details, though very possibly the increased luxuriance of the hedgerows, and the continual abstraction of
cheese and butter without any corresponding return of phosphates to the land, may have led to a perceptible decrease in its
annual produce. Water stagnates in the soil, the industry of
the farmer is paralysed, the energy of the labourer deadened,
— nothing seems to thrive but the gigantic trees, whose roots
in the smaller fields cover nearly their whole substratum like a
network.

A brief outline of the appearance of a Gloucestershire dairy
farm may serve to show the present state of the art in that
district. January is not the month most favourable for viewing the operations of a dairy, but the management of the
cows at this season is of much influence on their yield afterwards. An inconvenient road conducted us to the entrance
gate of a dilapidated farm-yard, one side of which was occupied by a huge barn and waggon-shed, and the other by the
farm-house, dairy, and piggeries. The farm-yard was divided
by a wall, and two lots of milch cows were accommodated in
the separate divisions. On one side of the first division was a
temporary shed, covered with bushes and straw. Beneath this
shed there was a comparatively dry lair for the stock; the yard
itself was wet, dirty, and uncomfortable. The other yard was
exactly the counterpart of this, except that it wanted even
the shelter shed. In these two yards are confined the dairy
stock of the farm during the winter months; they are supplied with hay in antique square hay-racks, ingeniously cap-
ped over, to protect the hay, with a thatched roof, very much
resembling the pictures of Robinson Crusoe's hat. In each

yard two of these are placed, round which the shivering animals station themselves as soon as the feeder gives them their
diurnal ration, and there they patiently ruminate the scanty contents. A dripping rain fell as we looked at them, from which
their heads were sheltered by the thatched roof of the hay-rack,
only to have it poured in a heavier stream on their necks and
shoulders. In the other yard, the cows had finished their
provender, and showed their dissatisfaction with its meagre
character by butting each other round the rack. The largest
and greediest having finished her own share, immediately dis-
lodges her neighbour, while she in her turn repeats the blow on
the next; and so the chase begins, the cows digging their horns
into each other's sides, and discontentedly pursuing one another
through the wet and miry yard. — Getting over an inner gate,
we came upon the piggeries, where a dozen well-fed and warmly
housed pigs showed by their sleek round sides the benefits of
food and shelter. Inquiring of the farmer whether he thought
his cows would not be bettered by equally comfortable accommodation, he said they would, but his landlord did not give that
accommodation, and he had no security by which at his own
cost he could safely make the outlay. — Leaving the yard, we
passed into the fields, sinking at every step in the sour wet
grass lands. Here little heaps of dung, the exhausted relics of
the hay from which the cows derive their only support in winter,
were being scattered thinly over the ground, to aid in the production of another crop of hay. But we need not continue the
picture farther. The management of the dairy itself did not
come under our observation, as nothing is done in it at this
season; but it is said to be conducted with great care, cleanliness, and attention.

It is well known that much benefit has attended the application of bones as manure to dairy farms in Cheshire, and this
benefit is accounted for by their replacing the annual abstraction of phosphates in the cheese sold off the farm. Bones are
not yet used to any extent in this district. The advantages of

drainage seem to be very little appreciated, the great bulk of the dairy farms of the county, even those which most require it, being inadequately drained.   Some instances were pointed out to us where this operation had been executed with great advantage, and which showed how much might be done by a liberal outlay for this object.   In these cases the landlord supplied the tiles, and the tenant was at the expense of opening the drains.   No winter food, except hay, is provided for the stock, roots being deemed injurious to them.   Under this management, even in the famed Vale of Gloucester, we were not surprised to learn that the annual produce of a dairy cow, on the average, does not exceed $3\frac{1}{2}$ cwt. of cheese, and that fully three acres of land are required for the annual support of each cow.

Can no improvement be made on this system?   We know there can.   Let the land be thoroughly drained, so that the manure the farmer puts on it may not be wasted; let the too numerous trees and hedgerows be removed, which at present rob the pastures, and whose roots (if the trees are left) would penetrate and obstruct the drains; let the farmer have suitable accommodation provided in which he may economically feed and shelter his stock; and then encourage him by security of tenure to increase the productiveness of his farm.   From the half of the land formerly devoted to hay, the same quantity will be got by the application of purchased manures; the other half will then produce him cabbages, kohl rabi, and mangold wurzel, which may be given to his stock in winter without injuriously affecting the taste of the milk.   He will thus be enabled to continue his sales of butter at the time when it is scarcest and sells best.   Manure of a richer quality, and in greatly additional quantity, will be accumulated, its application will improve the productiveness of the well-drained land, and both the annual produce of each animal will be more valuable, and the number which the land supports will also be materially increased.   Every one will be benefited by the change :

the landlord, in the increased capital invested by the tenant, will have better security for the regular payment of his rent; the tenant, in the increased produce, will have a better return for his capital; and the labourer, in increased employment, a better market for his labour.

The fall in the value of Gloucester dairy produce has been very considerable; a cow, which yielded cheese and butter worth 9*l.* last year, producing not more than 6*l.* 10*s.* this year to the farmer. As he has nothing else to recompense him, and cannot, without the co-operation of his landlord, adopt any system by which to increase his produce, he must of necessity suffer severely from the fall in prices. He employs very little labour, and that paid for at a low rate. There is not, therefore, any mode of managing land in which rent forms a more important item in the cost of production than in dairy farming.

The wages paid to labourers in the Vale have fallen 1*s.* since last year, 7*s.* and 8*s.* being now the average weekly wages of a man; 6*s.* is sometimes paid; but, as we were significantly informed, 6*s.* worth of work only is given in such cases.

Many of the dairy farmers of the Vale have a small portion of tillage land attached to their farms. This is commonly cultivated in a three-field course, viz., fallow or roots, beans, wheat; or a "four-field," viz., fallow or roots, barley, beans or clover, wheat. The average crops of wheat vary from twenty-four to twenty-eight bushels; barley, forty bushels; and beans, twenty to thirty bushels per acre.

It is worthy of remark, that the farmers on the Cotswolds seemed to be a superior class of men to those holding land in the Vale. Their farms are larger, and they are men of greater capital; but it appears somewhat anomalous that the better soil, with the better climate, and paying the higher rent and the higher rate of wages, should be held by the less intelligent class of men, and under the more backward mode of management, while the worse soil, with a cold climate and a low rent, is possessed by men of intelligence and energy, and cultivated with

considerable skill.   The small size of the Vale farms places
them within the reach of a greater number of competitors than
the large arable holdings on the Cotswolds, and they accordingly
bring a higher relative rent, while the "convertible " character of
the hill lands calls forth a greater exercise of skill, than the
monotonous routine of an old-fashioned dairy farm.   It may be
mentioned that in both districts model farms have been established — that of Whitfield in the Vale, and the College-farm
of Cirencester on the Wolds.   Whitfield example-farm has
exercised a beneficial influence on the national agriculture ; and
though it does not admit of being copied literally in its own
neighbourhood, on account of the dairy character of the district
and the numerous obstacles to arable farming, yet there can be
no doubt that by degrees many hints will be taken by the surrounding farmers, whose prejudices will yield to the satisfactory
evidence of success.

Before closing our notice of Gloucestershire, we shall state the
opinions most prevalent among the farmers themselves as to the
best mode of remedying their present distress.   If prices continue permanently low, the dairy farmers look to a reduction of
rent as their only remedy.   All parties seem to think that rents
have been run up too high by over-competition, and that the
landlord must either, by expending capital in improvements,
give his tenant a better article for his money, or content himself with less rent for the article he gives.   Repayment for unexhausted improvements was strongly urged as the foundation
of a better system of husbandry, in which the tenant was said to
have as much right to be secured by act of parliament as the
landlord has in the possession of his land.   The effect of this
was expected to limit the competition for farms, by the necessity it imposes of a larger capital on the part of the entering
tenant, while it would encourage a liberal expenditure by the
man of capital, and thus in a great measure supersede the want
of means on the part of the landlord himself, or his want of

interest in consequence of possessing, as in many cases, only a life
tenancy in his estate.   The effect of this increased investment
of tenants' capital in cultivation, amounting probably to not less
than two years' rent, would, it was thought, afford as much security to the landlord for the payment of his rent as if the same
sum had been previously lodged in his hands.

LETTER VII.

DEVONSHIRE.

EXETER, Feb. 5. 1850.

THE surface of Devonshire is of an undulating character. In the lower part of the county, hill and vale succeed each other in constant variety of outline, clothed with rich verdure waving gently under the influence of genial breezes, while the higher district slopes up into bare smooth wastes, dotted over with huge blocks of granite, and swept by the humid blasts of the Atlantic. With such various soil and climate, the Devonshire farmer, according to his locality, practices nearly every branch of agriculture. Dairy, tillage, orchards, irrigated meadows, the breeding and feeding of stock, and the reclamation of waste land, each engage his attention. And though, in every district of the county, the cumbersome and unskilful practices which rendered Devonshire farming a by-word in the estimation of the great corn farmers of the eastern counties are still too frequently to be met with, yet, of late years especially, the practice of agriculture has made great progress in this county.

That progress is not ascribed by the tenants in any considerable degree to encouragement from their landlords. Till within the last two years they, with few exceptions, are said to have done almost nothing towards the permanent improvement of their estates, though they have not been slow to avail themselves of every increase which the tenant's capital and skill, as well as the general progress of the country, have added to the value of their farms. From these circumstances the rent of land in the better parts of the county has increased one third within the last twenty years, excessive competition having been encouraged by the system of letting farms by private tender. And hitherto the landlord not only extracted by this means a full rent, but also grew his crops of timber in the farmer's hedgerows and at the farmer's expense.

There are two classes of farmers in the county, one consisting of men with small holdings, little elevated above the condition of the labourer, the other of educated agriculturists holding large farms, into which they have introduced improved methods of husbandry. By them draining has been introduced, and the levelling of hedgerows and enlargement of arable fields; the system of irrigated meadows has been extended, and the application of artificial manures practised. The improvement of the breed of Devon cattle, now one of the most shapely, graceful, and profitable breeds in Great Britain, has been by them brought to its present high state of perfection, — the names of Mr. George Turner of Barton, and the Messrs. Quartly of Molland, holding in that branch the most eminent position. To the exertions of this class is due much of the progress which the county has of late years made in agricultural improvement, the small farmers profiting by the example which their richer and more intelligent neighbours set before them. That this has been rapid we may show by mentioning that in one parish in North Devon there are now 800 acres of green crop raised, where, only eight years ago, there were not more than 80.

Circumstances have, however, in a considerable degree, prevented landlords from undertaking improvements which would have raised the value of their estates. Life leases were formerly very common, having been granted by landlords of settled estates as a mode of raising money partly at the expense of their successors. These were given at very low, or almost nominal rents, the tenant paying a fine or sum in hand equivalent to about eighteen years' purchase of the real value of the property. To raise this sum, the tenant was generally obliged to borrow from others on terms exorbitant in proportion to the uncertainty of the security, leaving himself without capital adequate to the management or improvement of his farm, and therefore incapable of developing its resources. During the life of the tenant the landlord of course had no interest in encouraging improvement, and, as many farms in Devonshire are still held on such unexpired leases, the evil effects of the system may be seen in the wretched management of such estates, and the poverty of their tenants. Life leases are now replaced as they fall out by leases of seven to ten years, for large farms, and of six years for small. Improved management has followed this change, and though the farmers consider a term of seven or ten years too short, they prefer that to the yearly tenures so common in other parts of England.

The rent of land appears high as compared with that of other counties in which the same competition is not encouraged, though the mildness and salubrity of the climate may render soil apparently of equal quality more productive to the farmer of this county. Within a circle of three miles round the city of Exeter, on a fine deep soil, well adapted for the growth of corn and green crops, rents vary from 30s. to 50s. an acre, exclusive of accommodation land, the local rates or "outgoings" amounting to about one third more. The poor rate varies exceedingly in amount, according as the labouring population happen to be conveniently situated for employment.

The Devonshire tenant being at once a dairy farmer, a breeder or feeder of cattle, of sheep, and of pigs, and a grower of corn and of cider, this variety of occupation, arising naturally from the character of the climate and soil of the county, as already described, has given to him a tone of intelligence and activity which one looks for in vain in a district like the Vale of Gloucester, where a monotonous routine narrows the intellect of the dairyman. Farms generally are of moderate, or even of small size, and, although individual farmers may hold 600 or 700 acres in several separate farms, the great majority run from 50 or 60 to 200 or 250 acres. Farm buildings are very often found collected in a village, the housing of four adjoining farms being placed at their point of junction, which, of course, must also be the extremity of each. In an earlier age of the country this arrangement may have been necessary from motives of self-defence or the pleasure of society, but now it is attended with inconvenience, and must increase the cost of production by waste of power in the cartage of produce. The buildings are of every variety of character, from the antique and dilapidated to the more modern and convenient form. On badly managed estates the farmer is frequently bound by covenant to uphold in repair the most rickety old mud and wooden houses at a cost to himself of as much as ten per cent. on the rental of a small farm; and we were assured by one respectable farmer, on the Bicton estate, that often in a windy night he got up and looked out with dread, lest his live stock should be destroyed by the whole fabric being blown down on them. Having ourselves inspected these buildings, we were at no loss to account for the apprehensions of the tenant. Accommodation of any proper kind for the accumulation and preservation of manure is in such cases out of the question, though where there is plenty of straw the cattle seemed warm and comfortable.

The better class of buildings are generally in the form of a square, close all round, and entered on the south side through a large arched door under the granary. Immediately opposite is the barn, cider-cellar, &c., which usually occupy one side of the square, having the rick yard behind. Two sides are for the accommodation of cattle, the back walls being built close up to the eaves; the front is in two stories, supported on strong

E 2

posts of timber, open from the ground to the eaves, the lower story occupied by cattle, the upper kept as a store for their provender.   The cows are usually kept in loose boxes; the fattening cattle are tied by the neck.   The fourth side of the square embraces the farm stable and waggon shed.

The soil, as already mentioned, is of very various character, good turnip and barley lands being intermixed with wet, stiff, unkindly soil, generally very imperfectly drained.   Tracts of fine green crop land, of deep friable texture, are met with in continuous succession, and there the cultivator reaps the best returns. The alternate system of husbandry is followed, varied by allowing the land to rest one or more years in grass, as may be thought best by the farmer.   The four or five field course would be an inappropriate term to use in this county, the number of hedgerows and fields on each farm being so great, that a forty or fifty field course would more truly designate its husbandry.   To give an idea of this, we may mention one case related to us by the tenant of a farm of 160 acres, from which seven miles of hedgerows were removed, and, on the ground being measured, it was found that thirteen acres of land were gained by their removal.   Nor is that farm by any means bereft of its hedgerows, the fields still being on the average not more than ten acres in size.   Where the ground is exclusively kept in pasture, the shade and shelter afforded by the trees may in some degree compensate for the injury they do.   But where the land is not adapted to continuous pasture, and is therefore let by the landlord for tillage, there can hardly be anything more injurious to the tenant than this multiplication of fields and hedgerow timber. Every operation of husbandry is impeded, a constant shifting of implements from field to field occasions waste of time, and does positive damage to the implements and the fields through which they are passed, while the corn in harvest is frequently injured by the difficulty, in this comparatively moist climate, of getting it aired in these small close fields.   But the direct injury sustained by the crops being robbed of their food by the hedgerow timber is most strongly shown in a turnip-field. "Look at these thieves," said a farmer to us, pointing to three stately elms; "I warrant there is not a root as large as an apple within many yards of them." On measuring the ground, we found that for forty feet out into the field from the hedgerow opposite these three elm trees the crop was quite diminutive, while immediately beyond their influence the size of the turnips was quadrupled. An idea of the positive injury thus inflicted on the tenant may be guessed at, when we mention further, that forty feet extended across about a fifth of the field, from the opposite side of which another line of trees spread their voracious roots. From one hedgerow pointed out to us, extending along one side of a five-acre field, sixty large elm trees were cut last season, and there still remained a superfluity. Till within the last two years very few landlords would listen to any complaint on this head (some are loth to do so still); but the value of the timber has fallen so much, and the tone of the farmer has become so stern, from the pressure of low prices, that less difficulty is now felt in getting leave to remove superfluous hedgerows. Their existence may be considered rather a fortunate circumstance at present, inasmuch as the value of the timber, and the amount of land gained in their removal, will amply compensate the expense, while the employment afforded in the operation will take up any superfluity of labour.

There is nothing particular to detail in the management of the arable land, which, however, is well and deeply tilled, not very heavily manured, but managed, on the whole, where the tenants have sufficient capital, with much sagacity and skill. Two-horse ploughs are universal, and light carts and waggons. Oxen are occasionally used in the plough, two young and two old ones being yoked together. They are fed very cheaply, and will plough an acre a day. Sixteen to twenty-four bushels of wheat may be reckoned an average produce for South Devon, and thirty-two bushels of barley. Stubble turnips are occasionally taken, but the general practice is a bare winter fallow in pre-

paration for a root crop. In many districts of South Devon
the soil and climate seem admirably adapted for crops of early
potatoes, to be followed by turnips; but we did not hear of
this as in any degree a systematic practice. The mildness and
moisture of the climate would also indicate the advantage of
taking crops of rye, winter vetches, &c., for spring food, but the
attention of the farmer does not appear to have been much
directed to this, very possibly on account of the irrigated
meadows having hitherto afforded such food in sufficient abundance. Nor does it seem that early lambs form much an object
with the farmer, though it might be expected that such towns
as Plymouth, Exeter, Torquay, Dawlish, and other watering-
places on the coast, occupied by rather a wealthy class of resi-
dents, would have afforded a good market for such delicacies.

Devonshire is justly celebrated for dairy management, the
perfect cleanliness and freshness of the dairies we examined
forming a marked contrast to what we saw in some other counties. Where the best butter is made, the cows are fed in winter
on fine meadow hay, seldom any roots. They are turned out
daily for an hour or two, generally to the orchard, where the
trees afford them shelter; their dung benefits the trees, and
they are kept off the pastures, which would be injured by their
feet in wet weather. Cabbage is seldom given for the winter
food of the cows; Mangold occasionally is. Dairies are
frequently let by the farmer to a dairyman, who pays for the
use of the cow, which is fed at the farmer's expense, and
managed by the dairyman. A sufficient supply of grass and
hay is provided, and with this the common rent paid by the
dairyman for each cow is 9l. a-year, having fallen 1l. since
last year. Fresh butter and clouted cream are the products of
a Devonshire dairy. Fatting cattle are supplied with 4 or
5 lbs. of cake daily, in addition to cut turnips. Sheep are bred
and fed on most farms, the young sheep being sold fat when a
year old. They are occasionally supplied with cake when
feeding on turnips. The lambing ewes are placed on the

watered meadows. Pigs are fed on every farm, and are well managed.

The value of watered meadows is highly appreciated by the Devonshire farmers, advantage being taken of every little stream to increase the produce of the land. The warmth of the numerous valleys in this county is highly favourable to rapid growth, and the declivity of their sides affords a cheap and most convenient means of laying on the water. The expense of cutting the gutters where the land is suited for "catch meadow" may be about 2*l.* per acre, and the annual cost of keeping open the watercourses and laying on the water about 5*s.* an acre. The increased produce is fully 100 per cent., but this depends mainly on the quality of the water applied, which is found to vary extremely. Its value is believed to depend partly on the warmth it affords to the soil over which it passes, and partly on the deposit it at the same time makes.

The cider orchard is another source of income to the Devonshire farmer, the value of which has decreased nearly a half within the last few years. An orchard produces ten to fifteen hogsheads an acre of cider, the selling price of which at present is 25*s.* to 30*s.* a hogshead, and the cost of preparing it 3*s.* to 5*s.* As much as 150 hogsheads are produced on some farms, the half of which is consumed by the farm labourers. On one extensive farm we were assured that 100 hogsheads of cider were annually drunk by the labourers, the consumption averaging three-and-a-half hogsheads per man! The wages of the labourer vary from 7*s.* to 8*s.* and 9*s.* a-week, with three pints to two quarts of cider daily, the men bringing in every morning their wooden bottle to receive their day's allowance. Taskwork is much encouraged, and under it better wages are earned. Little or no reduction has been made on the labourer's wages. Women are not employed at outdoor work.

In many parts of the county, where the land is reckoned too stiff for profitable green crops, it is necessary to have a bare summer fallow. That is usually made after grass, the land

E 4

being well ploughed early in winter, and four or five times turned over during the following summer, after which a dressing of lime is applied and the wheat sown. The produce does not average more than sixteen to twenty bushels. This is followed by oats, which are usually a good crop, probably fifty bushels an acre. With the oats the land is sown out to pasture.

In Devonshire we found among farmers an unanimous expression of opinion that prices must rise or rents be reduced. It was stated that many of the farmers were putting all the land they could under the plough for corn crops, and discontinuing the use of purchased manures, in order to get as much as possible out of their farms at the least expense, previous to giving up possession. This, however, was not the case where abatements were given, and it was generally admitted that if a fair deduction of rent was made, tenants in most cases would be able to withstand the present pressure. A general reduction of rents is, in point of fact, the great object which they are all now striving to accomplish. We found produce rents strongly advocated, several gentlemen giving it as their opinion that the relief afforded thereby would in most cases be sufficient to enable the farmers of Devon, who had sufficient capital and skill, to meet the times.

LETTER VIII.

DORSETSHIRE.

CONTRAST WITH DEVON — CHARACTER OF SOIL — VALUE OF DAIRY PRO-
DUCE — TENURE — CORN RENTS — LITTLE EXPENDITURE REQUIRED TO
BE MADE BY LANDLORD — MANAGEMENT OF ARABLE LAND DETAILED —
SINGULAR CUSTOM OF "WORKING" THE SHEEP — SYSTEM OF FARMING,
THOUGH APPARENTLY CAUSING FREQUENT REPETITION OF MANURE, REALLY
NOT ENRICHING — CHALKING — ITS EXPENSE — WATER MEADOWS —
SHEEP — MORTALITY AT LAMBING TIME ATTRIBUTED TO THE "WORKING"
SYSTEM — SUGGESTIONS.

DORCHESTER, Feb. 8. 1850.

PROCEEDING eastward from Devonshire along the south
coast, the change is rapidly made from the warm wooded valleys
of that genial county to the breezy open downs and bare
uplands of Dorset. The small inclosures, the green fields, the
numerous hedgerows, the frequent village, and the narrow
lanes, give place to a bare and undulating outline, where great
breadths of various crops are separated from each other by an ideal
line, running out at their highest points to the lofty downs still
untouched by the plough. The white flinty roads, marked here
and there by lines of cutting into the chalk substratum, can be
distinguished for miles, now dipping into a little valley bare of
trees, but clothed with the verdure produced by irrigation, then
winding over the long ascent of the smooth down, and again
descending on the rich, bare, open country which surrounds
Dorchester.

The contrast between the two counties is very great. Instead of tall and stately timber trees we find extensive copse-
woods, which serve the doubly useful purpose of furnishing the
farmer with hurdles for his stock, and the labourer with fuel.
The hedgerows, where they exist at all, are cultivated for their

septennial crop of hurdles and fire-wood, for at intervals of
seven years they yield a new crop.   The farms are very extensive, as may be seen by the large intervening space between
each homestead.   The soil is thin, the subjacent chalk frequently showing itself on the surface.   Beds of gravel and clay
are of frequent occurrence, but generally the soil in the central
parts of Dorset does not appear to have much natural fertility.
The rents, including rates and tithe, average from 16s. to 20s.
an acre, regulated in some measure by the extent of down
pasture attached to an arable farm.   That quantity varies
from one half to one third of the whole farm, and in recent
years has been decreasing, the downs, which yielded inferior
grass gradually undergoing a process of conversion into arable ; and the dairy stock, which was then combined with
sheep, now giving place to sheep altogether.   This change is
limited to the large holders of land, the smaller farmers of the
vales continuing to combine dairy with arable farming.   The
produce of the dairies being chiefly made into butter, the
Dorsetshire dairyman is suffering from the recent depression
less severely than the cheese-producing farmer of Gloucester,
the depreciation in the value of a cow's produce having been
stated to us in the latter case as from 9l. to 6l. 10s., while here
it has fallen only from 10l. to 9l.

The farms are chiefly held on yearly agreements, though
leases from seven and ten to twenty-one years are not uncommon.   Many of the largest and best farms are held under
lease, and though there is at present no anxiety on the part of
the tenants for long tenures, they generally express themselves
favourable to them, and attribute much of the improved agriculture of their district to the security thereby given to men of
capital and skill.   On the estate of Lord Orford the farms are
let on lease at corn rents, regulated every half year by the
market price at Dorchester for the preceding six months, of so
many bushels of wheat and barley.   This mode of adjusting the
question of rent has given great satisfaction to the tenants, but
the example has not been followed by any other proprietors in

the county. A temporary deduction of rent had been very
generally given.   Though some of the principal farmers are
men of property independent of their farms, yet too many have
been tempted to embark in holdings far too extensive for their
capital, and as low prices have crippled their means, they are
unable to continue that generous treatment without which these
hungry soils cannot be profitably cultivated.   Drainage being
unnecessary, and, owing to the present system of management,
extensive housing for cattle not being deemed requisite, the
landlords of this district are not called on for any heavy expenditure; and the tenant, though bound by his agreement to a
certain routine of cropping, is not much interfered with by his
landlord.   There is, therefore, a good understanding between
the two classes in this county.

The chief characteristics of the farming of Dorset are the
breeding of sheep, and the folding of them to enrich the ground
for the production of corn.   Its thin chalk lands yield naturally
a very scanty herbage, but when tilled and well manured, the
alternate crops being eaten on the ground by sheep, they bring
good returns to the cultivator.   The tillage farms, as has been
already mentioned, are extensive, and most of them include a
range of down or sheep-walk, which is not permitted to be
ploughed without the consent of the landlord.   On this the ewes
are fed during the day in summer, and driven to the arable
grounds, where they are folded, without food, at night; thus
carrying to the arable lands, and enriching them with the
fertilising droppings derived from the grass.   The points
chiefly considered are the production of wheat and barley, and
to promote these the breeding and management of sheep, and
the other operations of the farm, are subordinate.   The maxim
is, the greater the number of sheep the greater the quantity of
corn; so that the course of cropping, though uniform in principle, is varied in detail, according to the greater or less extent
of permanent grass or down land which a farm contains.   Thus,
a farm consisting one half of down land and one half of arable, is farmed on the four-field system; while another, with only

one third of down land and two thirds of arable may be farmed
on the five-field rotation, both systems being thus made to afford
nearly an equal proportion of grass land for the feeding of
sheep.   Each method is carried out much in the same way, only
the "seeds" are left a year longer unploughed in the five-field
than the four.   Beginning with the wheat crop, the land, after
the second crop of clover has been eaten by sheep, is dunged from
the farm-yard and then ploughed, pressed with the furrow presser,
and sown, the wheat seed falling into the hollows made by the
presser, where it is covered by the harrows.   The more common
plan, however, where the furrow presser is not used, is to harrow
the ground after it is ploughed, and drill in the seed with the
Suffolk drill.   To give it that degree of solidity which on chalk
lands is found advantageous (and which is accomplished on a
similar soil in the Wolds of Yorkshire by the use of Cambridge's
or Crosskill's rollers), 700 or 800 sheep are driven backwards
and forwards over the field, the shepherd beginning at five o'clock
in the morning, and keeping the sheep on the ground three
hours, by which time they are found to have gone sufficiently
over about ten acres.   This is repeated day after day till the
whole is accomplished.   When the wheat crop is removed,
one half of the ground is ploughed and sown, part with rye,
and part with winter vetches for spring food.   The other
half is manured during the winter by the flock of sheep being
brought from the land on which they feed during the day and
folded there at night.   This part is in the ensuing spring prepared for swedes, which are sown in June, and is followed by
the sowing of white turnips on that portion on which the rye
and vetches have just been eaten off.   A few of the swedes and
turnips are taken to the farm-yard for consumption by the cattle,
and the rest are fed off by the sheep.   This is followed by barley, one half of which is sown with clover seeds.   The barley,
when ripe, is mown with the scythe, and carried and stacked
without being tied in sheaves.   After the crop is removed, the
half of the land which was not sown with seeds, is harrowed

with a heavy drag, and part of it sown with scarlet trefoil and part with rye. The first crop of clover is usually cut, the second eaten off, as also the rye and trefoil. After the rye, which is first consumed, the land is sown either with mustard to be ploughed in for manure, or with rape to be eaten off. It is then ploughed for wheat, which again commences the course. When the five-field course is adopted, the clover is pastured the second year; or a portion of the land is dunged and ploughed in spring, and sown with pease, which are followed by wheat, as before.

During this course the ground is repeatedly manured. First, the clover is fed off and dunged for wheat; secondly, the wheat land is trodden by sheep after being sown; thirdly, the wheat stubble, after being ploughed, is folded over by sheep, and the rye and winter vetches are eaten on the ground as a preparation for the turnip crop. The swedes are manured with ashes and artificial manures, and the greater proportion of the whole turnip crop is afterwards fed off in preparation for barley. When pease are sown the land is previously " *muckled*," that is, straw is carted out and spread over the ground, the whole of which is then folded with sheep, and the straw and manure ploughed in together to enrich it for the pease. This constant manuring might be expected to force good corn crops, yet the average produce per acre of wheat is said not to exceed twenty to twenty-two bushels, and of barley thirty-two bushels. An explanation of this may, doubtless, be found in the fact that the sheep are a breeding stock, never very highly fed, and that year after year the land produces and parts with a crop of wheat, barley, lambs, aged ewes, and wool, for which the moderate quantity of purchased manures brought back to the farm is certainly not an equivalent. The folding of the sheep on the arable from the pasture and down land no doubt enriches the former at the expense of the latter, but does not in any degree, on the whole, make good what is thus annually abstracted.

It is usual to chalk the land once in twenty years, the sour description of soil being that to which it is found most advan-

tageous to apply it.   The chalk is dug out of pits in the field to
which it is applied, and it is laid on sometimes with barrows,
but more cheaply with the aid of donkeys.   The first method
costs 40s. an acre, the last 35s., where hired donkeys are used;
20s. to 25s. where the donkeys are the property of the farmer.
The chalk is laid on in large lumps, which soon break down by
the action of frost and exposure to the weather.   Chalk is occasionally burnt and applied as lime, in which state it is preferred
by many farmers, notwithstanding the additional cost of the
burning.

The hollows or valleys in most of the large farms are occupied
by water meadows, which seem to be carefully managed, and
furnish very useful early food for the ewes and lambs, and hay
for them in winter.

The sheep generally kept are South-downs, which have, in
most instances, superseded the old Dorset breed.   The ewes are
now dropping their lambs, and are driven every night into a
warm littered yard with sheds all round, where they are carefully watched and tended.   As they lamb they are drawn with
their lambs into the sheds, and separated from the rest of the
flock.   Next morning they are put into a well-sheltered pasture,
where they are supplied with swedes on the ground.   In a few
weeks, they will be placed in the fold on swedes, when the
lambs are strong, and afterwards on rye and vetches.   The lambs
are weaned in May, and shorn when six months old.   The
" pur " or wether lambs are sold in autumn, as also the four-
year-old ewes.   The number of lambs annually reared from the
flock of ewes varies exceedingly, but scarcely ever averages a
lamb for each ewe.   From a flock of 700 ewes shown to us, the
shepherd had on one occasion gained the prize of his district for
rearing 670 lambs; but this was regarded as quite uncommon,
500 to 550 being reckoned a good produce.

The losses by death, both of ewe and lamb, in lambing are in
some years very great, the owner of the flock above mentioned
having assured us that in one season he lost as many as 300,

stating at the same time that he was unable to account for this.
We think that the daily driving of the ewes while in lamb from
the feeding ground to the fold and back again—often a mile or
more each way, the close and crowded state in which they are
nightly confined, the perambulation of the wheat fields, and the
other peripatetic uses to which they are applied in Dorset, very reasonably account for the mortality and scanty
produce described.   Previous to the introduction of guano
and bones, and more recently of cheap feeding stuffs for stock,
the practice of folding enabled the farmer to produce fair crops
of corn on land which could not then have been otherwise kept
in tillage with advantage.   The high value of the corn made
the care of the flock a minor consideration.   Free trade has
changed these relative values, and there can be no doubt that
the sons of the men who introduced the practice of folding will
soon adapt their management with equal skill to the different
position in which they are now placed.   The increased use of
artificial manures and bought food, the general introduction of
the turnip cutter, a greater economy of straw, and its conversion
into rich dung, will gradually change the present system of
breeding sheep into that of feeding out their produce also,
thereby increasing the annual return from stock, and, by the consumption of better food, adding fertility to the corn land.   Folding as a system will probably be superseded by attention being
devoted to the feeding of the flock as the principal object, the
enriching of the land following as a matter of course, but not
forming, as heretofore, an object paramount even to the welfare
of the stock.   It may then be found that to enrich the soil by
wasting the substance and injuring the constitution of the sheep
in driving them to and fro, and confining them in a crowded
fold, is a more expensive and less effectual plan than the direct
application of those manures and food which science and commerce have placed within the reach of the modern farmer.

Nor is this the only mode of increasing his returns which the
Dorsetshire arable farmer fortunately has still to fall back upon.

On these thin chalk lands we were surprised to find that, in order to consume their straw, the farmers carried it out and spread it over the soil, where, after the sheep had been folded upon it, it is left exposed to the weather for weeks or months before being ploughed in. Its economical consumption by cattle in stalls, with the aid of roots and cake or corn, will no doubt ere long be found by the farmer a much more profitable mode of converting it into a source of gain to himself and of fertility to the soil.

LETTER IX.

DORSETSHIRE — *continued.*

Mr. Huxtable's Farms. — distribution of liquid manure by pipes — hill farm—yield of crops — buildings — machinery — preparation of food for cattle — pig feeding — accommodation of cattle — sheep house — pig house — calving house — liquid manure tank — results — condition of dorsetshire labourer — clothing club — complaint of low prices.

SHAFTESBURY, Feb. 1850.

An account of the agriculture of Dorset would be very incomplete without some description of the farms of Mr. Huxtable. This gentleman, the rector of Sutton Waldron, by his pamphlet on Present Prices, has raised such a storm of reprobation among the farmers of the several districts through which we passed, that we were anxious on every account to examine his system, and obtain accurate information for forming a correct judgment upon it. It may be necessary to premise that Mr. Huxtable is a self-taught farmer, and that in carrying out his plans he has intrusted them to the direction of the people he found on the land, without calling to his assistance the aid of a skilled bailiff. The outdoor work, therefore, wants that finish which a tasteful farmer likes to see in the management of his land.

The West Farm, about a mile from Sutton Waldron, is the first on which Mr. Huxtable commenced his improvements. It is very various in quality, but chiefly a rather wet and tenacious soil, now drained, and all superfluous hedgerows and timber removed from it. This is strictly a breeding farm, keeping a stock of milch cows, the calves of which, when reared, are removed to the Hill Farm to be fattened. There is nothing peculiar about the management

F

of the stock here which will not be detailed in the description of
the Hill Farm, so that it is only necessary to call attention to
the plan adopted by Mr. Huxtable for the cheap distribution of
liquid manure over the different divisions of this farm.   The
whole liquid is carefully collected in a series of tanks, from the
lowest of which it is discharged as required, by a force-pump,
into pipes, which carry it to the several fields in succession.   The
pipes are of well-burnt clay, an inch thick, their joints secured
with cement.   They cost 7d. a yard, and, inclusive of an upright discharge column every 200 yards, will not exceed 1l. an
acre.   The pipes and columns are now laid down for the accommodation of 60 acres of the West Farm.   When it is requisite
to apply the liquid to any portion of these 60 acres, the force-
pump is set to work, and a stop put on at the discharge column
nearest the place to be watered.   A hose is then attached to the
column and carried into a tub placed on a light broad-wheeled
water-cart, which, as soon as filled, is drawn off, and another of
the same kind put in its place.   The first is then emptied by a
man with a bucket, who scatters its contents over the land.   By
the time he has emptied the first tub the second is full, and he
repeats the same process with its contents, and so on, the man at
the forcing-pump being thus enabled to deliver a continuous
stream of manure at the distance of many hundred yards.   We
think no apology necessary for occupying some space in describing this process ; for the application of manure in a liquid form,
fitted for immediate absorption by growing plants, is a matter of
the highest importance to the farmer.   The time may come
when all manures will be first prepared in the dissolving tank,
and then carried, in a liquid form, without waste or expense,
and applied without injury to the surface at any stage in the
growth of the plant which may be deemed advisable.

The Hill Farm is the most interesting, for here Mr. Huxtable
has most elaborately carried his science into practice.   A few
years ago this was an open chalk down ; it is bare and barren,
high and windy, rising abruptly from the adjoining vale to an

elevation of 500 feet.   Its cultivation was undertaken chiefly
with the object of increasing the field of employment in the
parish ; and not only has that object been accomplished, but
problems, having an important bearing on our national agriculture, are here in course of being solved.   The farm consists of
280 acres of land, all of which bear every year alternate crops
of corn and cattle food.   The corn crops are wheat and barley,
the former of which will average this year thirty-six bushels an
acre ; the green crops comprise white and yellow turnips, swedes
and mangold, the latter of which especially was a heavy crop ;
and also rye, vetches, clover, and Italian rye grass, though the
last is not in great favour on this farm.   Implements of every
kind for economising and perfecting labour are in requisition —
cultivators, scarifiers, clodcrushers (the most approved of which
is the smooth-ringed, as not being so liable to choke in damp
weather), seed-drills, dibbling machines, and liquid-manure-and-
seed drill.

At this season the operations on the land are not so interesting as those going on in the buildings — the meat and
manure factory of the farm.   These comprise an extensive range,
not altogether on the most convenient plan, as additions and
alterations are constantly being made when more room is re
quired, and when experience and observation suggest a better
arrangement.   They are constructed, however, with a strict eye
to economy, both of expense and labour ; and there is not a nook
about them which the critical eye will discover as either unnecessary or very inconveniently situated.   The whole stock of the
farm, except the breeding ewes, are kept constantly housed
night and day, summer and winter, and no particle of their food
or manure is suffered to be wasted.

Beginning our description with the steam-engine : it thrashes
and winnows the corn, cuts the thrashed straw into chaff, turns
the stones for grinding the cattle-food into meal, and by a
separate belt, when requisite, works a bone-crusher, in which,
also, the hard American oilcake is broken down.   Over the

furnace is a drying-loft, where beans or damp corn are prepared for the better action of the millstones, by the waste heat
of the engine fire.   The strawchaff is carried to the root-house,
where, by Moody's machine, turnips, mangold, &c., may be described as ground down rather than cut, and the roots and chaff
are then mixed together in the proportion, by measure, of one
bushel of the former to two of the latter.   The cut straw sticks
to the juicy fresh-cut roots, the whole exudation of which it absorbs.   This mixture forms the staple winter food of the cattle
and sheep, cake and corn being added in such proportions as are
deemed necessary.   The cut straw is not, even in this state,
thought so soluble as it should be, and a large steaming-chest is
being erected, in which the steam from the engine-boiler will be
employed in preparing every substance used as food to afford its
entire nutritive powers to the animal.   The mess so prepared will
consist of cut straw-chaff, ground roots, meal, oilcake or bran, and
crushed furze ; for Mr. Huxtable turns nature to account in all
her productions, and the scrubby furze, which is, except in Wales,
generally looked on as a nuisance, is here enlisted into the
service of adding to the nation's food.   After due inquiry, he
satisfied himself that, properly used, this is a most nutritious
substance.   It becomes, therefore, an object of careful cultivation, and when crushed and steamed in conjunction with other
materials, adds a flavour to the whole which, besides its nutritive
qualities, makes an extremely palatable mess for any animal to
which it is given.   Some people, no doubt, will say that steaming reduces the bulk of the roots, and is, therefore, a wasteful
plan : but in the process as here carried on, nothing is wasted ;
the bulk which leaves the turnip, swells and melts the chaff, and,
carried into the paunch of the animal in this state, cannot produce that chilling effect which a mass of roots, containing 90 per
cent. of cold water, must necessarily cause.   Such is the winter
food of the cattle and sheep.

Pigs are treated differently.   They are kept as a manure
factory, from which a given expenditure in meal will be returned,

with the cost of attendance, in the increased value of the animals, and all the manure they leave be clear gain. When this is reduced to a certainty, our supply of the richest manure will be limited only by the means at our command for purchasing pigs and corn; and it is right to mention that, at the present prices of both, a handsome profit was this year made by Mr. Huxtable over and above the manure. The pig food is therefore all purchased exclusively on their account, partly in the market, and partly from the inferior corn of the farm. Cheap Egyptian beans, lentils, and barley are ground into meal, the proportion of beans being increased in cold weather, and barley in warm weather, as being then respectively most suitable to the constitution of the animal. The requisite quantity is steeped over night in cold water, to render it more palatable and soluble, but undergoes no other preparation.

Having described the machinery for the preparation of the crops and the management of food, we shall now proceed to the buildings in which the different kinds of live stock are accommodated. The average stock of cattle kept is thirty milch cows and their calves, the whole of which are constantly housed, the younger being promoted from stall to stall as their elders depart under the butcher's charge. From 90 to 100 head are thus regularly kept on the farm. The cattle are all tied up in stalls, occupying three parallel rows, have plenty of light and air, and exhibit, notwithstanding constant confinement, the greatest liveliness and contentment. To economise as much as possible every particle of straw, they are all placed on sparred boards raised six inches above the water-tight floor; by this arrangement the straw is kept dry, and fully half the usual supply of litter is saved. The liquid is collected in an underground drain, whence it passes off to the tank. The cattle-house is very cheaply constructed, the walls being of wattled furze, which admits air without producing a draught, and the roof is thatched with straw, as being not only more economical in first cost, but far better adapted than a slate or tile roof to insure an equable temperature, being warm in winter and cool in summer. This point is well worth consideration, as we were assured that under one part of the buildings which is slated, the stock are much annoyed in warm weather by flies, while they scarcely ever make their appearance under that portion which is covered with thatch.

The sheep-house comes next under observation. It is a light, cheap, thatched building, with a walk up the centre, and a double row of sheep, standing on sparred boards, and tied up by the neck, quietly feeding on either side. Mr. Huxtable first tried his sheep in small pens on boards, but they never became so quiet and docile as they do after being tied up for a day or two. The progress of the sheep is tested at intervals of a fortnight or more by placing them on the weighing machine; and it has been observed that, from some unexplained cause, they at certain periods seem to make little progress as compared with others. When this is noticed, the house is carefully washed with chloride of lime, which seems to have an invigorating effect on the animals. They are found to thrive and fatten rapidly under this system of house feeding. The solid manure is removed daily, and the liquid passes off by a covered drain. No litter whatever is required by the sheep.

The accommodation for pigs next demands attention. Two methods are adopted, the one most approved being a long low thatched building, with a walk up the centre, and divided into compartments on each side. The pigs all stand on sparred boards, beneath which the ground is shaped like the letter V, for the collection of manure. The solid portion is removed at convenience, the liquid passing off by a drain to the tank. Eighty or ninety pigs are kept on the farm, and seemed in a state of perfect contentment with their quarters.

It is unnecessary to describe the root-houses and farm stable, but, crossing the road, we enter a building isolated from the rest, and divided into loose boxes, to which the milch cows are removed when about to calve. This house is also used as a quarantine, in which are kept for a few weeks any animals that are purchased, and which must first pass through this ordeal before being admitted to free pratique in the rest of the farm buildings. A little way further down is the establishment for the collection and preparation of manure. It comprises two extensive water-tight tanks for liquid, and a house in two compartments for the different kinds of solid manure. Here every element that is not carried off in the substance of the cattle, the sheep, and the pigs, is carefully preserved, to be again in due time restored to the soil, and the slender stream which constantly runs into the liquid manure tank is directed into a box filled with gypsum, through which it is passed, in order to fix the ammonia, described by Mr. Huxtable as "the spirit-like essence of the farm, ever longing and struggling to fly off into boundless air." It would occupy too much space to detail all the other processes going on on this farm, the dissolving of bones, the extraction of ammonia from rags, the conversion of earth by fire into an absorbent of liquid manure, — all these must be examined to be thoroughly appreciated.

The practical farmer will ask, — Does all this pay ? For crop 1847 Mr. Huxtable's books show a balance in his favour. For crop 1848 he was a loser of 26l. 9s. 8d. after paying rent and all expenses, inclusive of interest of capital ; but that arose, not from a deficiency of crop, but from the ruinous harvest which so seriously damaged the quality of all corn crops in the south of England that year. For crop 1849 a very profitable return is expected.

Neither of Mr. Huxtable's farms, as will be observed from our description, enjoys any advantage of soil or climate. The Hill Farm is in both respects inferior to the average of the arable farms of this county. We have thought it necessary to enter into these particulars, in order to satisfy the public that Mr. Huxtable is no mere theorist, and we strongly advise such as doubt his conclusions to visit his farm before they condemn them as impossible. They may there acquire a knowledge of important facts in their business, elicited by a truly philosophic mind, guided by quick perception of detail and energetic action, and applied to the good purpose of increasing the food of the people and enlarging the field for their employment.

The condition of the Dorsetshire labourer has passed into a proverb, not altogether just, as compared with the counties adjoining. The large farmers are anxious to vindicate themselves from the imputation of underpaying their labourers. Exceptional cases, they affirmed, had been taken as examples of the whole, and from these they had been unfairly believed to be heartless grinders of the poor. The labour books we examined showed that on the large farms the usual rate of wages for a labourer is 8s. a week, a piece of potato ground, fuel, beer in harvest time, with extra wages, and in some cases the principal servants have a house rent free. The fuel is brushwood and turf, which each labourer prepares for use himself, and which the farmer's horses carry home for him. The allowance of beer is a gallon daily for each man, which is usually consumed in the following manner: — a quart to breakfast at ten o'clock, a pint at half-past eleven for luncheon, a quart during dinner between one and two o'clock, a pint at four, with something to eat at five, and the rest when the work is finished. On a large farm the consumption of beer occasions a cost of 70l. or 80l. for malt in a year. The supply commences with hay harvest, and ends when the corn crop is secured. Women are paid 6d. a day, and boys 2s. 6d. to 3s. 6d. a week. On the smaller farms, where the tenants are poorer, and the population in proportion to the means of employment denser, the weekly wages are as low as 7s., and even 6s. ; and we were told that even that small sum was in many cases paid partly in inferior wheat, charged at a price which the farmer could not realise in the market. Low as the rate of wages is, it has not fallen in the same proportion as the price of provisions, and the Dorsetshire labourer is therefore at present more content with his circumstances than he was in times when the farmers enjoyed a prosperity in which he did not participate.

A clothing club, which has been established in the district of country round Blandford, has been of much benefit to the labourers, and promoted good feeling between them and their employers. It is supported by the joint contributions of the labourers and their employers, the subscription being 1*d.* per week for each member of a family, a labourer with two children subscribing 3*d.* weekly, and his master an equal sum. At the end of the year the labourer gets clothing for his family to the amount of their united subscription, such as he chooses to select. For this purpose 3000*l.* are said to be collected annually in this locality.

The farmers of Dorset complain very bitterly of the present low prices, and see no relief except in a return to protection. Being seldom disturbed by their landlords in the possession of their farms, and paying moderate rents, they expect little benefit from matters which in other counties are frequent topics of discussion, such as compensation for unexhausted improvements, a readjustment of the burdens on land, produce rents, and security of tenure. The only saving they think possible is in a reduction of wages; but from that source there can in the end be no gain, as no labour is more unprofitable than that which is underpaid. Very possibly a great saving might be effected by a better application of labour, and a more economical distribution of it.

LETTER X.

WILTSHIRE.

Devizes, Feb. 1850.

In viewing the agriculture of Wiltshire, it will be most convenient to examine it in those natural divisions which have given a peculiar character to the modes of husbandry adopted. The north-west division of the county, with its fertile vales and rich pastures, presents a marked contrast to the chalk downs which occupy the south-eastern and larger portion of the county, stretching from its southern boundary through Salisbury Plain to the vicinity of Devizes. In the former, dairy farming and grazing are the pursuits of the husbandman; in the latter, tillage and sheep farming.

The northern and western districts, which we shall first endeavour to describe, have all the rich and luxuriant appearance for which English rural scenery is celebrated. Green fields, with lofty hedgerow trees, winding roads which lead ever and anon over modest bridges spanning the devious streams that drain the country to the Avon, succeed each other for miles along its fertile valley. To the critical eye the green fields are often seen to be wet and undrained, and the luxuriant hedgerows more numerous than useful, though not so injuriously so as in the dairy districts of either Devon or Gloucester. Commodious farm buildings are seldom to be seen, but the want of them is

everywhere discernible in the poached pasture fields, and the
young cattle huddled behind the hedges for shelter from the
February blasts.

The farmers are very sensible of these defects, and are beginning to clear out useless fences, and to adopt a more permanent mode of drainage than the wedge or turf. Tileries have
been introduced, and pipes supplied by the landlord; but, as
the tenant is usually left to put them into the ground, at
his own discretion and cost, the work is not efficiently done.
This is a work which requires the skill that is learned by practice, and every landlord would find it to his advantage to have it
done under the supervision of a practised drainer, in order
that so costly an improvement may be attended with its full
advantages.

The farms vary in size from 60 to 250 acres, held from year
to year by men who, on the smaller and more numerous class of
farms, manage, by the help of their families, to dispense with
hired labour. The rent of land ranges from 30s. to 60s. per
acre, with the "outgoings" or local burdens in addition, which
in this part of the country are seldom less than 10s., and in some
cases as much as 20s. per acre.

There is much complaint among the farmers of the severity
of the rates, which a variety of causes has produced. The
village inhabitants of these districts are principally a decayed
manufacturing population, among whom handloom weaving and
pillow-lace working still keep a languid existence. A man
who is a weaver himself, or descended from weavers, is not held
in much estimation as a farm labourer; and in this grazing
district, where the demand for labour is not great, the land
must bear the burden of a population not required for its cultivation,—those manufactures on which they formerly subsisted
having ceased to afford them support. The surplus labour is
not divided among the farmers by mutual agreement, as is
common in other districts, where the tenants are men of capital.
The fear of this pressure, aided by the present law of settle-

ment, has induced large proprietors to diminish the cottages on their estates, and thus the burden is increased on those open parishes to which the population is driven. In the union of Melksham some of the parishes have no paupers, having cast their labourers off upon their unfortunate neighbours, where property being more divided and cheap, cottages are run up on speculation, and the new comers are welcomed by a certain class whose property is thereby enhanced, to the heavy loss of the ratepayers. The tyranny practised on the poor labourer when he falls into arrear to his new landlord, is great. This man, often the keeper of a huckster shop where the labourer gets his various wants supplied, charges every article he sells at exorbitant prices, from which there is no appeal, as, if the labourer leaves his residence, he cannot get another.

The highway rates are another serious cause of complaint. An unusual number of turnpike trusts have been created; and the money having been recklessly expended, and without system, the country is covered with toll-gates, not so much to raise funds for keeping the roads in repair, as for the purpose of paying the interest of former debts. From Melksham to Bradford and back by Troubridge, a distance of only thirteen miles, the traveller must pay his way through seven gates ! The parish of Broughton Gifford is closed in on every side by gates, and yet, notwithstanding the number of toll-gates throughout this division of Wiltshire, the tolls collected hardly pay the mortgagees' interest, and the farmers are called upon for heavy highway rates over and above to keep the roads in repair. The enhanced expense of transit in an inland district, caused by this gross mismanagement, is a very serious hindrance to agricultural improvement, and the great landlords are much to blame for having permitted such an abuse to grow up without challenge. If the expenditure of the money had been under the control of those who paid it, the result would have been very different.

But in the relations of landlord and tenant in this part of Wiltshire, matters are permitted to take their own course. The

law says the tenant shall pay his rent and all rates; and the
landlord takes his rent, and gives himself little further concern
about the matter.   Fields are undrained; farms inadequately
housed; the tenant complains; his landlord tells him that he
has not raised his rent in good times, and will not lower it for
a temporary depression.   There is no progress, and no encouragement given by the landlord to a tenant so disposed.   To
this there are several exceptions, but these exceptions only
illustrate the rule.

The farm-yards have a few shelter sheds of the most rickety
description, and never ample enough to accommodate all the
stock of the farm.   Convenient arrangement has not been attempted; a rough wood and thatch shed being run up now and
then, as the necessities of the tenant most required.   The dairy
cows are kept during the winter in open yards, wading often to
the knees in filth, there being no straw to spare for litter.
Here they stand shivering at the old-fashioned racks where
their scanty provender is supplied.   The young cattle, having
no sheds provided for them, are wintered in the fields.   Manure
is looked upon as a troublesome nuisance, there being no means
taken to preserve it, and much of it is accordingly washed off
by the winter rains into the nearest ditch or watercourse.   Such
is the general rule, though on some estates a great improvement
is now taking place in the farm buildings.

The dairy farmers in the richest parts of the county do not
breed their own stock, but buy heifers in-calf from the Cotswolds and at Swindon fair, which they keep as milkers for four
years, and then sell them fat if possible.   The price of such
stock used to be 17*l.* or 18*l.*, and when sold lean 12*l.* or 13*l.*,
but it has lately fallen 15 to 20 per cent. from these rates.   Of
good grass land, two and a half acres are reckoned sufficient to
support a cow throughout the year; and to give an idea of the
quantity of stock actually kept, we found a milking stock
of forty cows on a dairy farm of 120 acres.   The cheese
made in this part of Wiltshire is of very superior quality,

and brings the highest price.   It is sold as "Double Gloucester," though really Wilts cheese.   Each cow yields from 3½ cwt.
to 4 cwt. of cheese in the year, and 1 lb. of whey butter per
week, which is at present worth 7d. per lb.   In the winter
skim-milk cheese is made, and the best quality of butter, now
selling at 11d. to 1s.   The best cheese sells at 50s. a cwt., which
is a fall of from 10s. to 15s. on the prices of 1849.   The dairies are
exceedingly clean and well managed, and do much credit to the
industry and skill of the farmers' wives, to whom exclusively
this important department is entrusted.   Full-milk cheese
begins to be made soon after the calving season, which is the
end of March and beginning of April.

   The richest grazings in the vales and along the rivers' banks
are stocked with fattening cattle, to the management of which
the principal farmers devote their attention.—On the arable land,
where the soil is strong enough, roots and wheat are taken alternately ; while on lighter soil, clover, wheat, and barley or oats are
taken in succession.   As the extent of crop on any one farm is not
considerable, it is common for the farmers to hire drills and
thrashing machines, which parties keep travelling about the
country for this purpose.   The thrashing machines are driven
by hand, and employ four men and a boy, who think it a good
day's work to thrash out six sacks or twenty-four bushels of
wheat !   A fanning or dressing machine, which we saw at
work on a considerable farm, was quite a curiosity, and
evinces, as much as any thing else, the very primitive state of
husbandry in this district.   It was simply a series of spars
nailed at each end to circular pieces of wood, and thus forming
an open cylinder.   To each spar a deep fringe of sacking was
attached, and the apparatus being turned round by a man at
each end, got up a breeze of wind, in which a third riddled the
corn, and so separated it from the chaff!   The same primitive
implement is still used in some parts of Surrey.

LETTER XI.

WILTSHIRE — *continued.*

SALISBURY PLAIN. — NATURE OF SOIL — CONVERSION OF PASTURE TO ARABLE ENCOURAGED BY COMMUTATION OF TITHE — CONSEQUENT INCREASE OF RENT — CHANGE FROM SHEEP TO CORN FARMING — LARGE FARMS — MODE OF MANAGEMENT — TASK-WORK — CONSUMPTION OF BOUGHT FOOD AND MANURE — FARM BUILDINGS — MACHINERY LITTLE USED FOR FEAR OF DISPLACING LABOUR — LOW WAGES — DIET OF LABOURER INSUFFICIENT — SUBDIVISION OF VERY LARGE FARMS BELIEVED TO BE INEVITABLE — FARMERS' OPINIONS AND PROSPECTS.

SALISBURY, Feb. 1850.

SOUTH Wiltshire includes the extensive district stretching from Salisbury to within a few miles of Devizes, called Salisbury Plain. This is a somewhat elevated expanse of chalk downs, of an undulating character. When seen from a distance, it appears a vast uninhabited tract, with slightly swelling slopes; but as the traveller passes through it he discovers that it breaks down into numerous valleys, nestling among the green meadows of which are the farm-houses and labourers' cottages, the parish church, and sometimes the well-wooded park and mansion of the lord of the surrounding manor. These are the sheltered spots of the district, for along the open downs the cutting blasts of winter meet nothing to intercept their severity. Here and there on the horizon a strip or clump of fir trees may be seen, but the face of the country is bare and unsheltered, with no fence dividing field from field, and no prominent landmark to guide the traveller, who must trust to the finger posts of the different roads which intersect the country.

The soil is of various character, all on a chalk substratum. On the hills it is mixed with flints; the sides of the hills are a chalky loam, the flatter parts a flinty loam, and the bottom of the valleys which drain the district consists of the *débris* of

the adjoining downs.   It is a thin, dry soil, well adapted for the
system of folding sheep, and hitherto kept in cultivation by a
diligent prosecution of that system.

The greater proportion of this extensive tract has been
brought under tillage since the passing of the act for the
commutation of tithes.   The fertility of the most of it is artificial, the result of capital and labour skilfully applied; and, as
the country is not fenced, requires no draining, and the sheep-
folding involves no expenditure in buildings, it appears that the
increased produce derived from the land is almost wholly the
result of the tenant's exertions.   The commutation of tithes was,
therefore, a great boon to the landlords, as their tenants then became desirous to plough up the down-lands, and obtained permission to do so on the condition that they should pay an
increased rent.   In this way down-lands not worth more in their
natural state than 3s. 6d. or 5s. an acre were at once raised to
15s. or more, and that without any outlay on the part of the
landlord.   The land was held in very large farms by men of
capital, whose chief dependence was on their sheep stock, and
who, occupying wide tracts as sheepwalks, became gradually very
extensive tillage farmers, willing to pay an increased rent for the
right of converting downs into arable, so long as they were encouraged to do so by a high price of corn.   This change of
system involved a greatly increased outlay of capital, for it is
obvious that a man with sufficient means to stock and carry on a
sheepwalk of 2000 acres would find that very inadequate for an
arable farm of the same extent.   It is to be feared that many were
tempted by high prices to embark a large amount of borrowed
capital, in the full expectation that those prices would be permanent, and the pressure upon such farmers at present is very
severe, as the landlord has abated only 10 per cent. of his greatly
increased rent, and the lender of money is very possibly pressing
the borrower, in the fear that his capital as well as the tenant's
may soon be absorbed.

The size of arable farms on Salisbury Plain varies from 800

up to 5000 acres, cultivated fields being often two or three
miles distant from the homestead.   A portion of this on most
farms consists of unbroken downs, till now undergoing an
annual diminution.   The extent of the operations here may be
indicated by the fact, that one farmer has 800 acres in wheat
annually, and that 400 or 500 acres of corn, and 200 or 300
acres of turnips, on one farm, are not uncommon.   The returns
are calculated on the gross; little items, which engage the
anxious attention of the small farmer, being here thrown over-
board altogether.   The farmers, in fact, who generally hold on
lease, are a superior class of men, renting the sporting of the
manor as well as the land, and in many cases occupying the
manor-house, and holding that position which, from the non-
residence of the owner, devolves upon them.

The sheepfold and artificial manures are looked upon as the
mainstay of the Wiltshire-down farmer.   The system of husbandry pursued is precisely similar to that already described by
us as practised in Dorsetshire.   When the downs are first broken
up, the land is invariably pared and burnt, and then sown with
wheat.   Barley is usually taken after the wheat, and this is
followed by turnips, eaten on the ground, and succeeded by
wheat.   The first three or four crops on the fresh land were
very remunerative, and formed a great temptation to the
farmer; but the soil is soon exhausted, and requires expensive management afterwards.   It then falls into the usual
four or five field course; a piece being laid out annually in
sainfoin, to rest for several years before being again broken
up.   The sheepfold is shifted daily until the whole space
required to be covered is gone over.   To economise labour,
much of the land is " raftered," or half-ploughed, one furrow
being turned over on an equal space of ground, the two surfaces
of which are thus partially rotted together.   Turnips and other
green crops are consumed where they grow, which saves the
labour of taking home the crop and fetching back the manure.
The sheep, as in Dorset, are made the manure-carriers for any

G

portion of the land on which it is thought desirable to apply it.
Much of the corn crop is stacked in the distant fields, as it
would be almost impossible to carry it home so far with the
despatch which is requisite in harvest operations.   In many
cases it is thrashed where stacked, a travelling steam thrashing-
machine being hired for the purpose.   The straw is then carried
out, and spread over the grass lands from which the clover-hay
had been cut the previous year.   Only a very small proportion
of the root crop is carried home for consumption by cattle, the
number of which on these large farms is quite inconsiderable.

According to the quality of the land, and the amount of
artificial manure and food expended on it, the corn crops vary
in yield from twenty to twenty-eight bushels of wheat, and
forty bushels of barley, per acre.   Wheat is sown from
October till the end of December; spring-sowing being dis-
approved of, as on these high lands a late crop is very subject
to blight in harvest.   The dry cold March winds, which
sweep over the unsheltered downs, often inflict serious injury
on the young wheat plant.   Barley is sown in March and
April, grass seeds being sown at the same time, and clover seeds
in the month of May.   If sown earlier, they would vegetate
too rapidly, and starve the barley crop.   Swedes give the best
crop when sown in the end of May, and white turnips succeed
winter vetches and rye, as soon as the ground can be got ready
in July.   Wheat harvest begins, generally, in the second week of
August.   It is reaped chiefly by strangers from the more populous districts in North Wilts, by task-work, the price varying,
according to season and crop, from 7s. 6d. to 10s. an acre.   Barley is mown by the regular labourers of the farm.   The turnip
crop is hoed by men at task-work— strangers who migrate into
the district every season, and work late and early, that they
may earn good wages.   They are paid from 7s. to 8s. an acre,
besides beer ; and expert hands can make 3s. to 4s. a-day at this
rate, beginning work, of their own accord, at three o'clock in
the morning, resting during the mid-day heat, and after resuming

labour in the afternoon, continuing till eight o'clock in the evening.

The artificial manure most in favour for the turnip crop is superphosphate of lime, which some farmers prepare for themselves by mixing the bones and sulphuric acid on their own farms. Three cwt. of superphosphate is reckoned a sufficient application for the turnip crop. It is usually drilled in along with the seed, the same machine sowing both. Guano does not appear to be much used on the downs. On farms of such great extent, the sum expended on artificial manures and oil-cake is necessarily large, though greatly contracted within the last two years. On a farm of 2000 acres the sum at one time expended on these substances reached 1100*l.* in one year; it has now fallen off about a half, the farmer fearing that a continued low range of prices, with no corresponding abatement of rent, will oblige him to leave the farm; and there being no custom here to pay an outgoing tenant for unexhausted improvements, he has of course no interest in maintaining the condition of the land for the benefit of his successor. The land is chalked once in twenty years.

The management of the sheep stock is much the same as that of Dorset, this being also a breeding country. The South Down is the favourite breed, and much attention is paid to the improvement of the flock. On a farm of 2000 acres of average land which we visited, 1400 ewes are kept, from which 1000 or 1200 lambs are reared, several hundred lambs being annually lost, and many ewes at lambing time. The quantity of cattle at present kept on this farm may be about thirty or forty in the straw-yards, and ten fatted oxen, with a few milch-cows for the use of the farm. There are also eighteen working oxen and twenty-two farm horses. About 500 acres of corn are grown annually, and 250 acres of turnips.

The farm-buildings comprise a stable for the horses, and, at a different part of the farm, loose boxes for the working oxen, an engine-house, one enormous barn, and two smaller ones, a

couple of sheds with yards, and a waggon-house. The steam-
engine is used to drive a small thrashing-machine, which merely
beats out the corn without separating it from the chaff. It is
also employed in cutting chaff and bruising corn for the stock
of the farm. The wheat only is thrashed by the machine, barley
being thrashed by the flail, partly to give increased employment,
and partly because the maltsters prefer it hand-thrashed. It is
then winnowed by hand-fanners, and the awns are knocked off
in another hand-machine, worked by two men. But this necessity, which the Poor Law imposes, of giving employment, is a
heavy tax on the farmer, as the whole of these processes could be
quite as effectually performed at one operation by a good thrashing-machine, and certainly at one-half the cost. The buildings,
being generally thatched and the walls of timber, are very expensive to keep in repair.

The wages of labour are lower on Salisbury Plain than in
Dorsetshire, and lower than in the dairy and arable districts of
North Wilts. An explanation of this may partly be found in
the fact, that the command of wages is altogether under the
control of the large farmers, some of whom employ the whole
labour of a parish. Six shillings a-week was the amount
given for ordinary labourers by the most extensive farmer in
South Wilts, who holds nearly 5000 acres of land, great part
of which is his own property ; 7s., however, is the more common rate, and out of that the labourer has to pay 1s. a-week
for the rent of his cottage. If prices continue low, it is said
that even these wages must be reduced. Where a man's family
can earn something at out-door work, this pittance is eked out
a little, but in cases where there is a numerous young family,
great pinching must be endured. We were curious to know
how the money was economised, and heard from a labourer the
following account of a day's diet. After doing up his horses
he takes breakfast, which is made of flour with a little butter,
and water "from the tea-kettle" poured over it. He takes
with him to the field a piece of bread and (if he has not a young

family, and can afford it) cheese to eat at mid-day. He returns home in the afternoon to a few potatoes, and possibly a little bacon, though only those who are better off can afford this. The supper very commonly consists of bread and water. The appearance of the labourers showed, as might be expected from such meagre diet, a want of that vigour and activity which mark the well-fed ploughmen of the northern and midland counties. Beer is given by the master in hay-time and harvest. Some farmers allow ground for planting potatoes to their labourers, and carry home their fuel — which, on the downs, where there is no wood, is a very expensive article in a labourer's family.

Both farmers and labourers suffer in this locality from the present over-supply of labour. The farmer is compelled to employ more men than his present mode of operations require, and, to save himself, he pays them a lower rate of wages than is sufficient to give that amount of physical power which is necessary for the performance of a fair day's work. His labour is, therefore, really more costly than where sufficient wages are paid ; and, accordingly, in all cases where task-work is done, the rates are higher here than in other counties in which the general condition of the labourer is better. We found a prevalent desire for emigration among the labourers themselves, as their only mode of benefitting those who go and those who remain behind.

A subdivision of the large farms on the downs would tend to increase the demand for labour, and, with a low range of prices, such a subdivision appears inevitable. These thin lands cannot be kept in cultivation except by a liberal expenditure of capital and the utmost economy in the consumption of the produce ; and this is scarcely compatible with a holding of 2000 acres under one management. Very few men, even if they possessed it, would risk a capital adequate for the thorough development of such a farm ; and where men of this class are to be found, they would probably get a better

return by dividing their land into four or five farms of 400 acres
each, with separate bailiffs vieing with each other in the care of
the land under their charge, and answerable separately to the
capitalist farmer, who would superintend and direct the whole.
In the dairy and grazing districts the wages are from 7s. to 8s.
a-week.

The opinions expressed by the farmers as to what is requisite
to be done under present circumstances, and with future prospects,
were of a much more practical character than those we heard in
Dorsetshire.   In the dairy districts the farmers ask for drainage
and better house accommodation, relief from the unequal pressure
of poor-rates caused by the present law of settlement, and the
consequent obligation to employ the whole labourers of a parish
whether their labour is needed or not.   The income-tax is also
much complained of, being arbitrarily exacted even when the farmer is actually losing money.   This is thought an act of great
injustice; and it is not easy to see why the farmers alone should
be subjected to an arbitrary assessment, as it is not more difficult
for them to strike a balance every year in their accounts than it
is for a merchant.*   Indeed the necessity for doing so would
introduce a business-like accuracy of accounts which could not
fail to be beneficial to the farmer himself.

On the corn farms a reduction of rent is considered indispensable, or a conversion of money into produce-rents.   The
idea of a return to protection appears to be abandoned; and,
in the dairy district especially, it is readily conceded that
free-trade has much less seriously affected the farmers than
their brethren in the corn districts, though they think it
right, nevertheless, as one man said to us, to " bear their share in
the general grumbling."   With the large corn farmers, however,
the suffering is very serious, and much individual loss is unavoid-

---

* This has been amended by the legislature since this letter was published,
and a farmer who can shew that he has not cleared 150l. will now be
exempted from assessment.

able before matters readjust themselves.   Their claims on the
justice of their landlords are of the strongest kind.   As the
landlords, in the manner already explained, without any outlay,
obtained a large increase to their rental, and by so doing had
in some degree become partners in the scheme of extensive corn
farming, they, when through unforeseen causes it becomes un-
successful, cannot honourably withdraw without bearing the
same share in the loss as they drew from the profits of the adventure.   A deduction of ten per cent. has in the meantime
been generally allowed.

LETTER XII.

HAMPSHIRE.

GENERAL DESCRIPTION — SIZE OF FARMS — DEFECTIVE DRAINAGE — LAND
MIGHT BE PROFITABLY RECLAIMED FROM SEA — FARM BUILDINGS, VERY
INSUFFICIENT — WATER-MEADOWS — RENT AND PRODUCE NEAR SOUTH-
AMPTON — DETAILS OF MANAGEMENT — RENT AND PRODUCE OF CHALK
DISTRICT — MANAGEMENT OF SHEEP — AND OF PIGS — SUGGESTIONS FOR
FARTHER DEVELOPING CAPABILITIES OF SOIL AND POSITION — PROPOSAL TO
HAVE SLAUGHTER-HOUSES NEAR RAILWAY STATIONS FOR SUPPLYING LONDON
WITH FRESH MEAT — WAGES — HIGH RATE OF COTTAGE RENTS — FARMERS'
COMPLAINTS.

BASINGSTOKE, Feb. 1850.

IN the short space of a single letter it is difficult to enter into
a description of all the varieties of soil which a county possesses;
and for general purposes the leading characteristics are all that
it is requisite to describe. Hampshire may thus be divided into
two districts, — the southern, in which a soil, varying in depth,
rests upon beds of clay and gravel; and the central and northern,
occupied by the chalk formation to near the borders of Berkshire,
where a tract of woodland country, with the clay and gravel
substrata of the southern district, again presents itself. In the
south and north the country is well wooded, the woodland
scenery of the New Forest, which occupies nearly the whole of
the south-west corner of the county, remaining to this day a specimen of wild sylvan beauty. The central parts of the county
exhibit the bare landscape common to the chalk districts, the
trees being scanty and of stunted growth, the arable-lands in
large fields frequently unenclosed, and the wide refreshing streams
which drain the lower country being exchanged for meagre

rivulets, which very imperfectly supply the wants of the inhabitants.

In the chalk districts the farms range in extent from 500 to 1000, 2000, and even 3000 acres, while in South and North Hampshire they are from 200 to 500 acres. In both divisions they are commonly held on yearly tenures. On the clay soils, where drainage is required, it is usual for the landlord to supply the tiles, and the tenant to put them into the ground. In very many cases, however, the landlord leaves the tenant to do all or nothing, as he thinks best; and much of the country, where nothing but drainage is required to render the soil abundantly fruitful, is accordingly either very imperfectly drained or not drained at all. The low grounds lying along the banks of the principal streams in South Hants are very liable to injury from sudden floods, which, falling on the extensive chalk uplands, collect with great velocity, and in their course to the sea overflow their channels, and injure the meadows and fields which they overspread.

Along the coast there are numerous inlets of the sea, which at low water present a wide expanse of mud, poisoning the air with noxious exhalations, and which, from the land-locked nature of such creeks, might be very profitably embanked and reclaimed for cultivation.

The farm-buildings consist of a huge barn, and a few sheds and yards; the barn and sheds being constructed of wood and thatch. The barn is generally large enough to contain a stack of 300 bushels of wheat, which is about the usual size of stacks on the larger class of farms. The sheds and cow-houses are very inadequate in extent, and entirely without plan or convenience for the economy either of labour or food. The tenants are bound to keep them in repair, however expensive it may be to do so. In every respect, the present state of the farm-buildings in Hampshire is unsatisfactory: insufficient in point of accommodation, placed here and there more by random than on any

definite principle, constructed of materials so frail as to be in
constant need of repair, they show that the landlords have given
little attention to the wants of their tenants, and that their
agents have failed to perform a most important part of their
duties, when these duties are rightly understood.

The agriculture of that part of South Hants including the
area drained by the rivers Test and Itchen, which fall into the
Southampton Water, is of a varied character. Much of the
land adjoining these streams is in water-meadow, the greater or
less proportion of meadow attached to each arable farm giving
the distinctive character to the mode of husbandry pursued on
it. Formerly these water-meadows were very valuable, as they
enabled the farmer to rear the earliest lambs for the London
market, and then yielded him a crop of hay. The use of artificial food for rearing stock in less early counties has deprived
the Hampshire farmer of the monopoly of the lamb-market, while
the introduction of railways, and the cessation of coaching and
posting, have greatly curtailed his market for hay. Farming is,
therefore, somewhat in a transition state here, some of the
holders of land turning their attention more to the dairy, some
to a mixed system of dairy and feeding, but all, more or less, to
the feeding of sheep in winter and the rearing of early lambs.
In many instances the water-meadows are not now made the
object of so much careful attention as formerly ; and their very
scanty produce (in some cases little more than a ton of hay per
acre) indicates that the meadow is either ill managed, or not well
adapted for the purpose, and that in such cases it might be
turned to more profitable account by being drained and converted to arable-land.

The soil of many of the meadows is a peat and peaty loam,
while the adjoining land is a rich loam, of variable depth, with
a few flints in it, lying on a bed of gravel. The depth of the
soil varies from a few inches to several feet, and generally it is
all naturally drained by the substratum of gravel. The farms

are divided into large fields by convenient hedges, with a few trees here and there, which add to the beauty of the landscape, without doing any material injury to the farmer.

Strong clay land is occasionally met with, and farming there is very backward, little having yet been done in drainage, though that is now beginning to be attended to.   The rising ground is of a more sandy character, well adapted to sheep-feeding, while some of it is good green crop land, but on a "burning gravel," and subject to serious injury by a dry hot summer.

The farms vary in size from 100 to 450 acres, 300 being the average extent.   The rent ranges from 20s. to 40s. an acre, according to quality of soil and locality, and the rates and tithes are about a fourth more.   Fine fertile dry loam, close to the line of railway, and within a few miles of Southampton, costs the farmer, for rent, tithe, and all rates, from 2l. 2s. to 2l. 12s. an acre. The average produce of such land is 34 to 36 bushels of wheat, and 40 bushels of barley, per acre.   Clay-land in the same locality costs the farmer 26s. to 28s. an acre for rent, tithe, and rates.

The four-field system is recognised as the custom of the country, and is generally followed.   But where land is let on lease, the farmer is sometimes allowed great latitude, no restriction being made except for the last two years.   In such a case, he usually takes wheat after a portion of his root crop, as well as after his "seeds," substituting it for barley where the ground is early cleared by sheep, or where their treading in wet weather causes the land to turn up "unkindly" for barley.   Occasionally, when the summer proves too dry for the small seed of the swede to vegetate, the land is bare fallowed, and sown with wheat in autumn, the wheat in such cases being followed by barley. The "seeds" are usually covered with farm-yard dung during the winter, or early in spring ; and this insures a good crop of hay, besides enriching the land for the succeeding crop of wheat. Immediately after the first crop of hay is cut from the more clayey description of land, the ground is ploughed, well dragged, then again ploughed in autumn; and after being reduced by the drag and harrows, the wheat seed is sown with the drill. Where the land is suitable, however, a second crop of clover is taken, and the wheat sown on one furrow. Early in November is reckoned the best time for sowing wheat, spring-sown crops being somewhat precarious, though occasionally very successful.

Artificial manures are never applied as a top-dressing for any corn crops. In the vale of the Itchen the swedes are the best crops of that kind we have yet seen in any of the southern counties, but they want that regularity of size which is the sure test of a heavy crop. Two cwt. of superphosphate per acre is the common application to this crop, some adding a few bushels of rough bones, and occasionally a few loads of rotted dung. A heavier application of manure would, in our opinion, be found in the end a more economical practice, as rent, rates, and labour are nearly the same in amount per acre, whether the crop is good or bad; and an increase of 6 or 8 tons of swedes is therefore cheaply purchased by an extra expenditure of 20s. on artificial manure.

On the chalk-lands from Winchester to Basingstoke the mode of farming is very much like what we have already described as practised on similar soil in other counties, only that the quality of the land being generally somewhat better, and the face of the country rather warmer here, the acreable produce is greater, and the style of farming more generous. The rent varies from 10s., 15s., 20s., to 30s. an acre, the rates and tithes adding about a third more. The country surrounding Basingstoke is a fine fertile tract of dry arable-land, of no great depth, but laid out in large well-fenced fields, suited for the growth of all descriptions of crops, intersected by excellent roads, and having the convenience of a railway-station within a couple of hours of London. For this the average rent per acre, including tithes and rates, does not exceed 30s., the average produce being 34 bushels of wheat and 40 of barley per acre, though 40 of wheat, and 50 or

60 of barley, are occasionally reached in favourable seasons. The loss sustained here by the cold summer and wet harvest of 1848, which rendered the scanty crop nearly unsaleable by the injury done to it before it could be carried, is telling heavily on the farmer now. The four field course is the rule of the chalk district, to which there are partial exceptions.

Sheep-feeding is the sheet-anchor of the farmers, and it is carried on in a better or worse style, according to the means at their disposal. A few use cake and corn extensively, in addition to roots and green food, both summer and winter; but the great proportion of occupiers cannot afford to do so, and continue to feed their flocks on the green crops produced by the land, without aiding them even by the use of the turnip-cutter. A sheep to the acre all round, or 500 sheep on a 500-acre farm, are considered to insure good farming. Very few beasts are kept; indeed, there is seldom anything in the yards but the milch-cows for the use of the farm establishment, the work-horses, and a few pigs. Only a small proportion of the turnip crop, therefore, is drawn for consumption in the yards, the farmer depending chiefly on his sheep stock for manure and profit. The sheep are managed differently on different farms, some keeping them for rearing fat lambs, others for wethers. In many cases the shep herd has a hut beside the fold, which is moved about with it, and in which he sleeps at night, ready to turn out and give assistance should occasion require. The bells on the necks of a few sheep in the flock give him notice when anything disturbs them.

Pig-feeding we found in favour with some farmers, as being at present the best paying stock. One farmer we visited has from 40 to 50 breeding sows, which he keeps in a very cheaply fitted-up yard and sheds, feeding them on swedes alone till they are nearly about to litter. They are then placed in separate pigsties, and supplied with more generous food. The progeny are kept till worth about 20s. each, when they are sold. A young sow pig can be bought for the same price as a ewe; the ewe produces only one lamb in the year, while the sow brings on an average two litters of seven or eight in each, or 14 to 16 pigs annually. Hence our informant considers the sow by much the more profitable investment for his money.

The soil, climate, and situation of Hampshire afford several sources of emolument to the farmer, of which he does not appear to have yet fully availed himself. The populous towns on the south coast, Portsmouth, Gosport, and Southampton, are excellent markets for the consumption of his vegetable as well as animal produce; yet we found that, in the article of potatoes, these ports are chiefly supplied from the coast of France. Considering the abundance of manure to be obtained from these towns, and the fitness of much of the soil for potatoes, we think the farmers very remiss in letting this trade out of their hands. The earliest crops in England might be produced here, which could be off the ground in ample time to be followed by a turnip crop the same season. The northern part of the county could in like manner, with the aid of the railway, send supplies to the London market; but, as the same sources of manure are not within reach of this division, the farmer would probably find it to his advantage to house-feed stock extensively for the purpose of increasing his dung-heaps. This might be united with the introduction of dairying, by which to send daily supplies of milk and butter to the London market. And from these sources, viz., the culture of edible vegetables, and the sale of milk and butter, we have a strong impression that the difference in the price of his corn would be amply made good to the farmer.

There is another branch of industry which the convenience of railway accommodation and the movement for sanitary reform, are likely to introduce in this and other counties at a moderate distance from the large consuming towns of the kingdom. The rapid transit now afforded has put an end to the necessity which formerly existed, of sending cattle up to town, to be driven through the crowded thoroughfares, to the inconvenience and

danger of the passengers, and afterwards slaughtered, to the pol-
lution of the atmosphere, and the absolute waste of one of the
most valuable sources of reproductiveness. Should Smithfield
be abolished, the farmer's market might be brought nearer his
own door, if the carcass butchers would remove their establishments to convenient points on the different lines of railway, to
which cattle could be driven without unnecessary cruelty, and
whence the meat might be delivered in London with nearly as
much despatch, and certainly in a more wholesome state than
from the city slaughter-houses ; while the blood and other offen-
sive matter, instead of going to waste, would be carefully
retained for the benefit of the soil.

The rate of wages for labour in Hampshire is at present from
8s. to 9s. and 10s. per week, the higher scale prevailing in the
southern districts and the lower on the chalk-lands. Task-work
is extensively resorted to by the farmers, and, on the whole, the
labourer is better paid than in the adjoining counties. In many
cases, however, he is not better off or more comfortable on that
account than elsewhere, having to pay an increased rent for his
dwelling-house, from the scarcity of cottage accommodation. This
scarcity arises from the effect of the law of settlement, which has
induced the landlords in some parishes to pull down, whenever
they had the opportunity of doing so, all buildings that were
likely to afford a harbour for the poor. The consequence has
been that the labouring-classes in Hampshire have had their families crowded together, to the great detriment of their morals.
The rent of cottages, which in many of the surrounding counties
does not exceed 3l. per annum, here rises to 5l. and 6l., and in
some cases even to 10l. a year.

The farmers complain greatly of the injury they allege themselves to have sustained by free-trade, but they do not seem to
have any strong hope that protection will be restored to them.
Leases, produce-rents, compensation for unexhausted improvements, reductions of rent, the abolition of the law of distress, are

severally looked upon as measures which would contribute to
relieve them.   But effectual aid, they conceive, can only be
rendered by higher prices.   They are unwilling to admit that
their mode of farming can be profitably altered for the better;
and, even if it could, there is said to be a general want of capital with which to effect improvements.   Some farms had been
given up in despair to the landlords, and remained unoccupied,
while others had been relet at diminished rents; but the number
of these was not considerable.

LETTER XIII.

NORTH HANTS.—BERKSHIRE.

STRATFIELDSAYE — DUKE OF WELLINGTON AS A LANDLORD — RENT AND RATES — SYSTEM OF AGRICULTURAL MANAGEMENT.——BERKSHIRE.— DIVISION OF SOILS — WANT OF CAPITAL BY FARMERS — SIR JOHN CONROY'S FARMING — DRAINAGE AND TRENCHING THE FOUNDATION OF HIS SUCCESS — PROCESS OF IMPROVEMENT — MANAGEMENT OF EARLY LAMBS — CORN CROPS — FARM GARDEN — BUILDINGS — MACHINERY AND IMPLEMENTS — PIG MANAGEMENT — THIN SEEDING, WIDE DRILLING, AND FREQUENT HORSE HOEING — LABOURERS.

READING, March, 1850.

STRATFIELDSAYE, the gift of the country to the Duke of Wellington, lies on the northern border of Hampshire, near the line of railway from Reading to Basingstoke. The estate is chiefly a strong retentive clay, naturally wet, and requiring very delicate management to render it productive. Drainage, which is the foundation of all improvement on this description of soil, is being carried on very extensively at the Duke's expense. Chalking, which is second only in its importance to drainage, has hitherto been difficult to accomplish, from the great expense of carting for several miles so heavy a material, 20 tons being the usual quantity applied to each acre. The opening of the railway has facilitated this improvement, the article being now carried from the edge of the railway cutting, and conveyed to various points on the line, at less than half the former cost of cartage. His Grace takes the principal share in the expense of this improvement also, chalking the lands of some of his tenants at his own cost. The farm-buildings moreover present a striking contrast to the general style of accommodation provided by the landlords of Hampshire for their tenants. In this branch, the Duke has been at an immense outlay, substituting, whenever an opportunity arose, substantial buildings of brick and slate

H

for the wretched old wood and thatch hovels common in the
country.   The farm-houses have also been renewed or rebuilt.
and the labourers' cottages have equally shared the benefit of
his Grace's improvements.   The cottages have been fitted up so
as to afford comfortable accommodation for their occupiers, and
are held directly from himself, that there may be no exaction in
the matter of rent.   To each cottage about a quarter of an
acre of garden ground is attached, and for the cottage and
garden the labourer pays 1s. a week.

The mansion, which was formerly the seat of Lord Rivers, is
of moderate size, and in rather a low situation, but the park
which surrounds it is extensive and well wooded.   The stream
which flows through the grounds and the extent of woodland
scenery make it very picturesque.   But the stubborn nature of
the soil renders this estate, as an agricultural property, expensive
to improve.   For many years his Grace has laid out on its
improvement nearly the whole amount of its rental.   The same
liberal expenditure on a kindly soil would have been tenfold more
productive, but the true spirit of a benevolent landlord is the
more strikingly displayed on a field where there can be so little
return for it.   It is delightful to his countrymen, among all
classes of whom his Grace is, and ever will be, distinguished, as
emphatically " The Duke," to find that in the more private
capacity of a landlord his duties are performed with the same
wisdom, attention, and unswerving faithfulness, which have
rendered his public character so exalted.

The rent of land on the Stratfieldsaye estate is about 20s.
an acre, tithes about 7s., and poor and other rates 2s. 6d. to
3s. 6d.   It is strictly corn land, wheat and beans being the
chief produce.   The system of cultivation pursued is to plough
up the clover lea, after the second crop is consumed in autumn,
that the furrow may be exposed to the pulverising effects of
the frosts and thaws of winter; after which it receives a clean
summer fallow, being repeatedly ploughed and harrowed until it
is brought into fine condition, when it is sown with wheat in

October. After the wheat is reaped, the land lies untouched
during the winter, and as soon as it is dry enough in spring, a
heavy dose of dung is spread upon it, which is immediately
ploughed in, and the ground planted with beans ; the beans are
dibbled in by women, who are employed by task-work, and who
set the seed in rows, marked by a garden-line. During the
summer the land is carefully hoed between the rows, and after
the bean crop has been removed it is ploughed and sown with
wheat. After wheat follows barley, a portion of which is laid
down with clover, the rest being reserved to be sown in the fol
lowing spring with peas, of which an excellent variety, called
the " Victoria marrowfat," is in great favour, selling at 40s.
a-quarter. The average produce of wheat is from 26 to 30
bushels per acre. From the nature of the land it is found very
injurious to work it when wet, and a great number of horses are
therefore kept to push forward the work in favourable weather,
a farm of 300 acres having as many as 16 work-horses upon it.
The only other stock consists of a few milch cows, some colts,
and a number of pigs, which go loose in the yards. Stall-feeding is little practised, and when tried has been found very unprofitable : but this is not surprising, as fattening oxen are
fed on cake and other substances, costing 10s. 6d. a-week for
each animal.

On leaving Stratfieldsaye we enter BERKSHIRE, which, from
its extent and variety of soil, exhibits many modes of agricultural management. Along the Isis, in the vale of White Horse,
and on the banks of the Kennet, dairy farming predominates.
On the richer pastures sloping to the Thames, the fattening of
stock is practised. The range of chalk hills, which, entering the
county from Oxfordshire, crosses it in a westerly direction, are
employed in the rearing and feeding of sheep, combined with corn
farming. The mixed soils of clay, gravel, and sand, in the
district to the south of Reading, are chiefly under tillage, as is
the rich tract of corn land to the east of Wantage, which is an
eminently fruitful and fertile country.

On the stronger lands, where sheep are unsuitable, from the impossibility of folding them on the ground, pigs are fed in yards, in great numbers, on account both of the value of their manure, and of the profits arising from the excellence of the breed.

On soils suitable for turnip culture and sheep, it is no infrequent practice in the eastern parts of Berks for one farmer to give his turnip crop to another without any charge, on condition that the crop is to be consumed by sheep on the ground where it is grown. This practice infers a want of capital on the part of the farmer, who could turn his crop to much better account by putting his own stock upon it, and thus keep to himself the profit, without which his wealthier neighbour would not buy stock for such a purpose. But it also shows that the value of the crop is little appreciated, when within thirty miles of London it is ever turned to so unprofitable an account.

In farm buildings, roads, and drainage, the eastern part of Berks is generally very deficient. To this there are many exceptions, and the most instructive of these merit a full description.

The farm of Sir John Conroy, at Arborfield Hall, about four miles to the south-east of Reading, consists of various soils, but principally fair stock land, not very deep, some of which lies on a retentive substratum of clay, and some on an open gravel. Four years ago (for Sir John is a farmer of only four years' standing) the whole of the arable farm, comprising about 320 acres, exclusive of the park surrounding the mansion, was divided into numerous small fields by high wooded banks, every one of which has been removed, the soil in them being scattered over the adjoining land, while a sufficient number of the best trees were left to give variety and charm to the landscape. Every acre of the land was then drained with inch pipes laid four feet deep, the drains being 15 feet apart in the stiffer lands, and 30 feet apart in those which were of a drier character. Further experience leads Sir John to think that a greater distance would have sufficed on the latter, much of which is of a kind which

most farmers would consider a waste of money to drain at all;
and probably nothing could better convince them of the incor-
rectness of such an opinion than a peep into what Sir John calls
" the bigot's hole," a square box, in which at a depth of between
four and five feet, two main pipes are to be seen constantly
pouring out the drainage of 40 acres of this description of soil.
Immediately following the drainers, the whole farm was trenched
by forks to a depth of 22 inches; the surface being carefully
retained uppermost by being thrown forward to cover the previously trenched portion of subsoil.   The cost of both operations,
drainage and trenching, was nearly 12$l.$ an acre; so that, if a
great improvement has been effected, it must not be overlooked
that it has been done by a large outlay of capital.   Farm roads
were at the same time made, which serve the double purpose of
accommodating the different fields, and of separating the one
from the other.   Commodious farm buildings were also erected.
The land is managed strictly on the four-course system; every
modern improvement which is applicable to this system, and
has been previously proved to be profitable, being adopted.
The swedes, of which we saw an excellent crop, are manured
with yard dung of the richest kind, and 4 cwt. of superphosphate to the acre.   A portion is drawn for consumption in the
stalls, and the rest are eaten on the ground by sheep; that
part which in March is still to be eaten, having been laid in
heaps, and covered with a little earth to shield them from the
changes of weather, and to prevent them exhausting themselves and injuring the ground, by running to seed.

   The sheep are a Southdown ewe stock, crossed with a short-
woolled Leicester, the produce being a half-bred lamb which
grows and fattens very rapidly.   At the end of the fold, next
the untouched turnips, spaces are left through which the lambs
only can pass out and in, and here boxes are placed containing
an unlimited supply of bruised beans, peas, and oilcake, of
which they partake liberally, besides nibbling at the green
turnip tops, and at the turnips also as their mouths get strong.

They are ready for the London market by Good Friday, or as
soon after that as possible; and the ewes are immediately afterwards put on the best feeding, to fit them for the market with
the utmost despatch, the great object being to turn the capital
over in the shortest possible time in which a profit can be
secured.   A new stock is again purchased at the first favourable
opportunity after the old has been disposed of.

As the turnips are consumed, the land is ploughed and sown
with barley, drilled in rows six inches apart, by Garrett's drill;
two bushels of seed having been last year used to the acre,
though, considering the high condition of the land, and its perfect drainage, that quantity is thought too much, and one
bushel and a half is to be tried this season.   The produce last
year was seven quarters an acre.   The barley is followed by
clover and seeds, part of which is fed and part mown.   After
the second crop is fed or mown the ground is ploughed (Howard's
Bedford two-horse plough being in all cases used, and much approved), and then rolled by Crosskill's clodcrusher.   The ground
is, next harrowed, and one bushel of wheat per acre drilled in, in
rows $12\frac{1}{2}$ inches apart, and nothing can exceed the regularity
and beauty of the plant at this moment.   But those who may
wish to imitate Sir John in the economical use of seed must
not forget the important adjuncts of that system already described, — the perfect drainage and disintegration of surface and
subsoil, the subsequent manuring of the turnip crop, the corn
and cake fed sheep, and the final consolidation of the furrows, by
all of which, as far as possible, security is taken that nearly every
grain sown shall vegetate.   As soon as necessary in spring the
crop is hoed by Garrett's horse-hoe, two of which are used on
this farm, and with them 20 acres a day can be got over.   Sir
John thinks very highly of this implement, not more from the
speed and economy, than from the efficiency with which it does
its work.   The wheat crop last year averaged six quarters an
acre, and was of very superior quality.

Adjoining the farm buildings is the farm garden, a plot of

6 acres, where vegetables, such as cabbages, mangold, potatoes,
&c., are cultivated, chiefly by manual labour, for the consumption of the house-fed stock and pigs; and where experiments
are tried with various seeds, to discover the most valuable kinds
of corn for cultivation.   Near this is a pump, communicating
with the liquid manure tank, from which, by a hose attached to
the pump, the surrounding land, to the extent of 20 acres, can
be watered.   The hose has been in use for the last two years
without being much worn out.   It is manufactured by Paterson
of Manchester, and costs 5d. a foot.

A new space, as large as the original rickyard, has been
cleared in order to afford room for the increasing bulk of the
crops. — Resting on the outer wall of the rickyard is a light
thatched sheepshed, with well-littered yards, into which the
ewes are brought for shelter to drop their lambs, and where
they remain for a few days till the lambs are strong enough to
follow them to the turnip-fold.   When the whole of the ewes
have lambed, the hurdles enclosing the yard are removed, the
solid dung is carted out, and the land beneath is then dug and
planted with potatoes, for which it is sufficiently manured by the
liquid which has penetrated it.   The shed of course remains for
use in the same way in subsequent years.

Next the rickyard is the barn, the whole machinery of which
is driven by a 10-horse steam-engine, the cost of working which
is 2d. an hour for labour and 10d. for coals, or 1s. an hour
altogether.   A covered gallery extends along the back of the
barn, through the length of which the driving-shaft of the
engine passes, with pullies and belts at intervals for attaching
the power to the several machines as they are wanted.   In this
gallery Sir John proposes to erect steaming chests, should he
adopt the plan of giving his stock cooked food.   The thrashing
machine is an inferior one *, but it is fitted with an excellent shaker
by Garrett, the double motion of which at once separates the

---

* This machine has since been replaced by a very efficient one from the
manufactory of Messrs. Garrett.

grain from the straw and causes it to fall lengthways from the machine. It is then passed to a straw-cutter by which it is cut as required, either into 4-inch lengths for litter, in which state all the litter of the farm is used, or into $\frac{1}{2}$-inch lengths for food, a hopper from each side of the machine carrying the respective kinds into separate compartments of the building beneath. A corn and cake bruiser and turnip-cutter are all attached, when requisite, to the shaft of the steam-engine. Immediately opposite to the barn door is a high, open shed, in which straw is stored dry.

One side of the square is the implement shed, in which every implement not in use is kept under cover; and it is proper to mention here that Sir J. Conroy has disposed of all his waggons, and adopted Crosskill's one-horse carts, having fully persuaded himself, after trial of both, that there is no comparison in point of economy of labour between the two, this being a level part of the country, with good roads through the farm. The more intricate machines are kept under lock, the house being provided with shelves on which the different parts of the machines not required for the particular work in hand, are carefully arranged. Adjoining this is a carpenter's shop, and, at another part of the buildings, a smithy.

The farm stable has at one end of it a harness room, well lighted, where all the cart and plough harness is kept, and which is cleaned every Saturday afternoon. Water is supplied by pipes to the stable and harness room, and indeed to every part of the farm buildings. The fattening oxen are kept in loose boxes under cover, with a passage before them for the convenience of the feeder. Each box is supplied with water, *all soft water*, which Sir John rightly considers of much importance to the thriving condition of his stock. The centre compartment of the building is occupied as a store for preparing the food, cutting turnips, mixing meal, cake, &c. — the "kitchen," as it is termed.

The next yard is the great feature of the in-door management — the pig establishment — upon which great attention is bestowed. It should have been mentioned before that the only things sold off this farm are live stock and wheat, everything else, including the barley, being consumed on it. Of course, when a fine malting sample of barley is produced it is sold, and an equivalent quantity of feeding barley bought to replace it. Besides this Sir John buys a large quantity of cheap grain and oilcake for his stock. He endeavours to fatten, every year, as many hogs as he has acres, and has therefore always on the farm between 300 and 400, 80 of which are in the fattening pens to be finished. The only food they receive is barleymeal and water, a " kitchen " being conveniently placed for each pig-yard, with a trough sunk·in the ground into which the requisite quantity of meal is put among water, in the evening, for the morning's feed, and the trough again filled in the morning for the evening's meal. Each pig is calculated to consume about ten bushels of barley in the course of feeding. The 80 fattening pigs are kept in three yards, with a shed, all well littered with cut straw.*

---

* Since our visit in 1850, Sir John Conroy has steadily persevered in his course of agricultural improvement. To deep drainage and trenching, he attributes the foundation of all his after success. He has now covered in his two farm yards, one of which is 2500 square feet, the manure pit below, and a sparred floor above, 7 feet from the heap, in which 300 sheep are fattened at a time. The other covered yard is a rectangle of 1100 square feet, in which 100 pigs are fattened in a chamber aloft, with a sparred floor. The manure falls through the open boards upon cut straw chaff, which is laid in every morning, and is ready to plough in when wanted. The first building cost 80l., the second 45l., and by their aid the stock are fattened quicker and with less food than before, and the manure is preserved from the injurious action of the air, and sun, or rain. Besides other stock, 500 pigs are fattened annually, and their manure, mixed with ashes, is very valuable. Additional experience has also enabled him to feed cattle quicker and better. He can now turn out in six months oxen as fat as he at first took a year to do.

He is more than ever convinced of the advantage of thin seeding and wide drilling. Of wheat he sows 3 pecks, and never above 4 ; of barley 1 bushel ; and of oats 2 pecks to the acre, — and all are drilled in at 13 inches apart ; vetches for soiling are drilled at the same distance, and at the rate of 1 bushel an acre ; beans 1 bushel an acre, and 2 feet apart in the rows ; mangold and swedes in rows 31 inches apart. This wide drilling admits the horse hoes of all sorts to be constantly working among the growing crops as long as possible. The produce realised from this management

There are many other interesting points of detail which we
have not space to enumerate. The houses are all spouted to
carry off rainwater, and every particle of liquid, as it escapes
from the feeding-houses, is secured in drains and carried to the
tank. An eating-room, with benches and a table, is provided
for the people to eat their midday meal. Here there is a fire
for cooking, and a washhand basin in the corner, with water laid
on, which is regularly used at night after work is over. The
same orderly precision which regulates all the departments of
the farm is pre-eminently displayed in the management of the
farm servants. They are engaged by the week, the present
rate of wages being 10s. They are paid every Saturday in
small silver, so that they may have no necessity to go for change
to the public house. A serious fault is never passed over (about
which rule, however, we desire to express no opinion) ; no abusive
language or high words are permitted to be used to any person
engaged on the farm ; and, should any misconduct occur, or any
serious neglect of duty, the offender receives with his pay on
Saturday a notice that his further services are dispensed with.
All extra time is paid for, and every man made to feel that,
while the exact performance of his duty is required, he is at the
same time treated with perfect fairness. We can testify to the
intelligent appearance of the men, and the cheerful *esprit* with
which they seemed to be animated.

Such is the style of farming adopted by a gentleman bred in
the camp and the Court—a farmer of four years' practice, but
of many years' observation ; who, notwithstanding all the outlay
he has made, finds the business remunerative. We have been
thus minute in our description in the hope that other country
gentlemen, now compelled by necessity to look strictly to their
own business, may be tempted to take a lesson from Sir John
Conroy, and to learn from him how much healthful excitement is
to be obtained by personal attention to the business of farming.

may challenge comparison with that of any other system practised on similar
soil in this country.

LETTER XIV.

BERKSHIRE.

NORTHERN DIVISION. — FARM OF MR. PUSEY M.P. — LAND ALWAYS UNDER
CROP — DETAILS OF MANAGEMENT — SHEEP STOCK FED ON RAPE CAKE AND
BARLEY IN ADDITION TO GREEN FOOD — EXPENSIVE MODE OF FEEDING
OXEN — WATER MEADOWS — CAUSE OF FAILURE THE FIRST YEAR — EX-
TRAORDINARY FERTILITY AFTERWARDS — LETTER FROM MR. PUSEY —
BENEFITS OF HIS EXAMPLE — WAGES AND COTTAGE RENTS. — RENT AND
PRODUCE — VALE OF THE ISIS — FERTILE DISTRICT NEAR WANTAGE —
CHALK DISTRICT ROUND ILSLEY WELL CULTIVATED — INCONVENIENCES OF
" COMMON FIELD " — TENURE — FARMERS' OPINIONS — IMPROVEMENTS AT
BEARWOOD.

PUSEY FURZE, BERKS, March, 1850.

IN the north-western division of the county, between four and
five miles north of the Farringdon-road station on the Great
Western Railway, is situated the estate of Mr. PUSEY, M. P.
for Berks. The prominent position held by this gentleman in
our agricultural literature, and the many aids he has given by
his writings to the general diffusion of enlightened agricultural
practice, determined us not to pass through Berkshire without
examining the system of farming followed by him. The outline
of the country, and the soil of which it is composed, are eminently favourable to economical and remunerative farming.
Large open level fields, sufficiently sheltered by lofty timber to
break the rigours of winter and to afford a shade from the heats
of summer, offer a refreshing landscape to the eye of the
practical farmer. No rocks or stones obstruct the operations of
his implements, but a fine dry easy-working soil presents to him
a field on which his skill, capital, and enterprise may be em-
barked with every reasonable hope of success.

There are several varieties of soil on Mr. Pusey's farm, which
contains between 300 and 400 acres, part of it being stone-brash,

part fine loam, and part an inferior and somewhat moory soil,
mingled with peat. The first two are excellent corn and turnip
land, the last is being chiefly devoted to water meadow. The
breeding and feeding of sheep is the point on which everything
else on this farm is made to hinge, and large quantities of artificial food are bought in order to increase the capacity of the
farm for sheep, of which there has been for the last year a very
large stock kept. Corn crops, consisting of wheat, barley,
and oats, are taken alternately with green crops, which are consumed on the ground. No clover or seeds are sown, as the water
meadows, which will be afterwards described, supply all the
summer food that is considered requisite.

As soon as the corn is removed, the stubble and a thin surface
are turned over by Glover's skim plough, with which implement
(said to be a most efficient one) two horses can go over two acres
a day. This is well knocked about by the harrows, and white
turnips are then drilled in, on the flat, with 2 to 3 cwt. of superphosphate. Another part is sown with winter vetches and other
spring feed. These are eaten on the ground, and followed in
May and June by swedes and mangold, the former of which are
also eaten on the ground, the latter drawn home for consumption
in the yards. The swedes are manured with 3 cwt. of superphosphate per acre, and sometimes yard dung also, in which case
they are sown on ridges, in which the manure has been previously
laid and covered ; and this mode of mixed manuring, with dung
and superphosphate, invariably brings the best crop. As the
swedes are eaten off, the land is either lightly ploughed, and the
wheat, barley, or oats, as may be, drilled in with Hornsby's drill,
drawn by four horses, and covered by one stroke of the harrows, or it is breast-ploughed by men, about one inch in depth,
at a cost of 5s. per acre, and the seed then drilled in and covered
as before. This last is considered the best operation, as offering
the firmest and surest seed bed for the wheat. For barley and
oats the land is lightly ploughed. All the corn crops are sown
in rows, and hoed by Garrett's horse hoe. The land, being in

constant tillage and highly manured, is very clean and free from
weeds.

The sheep stock kept on the farm averages 800 in number,
the half of which are breeding ewes. During winter they are
folded regularly over the rape, turnips, and swedes, the ewes
getting no other food except hay-chaff. The " tegs " receive a
little rape-cake and barley besides, the quantity being gradually
increased to three-quarters of a pound of the former and a pint of
the latter daily to each, for six weeks or two months before they
are sent off fat to London. Rape-cake is given by Mr. Pusey,
as he has found it as good an article for feeding as oil-cake, and
much less expensive. As soon as the " tegs " are ready they
are shorn before being sent to market. When the winter food
is consumed, the sheep are folded on the water meadows, on
which the whole stock is kept for five months in summer.

Twenty to twenty-four oxen are purchased annually, more
for the purpose of making the straw into manure than anything
else. They are kept in a yard with an open shed, in which
each is tied up to a stake to be fed three times a day, being
loosed again as soon as they finish their bait. That consists at
present of 7lb. of oil-cake and a peck of barley-meal mixed with
hay-chaff for each animal, and cannot cost less than 10s. a week
exclusive of attendance. They get no roots or green food whatever, but are allowed to wallow among straw. At the bottom
of the yard the stream for irrigating the meadows passes through,
supplying water to the cattle, and carrying off the liquid of the
yard to enrich the meadows.

These water meadows form the great feature of Mr. Pusey's
management. He introduced the system two or three years ago
from Devonshire, having entered into a contract, for laying out
an experimental portion, with an experienced irrigator from that
county. The whole cost of levelling the ground, making the
gutters, and the further charge for carriers to bring the water
from the brook to the meadow, was 5l. 10s. an acre. The first
year the experiment proved a total failure; in one case the

ground seemed positively injured, but this arose from the water
having extirpated the moss which previously overspread the
meadow. It was suggested that the failure arose from the
poverty and low condition of the land; and it was noticed that
on a portion where some burnt ashes had been spread, the action
of the water had produced a luxuriant growth. Next spring,
therefore, the whole meadow got a dressing of burnt peat ashes.
The water now had its full effect, and so great was the growth
produced, that one meadow, 20 acres in extent, was fed four
times with a flock of sheep during the summer, the water being
let on immediately after the fold was removed, thus washing
down to the roots of the grass the whole enriching substance,
before there was time for it to be lost by evaporation in the
heat of the sun. This meadow afforded keep to a flock of 400
sheep for five months of summer; and Mr. Pusey states that
a smaller one, of two acres, yielded keep for 73 sheep, or 36
sheep on one acre, for five months. But we are bound to say
that some of the neighbouring farmers allege that the sheep
were *kept*, not *fed*, and that it was marvellous to them how
Mr. Pusey had managed to keep so many sheep, even alive, on
this small space during the whole summer.* So well satisfied is

* With reference to this, Mr. Pusey sent the following letter to the Editor
of " The Times : "

To the Editor of " The Times."

" Sir,—I beg permission to advert to a single point in your Commissioner's
account of my farm ; but in so doing I ought first to say that nothing can
be more fair than that report, and, indeed, it surprised me that he should be
able to collect so accurate a statement during my absence, which I regret
the more as it prevented me from making his acquaintance.

" The point, however, is simply a rumour, which he felt bound to allude
to, that on a field of two acres which had supported, in consequence of irrigation, 36 sheep per acre for five months, or rather had supported sheep at
that rate, the sheep were ' kept alive, but not fed.'

" This rumour, I beg to assure you, is utterly without foundation. The
sheep left the field, after each time of eating it off, in thriving condition.
The whole of my last year's lambs will at the end of seven days have left the
farm for Smithfield to be sold, though but a year old, as mutton, and are
this year unusually fat. Your readers, who like myself remember that, in

Mr. Pusey himself with the results of irrigation that he is extending his meadows on his own farm, and, it is said, intends to lay one out for each of his tenants where the necessary supply of water can be found. On his own farm he has no other summer keep, as he now dispenses altogether with laying any part of his arable farm into clover or seeds. As this experiment has proved so successful in one of the drier counties of England, it may be useful to mention that a detailed account of the whole, by Mr. Pusey, is to be found in the 24th number of the Journal of the Royal Agricultural Society.

The benefit which Mr. Pusey does to the district around him,

their youth, butchers were required to furnish five-year old mutton, may not like this rapid production; but it is required by the increase of our population. It clearly cannot be accomplished without plenty of food. My flock of ewes also were never at any former lambing season in better order than now, after being kept on these catch-meadows. The rumour therefore is one of those by which men endeavour to account for things which exceed their powers of belief, and appearing to them fabulous or mythical, seem to require a rational explanation. Your Commissioner was quite right to mention it, but the sceptics would do well to inquire whether irrigation does not at least double the yield of grass-land.

"I should hardly have troubled you on this matter if it had been merely agricultural, but have done so because I should be sorry to be thought a hard flockmaster.

"It has always been an agreeable thought to me that the improvement of farming tends greatly to increase the comfort of all the animals usually found on a farm. Under the old system there was, and still is where it lingers, a great deal of unreflecting cruelty. The sheep, when kept for wool only, is even yet, on some of our moorlands, left to his fate in the winter, and not uncommonly dies of starvation.

"By the improved system the farmer is taught to keep his animals in a thriving state steadily from their birth. Even horses, though not meant to be eaten, should not be stinted of food. Railway contractors hardly measure their horses' oats, and two well-fed horses can do as much work or more, for the same provender, which on the old system enabled three horses barely to crawl.

"We have now learnt that, for our own interest, every animal on a farm should live well, and that a hard stockmaster is a bad farmer.

"With sincere respect for your Commissioner's ability and fairness,
                        "I remain, Sir, yours faithfully,
"Pusey, March 29."   "PHILIP PUSEY.

by introducing new agricultural implements, is readily recognized
by the farmers, who profit by adopting those which he finds
successful, while they, of course, avoid his failures.   Hornsby's
drill, Garrett's horse-hoe, Glover's skim plough are the most efficient of these.   The great increase of sheep stock now kept
on the farm, and the larger annual produce in corn, are readily
admitted by the farmers, who say, however, that they are
gained at a greater cost of artificial food and manure than they
are worth.   This we think is very unlikely; but it might be
worth Mr. Pusey's consideration so to systematize his management (which, hitherto, has necessarily been irregular from being
in some degree experimental) as that it would be readily understood by his neighbours; and we would add that he should not
be content with a mode of feeding oxen which *must* entail loss,
besides a great waste of straw, when there can be no doubt that
a judicious mixture of roots with the more expensive food, corn
and cake, given to house-fed cattle, might be adopted with a
profitable result, and a much more economical consumption of
straw.   With the improved farming now carried on, we were
informed that Mr. Pusey has quadrupled the sheep stock and
doubled the corn annually maintained and produced, as compared with those of the tenant who previously occupied the
farm.

Labourers are here paid 8*s*. a week, working nine hours a
day.   Cottage rents are 1*s*. a week, and each labourer has an
allotment of a quarter of an acre of excellent land adjoining his
cottage let to him at the rate of 40*s*. an acre, which the land
would readily bring from a renting farmer.   The soil is so easily
wrought that three, and occasionally four, pairs of horses and a
yoke of oxen are found sufficient for Mr. Pusey's farm of between
300 and 400 acres, but the usual depth of furrow in ploughing
does not exceed three inches, and nearly the whole green crops
are consumed on the ground where they grow.   Horses are fed
on hay and two bushels of oats a week to each.

The surrounding country is very much of the same description of land as that occupied by Mr. Pusey. The rent and rates amount on an average to 30s. an acre. The four-course husbandry is the rule of the district, and in many instances it is carried out with much skill and spirit. The details are very similar to those described by us as practised in the southern division of Oxfordshire, though we think the farming on the whole is not so good as in that part of the country. In many instances turnips are not cut for the sheep, nor is the system of winter green crops so diligently pursued. Twenty-eight to thirty bushels of wheat an acre may be reckoned an average crop.

Towards the Vale of the Isis the land becomes very stiff and worthless, "too strong for cultivation, and too weak to carry crops," being the terms in which it was spoken of to us. Adjoining this tract is a deep sandy loam and a light sand, and each farm generally contains a proportion of all, so that the bad is kept going with the good. This variety of soil likewise gives rise to a mixed system, the farmers having a portion under dairy, as well as sheep and corn. The fall in prices has, therefore, not so much affected them, the dairy and the sheep stock still continuing to bring in regular returns. — Farms are never let by tender. Poor-rates are said to have increased in some parishes from 2s. 6d. (a few years ago) to 4s. per pound. Labourers' wages are from 7s. to 8s. a week, and cottage rents 2l. 10s. to 3l. Twelve horses are requisite for the cultivation of a 400-acre arable farm, managed in the four-field course. Chalk is not used in this part of the county, as the distance is considered too great to fetch it. Turnips are frequently destroyed in winter by rapid changes from frost to thaw ; and it is somewhat singular that the warmest parts of the county suffer more from this than the more exposed — the northern side of the Wantage hills, for instance, where the frost is more intense, but, by its aspect, protected from the sudden alternations produced by the rays of the sun, suffering less than the southern and warmer slopes of the same range. Excellent Southdown sheep and Berkshire pigs are bred in this part of the county.

I

Eastward from Wantage is a tract of very fertile corn and bean land, not too strong for swedes and other green crops. Corn and leguminous crops here follow each other in succession; very little stock is kept, the farmer's sole dependence having hitherto been on corn. The country is open, there are no fences along the public roads, and none dividing the different kinds of crop. Rent in some cases reaches 40s. an acre, while the tithe, which is greatly complained of, is as high as 12s. and 15s. an acre, and the poor and other rates about 5s. per pound more. The land seemed to be cleanly, but not richly farmed. The farmers are busy sowing peas; a variety called the blue pea, said to be excellent for boiling, being sown at the rate of four bushels an acre. They are put in by a one-horse drill, in rows about a foot apart, the machine (which has been long in use in the county) acting like a light plough with a seed-box fixed to it, the seed falling into the rut just as it is made, and being covered by the crumbling mould which falls in of itself upon it. The depression in the price of corn tells very severely in a high-rented district like this, which is altogether dependent on corn.

Southwards of this tract, we get on the chalk downs round Ilsley, which are of superior quality. Labourers' wages were here lowered last week from 8s. to 7s. Cottage rents are 1s. to 1s. 3d. a week.

The land is cultivated on the four-course system, with some variations peculiar to this place, barley being taken after clover lea, and wheat following turnips and rape, which are eaten off early. About a tenth part of the land is kept under sainfoin, in which it remains for four years, being each year cut for hay, of which it gives an excellent crop. A farmer having 40 acres of sainfoin, sows out 10 acres and breaks up 10 acres annually. This goes regularly over the whole farm, the sainfoin not returning on the same field for considerable intervals, and when its turn comes round the field receives a rest of four years from the routine of cultivation. It is then ploughed up in spring and sown with oats on one furrow, the crop of which is generally excellent, as much as 80 bushels an acre not being uncommon. The average yield of wheat on the better class of down land here is 30 bushels, and of barley 40 to 48 bushels an acre.   On a farm of 380 acres a stock of 300 ewes is kept, and their produce fattened off, the farm maintaining about 700 sheep during the year.   This is very much greater than we found common on the down lands of Dorset or Wilts.   But here all the beans and peas, and part of the barley, grown on the farm are " spent " in feeding the stock.   Artificial manure is not used to a great extent.   The rent, tithe, and rates are about 30s. an acre.

In one parish here the land is " common field," one farmer's fields being intermixed with those of another, and thus producing great inconvenience and expense in management.   One consequence to which this leads is, that a man who farms much better than his neighbour, expending more capital and getting his land into higher condition, reasonably objects to what would otherwise be a most desirable improvement, an enclosure and new distribution of the land, as he must suffer in being compelled to take what is out of condition in lieu of that which he has at so much cost put into condition.   The rector of the parish, in the case referred to, has between 50 and 60 acres of land, which he cultivates on his own account, scattered in 50 different places among the fields of his parishioners, through any of which he has of course right of access !

Farms in Berkshire are generally held from year to year. The farm buildings are old and insufficient, though there are, of course, many exceptions.   Some farmers have abundance of capital; but it is too common here, as in other parts of the country, for farmers to take farms too large for the means at their disposal.   Compensation for unexhausted improvements, it was urged, would tend to prevent this, by the greater capital in hand which an entering tenant would then be obliged to be possessed of.   And the fact that the landlord would be liable, in the first instance, for this claim, would, it was thought, prevent him giving a tenant notice to quit, except on grave necessity.   The want of such right is felt as a great bar to the free investment of tenant's capital, and the full cultivation of the soil.   The arbitrary exaction of the income-tax is another grievance much dwelt upon.   Corn rents are advocated by some. Two years' notice, with compensation for unexhausted improvements, are regarded by others as a better tenure than a lease. The malt-tax was not much complained of.   There can be no doubt that in many parts of this county farmers are diminishing their expenditure, and that the small country tradesmen are now suffering on that account.

Labourers on the whole are considered to be better off than before.   The rate of wages in Berkshire contrasted favourably with some other counties we have lately been in, while the rent of cottages is moderate.   In many parts of the county the execution of drainage and other permanent improvements afford employment.   On the estate of Mr. Walter, of Bearwood, which we visited, an extensive system of drainage was going on, and very substantial farm-buildings were being erected.   In the erection of farm-buildings we may remark that it is important they should not be executed on a scale more expensive than is requisite for the purpose in view, as in that case the interest of the outlay becomes a permanent dead weight which can never be remunerative.   And we think it also injudicious to erect costly buildings (as we have seen instances in this county) for the use of farmers who are unable, from want of capital or want of skill, to turn them to a profitable account.

LETTER XV.

SURREY.

DESCRIPTION OF COUNTY — BACKWARD STATE OF AGRICULTURE — TENURE — "CUSTOM" OF THE COUNTY — SAID TO PROMOTE FRAUD AMONG THE FARMERS—NUMEROUS BODY OF LAND-VALUERS UNFAVOURABLE TO MUTUAL CONFIDENCE BETWEEN LANDLORD AND TENANT — STATE OF AGRICULTURE NEAR GUILDFORD — VALLEY OF THE WEY — ALBURY — PREJUDICES OF FARMERS NEAR REIGATE — EXCELLENT MANAGEMENT OF A BUTTER-DAIRY — WEALD FARMING, MEAGRE RESULTS — PRIMITIVE BARN IMPLEMENT — — SUGGESTIONS FOR IMPROVEMENT — EXTENT AND RENT OF FARMS — WANT OF INTELLIGENCE AMONG FARMERS — WAGES — INFLUENCE OF RAILWAYS IN LESSENING PRESSURE OF RATES — EFFECT OF TENANT RIGHT IN DEPRESSING RENTS.

REIGATE, March, 1850.

SURREY, described by Cobbett as on the "sunny side of London," is one of the warmest and driest counties in England. With many varieties of soil, and immediate contiguity to London, and with every facility which railway or road can offer, the farmers of this county possess advantages of no common kind. Excepting the Weald, the face of the county presents a pleasing variety of surface. In the vales, the deep lanes and lofty hedgerow trees remind one of Devonshire; the bare uplands of the chalk hills recall the open downs of Dorset; while the rich woodlands match those of Berks or Hampshire. From Guildford to Dorking we pass along a picturesque road, hills rising on either hand wooded along their summits, and with frequent hedgerows dividing their sunny slopes; large sombre yew trees in great numbers interspersed through the fields giving a peculiar aspect to the scene.

The soils include clay, loam, chalk and heath. The Weald of Surrey, occupying the whole of the flat district on the southern boundary of the county, and forming part of the extensive

Wealden tract which stretches over the adjoining counties of Sussex and Kent, is a cold retentive clay on a clay subsoil. To the north of this is a district of sandy loam, on the green-sand formation, with blowing sands on the hill tops. The chalk hills stretch from east to west through the centre of the county, with a breadth of some miles on the Kentish side, gradually diminishing towards Hampshire. Approaching the Thames the soil is sandy, with loam and clay intermixed. The north-western corner to Bagshot is a moorish soil, with a considerable extent of barren heath.

Near the points of junction of these different tracts, the soil varies so considerably that, on the same farm and in contiguous fields, the systems of management are very different. In the immediate neighbourhood of Guildford there are clay, chalk, moor, and sandy soils, some very superior and some very indifferent in quality. However various the soil, its cultivation exhibits too great uniformity in one respect — the absence of enterprise. Throughout the county, neglect and mismanagement are apparent; and the general features of its agriculture betray a low scale of intelligence and a small amount of capital and industry. The denizen of the metropolis, if in quest of rural scenery untouched by the hand of modern improvement, need not journey for it to the remote parts of the kingdom. An hour and a half's ride from London will set him down at the Gompsal station of the Reigate and Guildford Railway, where a short half-hour's walk will exhibit to him a state of rural management as completely neglected as he is likely to meet with in the remotest parts of the island. He will there see undrained marshes, ill-kept roads, untrimmed hedges, rickety farm buildings, shabby-looking cows of various breeds, dirty cottages — nothing indeed exhibiting care or attention, except covered drains from the farmyards, which ostentatiously discharge the richest part of the manure into the open ditches by the wayside.

The relations subsisting between landlord and tenant will be found to explain, in some degree, the backward state of agriculture in Surrey.   Farms are principally held on yearly tenures, though leases of 7 to 14 years' duration are not uncommon. The landlords are not the parties who object to leases, but the tenants, from the "custom" of the county presently to be described, have a practical security of possession not inferior to a lease.   This custom is somewhat of the nature of " com pensation for unexhausted improvements," with this difference, that it embraces also large payments for imaginary improvements and alleged operations, which, even if they had ever been per- formed, would be more injurious than beneficial.   Under this custom the outgoing tenant receives from his successor the amount of a valuation, which includes " dressings and half dressings of dung and lime, and sheep foldings, the expense of ploughings and fallows, including the rent and taxes of the fallows, half fallows and lays, the value of ' seeds,' the under- woods down to the stem, hay and straw at a feeding price," and other items greater or less in proportion to the expertness of the out-going tenant's appraiser.   This practice is described before the Parliamentary Committee of 1848 by Mr. Robert Clutton, an experienced land agent in Surrey, as " promoting an extensive system of fraud and falsehood among the farmers."   He says

" Where manure has been put on at a distance of time, it is ex- ceedingly difficult to check the quantity or quality of the dressings ; and we find that very false returns are made of it, both in respect to quantity and quality.   Outgoing tenants ' work up to a quitting, —that is, they work out the farm, and put in inferior manure, in order to receive payment for it as if it were of good quality.   Having been so imposed upon in starting, they feel justified in playing the same tricks upon quitting.   There is not much difficulty in ascertaining the value of the manure while it is in the yard, but there is a great deal of difficulty in ascertaining its value after it has been carried out and mixed with the soil.   Even when no crop has been taken this is the case ; and the difficulty is increased, of course, with half-dressings. A disposition has arisen among the tenantry to lessen their payments in this respect by getting their landlord to buy up their dressings and half-dressings.   I have found that appraisers are appointed by farmers to go over their farms and tell them how to make a high valuation, and this has been found practically to limit the choice of tenants and to lock up their capital.   The tendency in Surrey has been to lower the rent of farms, as compared with other parts of England, and to have the same money paid for bad as for good farming."

These objections, it will be observed, apply more to the manner in which the custom is exercised, than to the justice of the principle of compensation for unexhausted improvements. The information we received confirms this evidence, and the demoralising influence of such a practice on the conduct of the farmers, in their relations with their landlords and each other, is just what might be naturally expected.   In every little town in the county, the brass plates on the doors which are brightest and most numerous, are those of the land-valuers and appraisers; the rapid increase of which class is deprecated by the most intelligent farmers as equally injurious to the owner and occupier. With a business which can thrive only by promoting constant changes from farm to farm, which encourages an involvement of claims, having a tendency from their embarrassing character to destroy confidence between landlord and tenant, preying upon the capital of the entering farmer, and rendering it necessary for the landlord in self-defence to commit his interests to their charge, they interpose injuriously between the landlord and his tenants, and close the door against that individual responsibility and personal communication which a proprietor can never neglect without injury to his estate.

The neighbourhood of Guildford supplies various examples of husbandry.   The clay lands on the hill-sides, in many cases, still undergo the process of naked fallow, tile drainage not yet having been so extensively adopted as its importance on such soils renders necessary.   On the chalk lands the usual husbandry described in other counties is here adopted; but, as the farms seldom exceed 500 acres, and generally run from 200 to 300, more care and minute attention to details secure better returns. Oilcake is used for feeding the stock to some extent, and artificial manures for increasing the green crops; so that the returns of wheat may be reckoned on the average at nearly 28 bushels, and of barley 40 bushels, while the average of sheep stock is $1\frac{1}{2}$ per acre.

On the sides of the valley sloping to the Wey, the operations of the farmer are much impeded by small enclosures and hedge-row timber, though on all sides indications are here afforded that landlords are now giving way on this point. Generally speaking, these sloping fields are greatly injured by water, and there did not appear to be much drainage going on. The style of agriculture is, therefore, very defective, when the quality of the soil and the conveniences of the situation are taken into account. Along the Wey the land is a deep sandy loam, much of it in pasture, but much also under tillage. The foul appear-ance of many of the winter fallows, the paltry green crops, and the old-fashioned plans of ploughing so generally adhered to, indicate a very backward state of husbandry; while the neglected state of the farm-roads and farm-buildings is in perfect keeping with the implements and the stock. Draining, generally too shallow, is here followed to some extent on most farms; but very seldom does there seem to be proper accommodation pro-vided for the milch cows or their produce.

At Albury we turned into the farmyard of Mr. Drummond, M. P., of whose agricultural improvements we had heard at Guildford. The buildings are constructed somewhat on Mr. Huxtable's plan, all the animals being stall-fed and placed upon boards. Covered houses for dung, and tanks for storing the liquid, are provided. The houses were in good order, and the animals seemed to be very healthy and thriving, but we should fear that in the heat of summer the houses, which present an enormous surface of dull black roofing, would be unwholesomely warm. We would venture to suggest to Mr. Drummond that he should try the effects of a thin lining of thatch straw (which Mr. Huxtable finds the best equalizer of temperature), or even a good outside coating of whitewash, which would reflect, instead of absorbing, the piercing rays of the Surrey sun. The soil here is a light sand, wearing probably its best aspect at this season, as it must be very subject to injury by drought in summer.

In the neighbourhood of Reigate, along the valley, the land is of a friable texture, fairly cultivated in some instances, but not, in any one that came under our observation, with that energy and skill which are to be met with in districts of the country which have very few of the advantages enjoyed by this. Four and five horses are frequently used in a plough. The plough itself is of very antiquated construction. As illustrating the prejudices of some of the farmers, we were told of an instance in which a farmer coming here from a "two-horse country," introduced the two-horse plough, continued to use it for 30 years, turning over on an average an acre a day; and yet his neighbours on both sides of him, at the end of that long probation, still insist that it is impossible to plough land with only two horses! The course of husbandry followed, where any is adhered to, is the four field; but we were assured on very competent authority — that of an intelligent farmer long resident in the district — that the ordinary farmers have no plan, but usually decide as to the next crop of a particular field according to the opinion of one or two neighbours, at their weekly consultations in the alehouse on market-days! The stock kept on the different farms varies with the character of each, some rearing early lambs for the London market, some keeping also a few cattle, and some dairying.

On one farm we found a butter dairy of 40 cows, from which the farmer derived a larger and less fluctuating return than from any other branch of his business. In this case, however, the cows are not suffered to stand exposed among filth and wet in an open yard, as is usual in this county, the farmer having, in default of his landlord making the fair and necessary outlay, built substantial cow-houses at his own cost. In these the cows are fed in stalls, each animal receiving, besides hay, three fourths of a bushel of brewers' grains and a supply of mangold daily; and it may be instructive to the dairy farmers of the south-

western counties, who despair of producing a marketable article
with such feeding, to know that the butter produced on this
farm is supplied by contract to one of the first hotels in Brighton
and to another in London, the tastes of the frequenters of
which are likely to be sufficiently fastidious. The contract
price is 1s. 4d. per pound in winter, and 1s. 2d. in summer.

That portion of the Weald which we have examined in Surrey
is for the most part a stiff wet clay, becoming at intervals more
loamy and friable, and rising in some instances to good stock and
green crop farms. Being naturally very difficult to manage profitably, it has for a series of years been gradually deteriorating
under the present management, and while it yields scarcely a
subsistence to the cultivator, it affords a scanty rent to the owner
and a niggard supply of work to the labourer. The system of
cultivation is begun by a bare summer fallow, the ground being
as carefully managed as its undrained state admits, and then
dunged with such manure as the farm produces, and limed, if the
farmer can afford the expense. The wheat is then sown, the
field being ploughed in "lands," so as to admit the horses in
drawing the harrows to pass up the open furrows without trampling the rest of the land. The crop reaped after this preparation
varies from 12 to 20 bushels an acre. Four or five crops then
follow, according to the taste of the cultivator, whose study
is how to get from the soil, at the least expense, the different
qualities it may have imbibed or accumulated during the year
of bare fallow. When it is clearly ascertained that these are
thoroughly exhausted, the land is again bare fallowed. Scarcely
any stock worth mentioning is kept on these farms. The implements used are of the rudest kind; the barn implements in an
especial degree, the use of the common barn winnowing machine
being frequently unknown. Its place is supplied by sacks nailed
to four horizontal spars, which are fixed on a pivot at both ends,
and when turned briskly round get up a breeze of wind, in which
the corn is riddled by hand, and the chaff blown away !—Under
such a system it is quite impossible that this land can long

continue in cultivation. The first improvement necessary is thorough drainage, and after that is accomplished we should expect much assistance in the further development of its resources by the facilities of communication afforded by the several lines of railway which traverse it. We should anticipate great benefit to the texture of the soil by heavy applications of chalk, which might be brought along the line from the nearest chalk cuttings, and if the railway companies would co-operate with the farmer, it might be worth his while to bring down from London large quantities of the cheapest manure, — coal ashes and street sweepings, to be laid on in heavy doses, in the hope that by this management the soil might gradually be rendered friable, and suitable for the production of green crops as well as corn. This no doubt contemplates much outlay of capital; but when regard is had to the impossibility of things going on as they are at present, and to the advantages this tract enjoys in being little more than an hour distant from London, we have no doubt the experiment, in good hands, would prove successful. This soil, if dry, and if its texture can be altered so as to admit of being kept clean under constant tillage, possesses a strength and depth of staple which could not be easily exhausted.

The farms are from 50 to 200 acres in extent, and are let at from 5s. to 15s. an acre, of rent, to a class of men whose families, though they may shift from farm to farm, have been located in the district for many generations. In intelligence and education, they are extremely deficient; many of them, as we were told, being scarcely able to sign their own names. The efforts of their landlords, some of whom are anxious to promote drainage and other fundamental improvements, are greatly frustrated by the prejudices of such a class of tenantry. Not a few of them are now two years in arrear of rent, and all are every day becoming less able to meet those increased outlays by which alone larger crops can be produced, and diminished prices compensated.

Labourers' wages in Surrey are from 9s. to 10s. Taskwork

is very common, and 12s. a-week is often earned.   Cottage rents
are high, varying from 1s. 6d. to 3s. and 3s. 6d. a-week, with
very little garden ground.   The cottages on farms are sometimes held by the labourers direct from the landlord, in others,
they go with the farm.   Beer is generally given in hay and
harvest-time, but there is no rule on the subject.   Many farmers
are reverting to the custom of keeping the farm servants more in
the farmhouse, the low price of corn and meat rendering this the
cheapest plan they can now adopt.   Besides the facilities which
they afford, the railways, by sharing the burden, have exercised a
very beneficial influence on the "rates" of the parishes through
which they pass.   Poor-rates and highway-rates in some parishes
are, from this cause, extremely moderate.

The chief complaint among the farmers themselves, apart from
that of low prices, was the heavy burden of the tithe.   The unfair character of some of the payments claimed by the out-going
tenant from his successor, which have already been referred to,
was also mentioned as a heavy tax on a farmer's capital.   One
fact arising from this "tenant right" in Surrey is that there is
less competition for farms and a more moderate scale of rent
than we have met with in other counties; but we are bound to
add that these advantages have not contributed to better cultivation, as we should have anticipated.   Incapable of appreciat-
ing the advantages of their proximity to the best market in the
world, within a distance varying from 10 to 30 miles of London,
with railway accommodation if they choose, with a soil and
climate adapted for the production of the earliest vegetables of
every kind for the use of the table, the great body of the Surrey
farmers follow a system suited to farms 500 miles distant from
the metropolis, where it is necessary to convert every thing the
land produces into the least bulky form for cheap transit, so that
the produce of two acres of wheat may be condensed into a ton
weight, and the whole green crop of the farm be packed up and
borne to market, after being digested, in the living bodies of the
sheep stock.

LETTER XVI.

SUSSEX.

LEWES, SUSSEX, March, 1850.

THE county of Sussex possesses soils of chalk, clay, sand, loam, and gravel. From Beachy Head, on the English Channel, the chalk hills, called the South Downs, stretch westward past Brighton, touching Arundel, through the county to Hampshire, the elevated parts with a south-western aspect being exposed to the injurious influence of very boisterous winds. Along the whole northern boundary of the chalk, a strip of green-sand intervenes between it and the clays and sands of the Weald, which comprise the largest portion of the county. An extensive tract of marsh land extends along the coast towards its boundary with Kent. In the western part along the coast, the climate is mild; and as the roads throughout the county are good, and the convenience of railway accommodation very general, the agricultural management might be expected to be fully developed.

The husbandry of the Weald district is very similar to that of Surrey, the farms being small, the land ill-drained, half cultivated, and inadequately stocked; while the face of the country is too much occupied by wood, and cut up by over-grown hedgerows. The farmers as a class are unskilful and prejudiced in their methods of cultivation, and usually hold their farms on yearly tenures.

In the eastern districts of the county between 10,000 and 12,000 acres are annually employed in the cultivation of hops This plant requires the richest soil of the farm, and receives nearly all the manure produced, robbing the corn and root crops of the share which rightly belongs to them.   The farmer's attention is concentrated on his hop garden, and the rest of his farm receives very little of his regard, and hardly any of his capital.   The operation of the excise duties gives the business a gambling character.   A favourable season with a large yield of hops is disastrous to the farmer, as the market value of the article falls, while the duty swells in proportion to the bulky character of the crop: when the crop is a short one the farmer prospers, as the price of the hops rises, and the total amount of duty declines.   There is thus a constant succession of chances, extraordinary profits being sometimes realised, which tempt men to farther adventures, and withdraw them from that steady persevering industry without which agriculture cannot be profitably carried on.   The uncertainty of prices and crops, and the peculiar bearing of the duty, are such that very few of the hop farmers are enriched by it; while many are ruined, and still more kept on the verge of bankruptcy.   It is very probable, therefore, that if the cultivation of hops were to cease, it would in the end be no loss to the Sussex farmer, as his richest land would then be released for the growth of crops of a less hazardous kind, and the rest of his farm receive its fair share of manure and cultivation.

On the Sussex Downs the cultivation of the soil and the management of stock differ in some points from what we have hitherto met with in the chalk country.   On the better lands the four-field course is adopted, and this is extended to a five or even a six field (being laid one to two years to " rest "), where the land becomes thinner and less valuable.   Very old-fashioned clumsy ploughs are used, made of wood, with a bit of flat wood for a mould-board, which is shifted from side to side at each turning; the beam, a thick, strong, straight piece of wood, set on to the head of the plough at an angle of 45 degrees, and borne up in front in a very solid and substantial manner on a pair of wheels from two and a-half to three feet in diameter. This implement is drawn by three or four horses, or six bullocks. Within a couple of miles of Brighton these ploughs may be seen in use every day; and we saw in that neighbourhood a working team, which, for waste of opportunity, of power, and of time, could probably not be matched in any other county in the united kingdom. At the end of a ploughed field were a lot of bullocks, all crowded together, but which we presently perceived were in the yoke, and being turned round. Slowly the crowd separated, each team wheeling about; and steadily advancing up the hill came 18 heavy bullocks, two and two abreast (six oxen in each plough), drawing three ploughs following each other, one man guiding each plough, while another, armed with a long pliable stick, like a fishing-rod, kept the team under his charge at their duty. The furrow was of an ordinary depth, and the land by no means very steep or heavy to cultivate. On the next farm to this we found that a well-managed dairy cow produced upwards of 20*l.* a-year, the milk being sold in Brighton, one of the best markets in England ; and here, with the command of the same market, on precisely similar soil, was the keep of six oxen lavishly expended on an operation which could have been infinitely more cheaply executed by one man and two good horses.

Oats are grown extensively, the soil being found better suited for them than barley. The wheat chiefly cultivated is a brown species, less in value by 6*d.* a bushel than white wheat, but a third more prolific : 26 bushels an acre may be reckoned an average produce.

Sheep flocks are the principal dependence of the Down farmer; and on a farm of 1,000 acres, part sheep-walk and part arable, 800 ewes are considered a fair stock to be kept. They are all

of the pure South Down breed, this being the county where
that celebrated stock originated.   Breeding flocks are kept on
the downs, the lambs being sold every year, in August and
September, to be fed in richer parts of the country.   For this
purpose many go to West Sussex, the price of lambs averaging
17s. each.   Old ewes are sold for early lambing in warmer
districts; the price last year was 28s.   A lamb reared for each
ewe is reckoned a good produce, and not often realised.   The
wool yielded by each sheep averages 3 lb.   Folding is regularly
practised, much in the same way as in Dorset and Wilts, the
sheep being " worked" harder in some seasons than in others.
We cannot reconcile this system of " working " the sheep (such
is the phrase) with that economy of food and full development
of the substance of the sheep which, in other districts, is regarded as necessary to profitable returns.   If manual labour
was scarce and dear, and the price of mutton extremely cheap,
there might be some wisdom in the plan.

In the immediate neighbourhood of Brighton we visited a
farm 450 acres in extent, comprising 300 acres arable, 50 meadow,
and 100 down.   This farm keeps a stock of 40 milch cows and
400 ewes.   The arable is managed in the four-field course,
wheat and oats being grown alternately with green crops and
grass.   Very little barley and no potatoes are grown on the
farm.   The buildings contain comfortable stalls for the milch
cows, well ventilated, and provided with the necessary means of
economizing food and manure.   The stock are kept in excellent
order, and yield great returns.   The cows are house-fed during
the winter on carrots, mangold, swedes, and grain.   They are
housed during the night in summer, and tethered (!) by the head
on the grass-land during the day, both to make them consume
the grass more regularly, and because there are no fences.   This
is not without inconvenience, as may be easily supposed when
one thinks of 40 milch cows tethered within short distances of
each other, under a burning sun, without the shade of a tree,
and tormented by flies!   On a farm where the other opera-

K

tions are conducted with so much prudence and skill we are at a loss to account for such anomalous management. The sheep are folded and "worked" in the usual manner. Two horses are found sufficient for a plough; and their draught would probably be much lessened by the substitution of a light improved wheel-plough for the cumbrous machine of the country, already described. Much care and attention is paid to the economy and accumulation of manure. It has been proved by measurement on this farm, that, the litter included, each milch cow leaves a cubic yard of dung a week in winter, and the half of that quantity in summer. The liquid is carefully collected in a tank, and pumped regularly over the heap to moisten and enrich it.

Great accuracy and attention are carried into the different operations on this farm, all of which are checked by a system of bookkeeping, which shows at once the loss or gain attending any particular practice. For last year the books show a very handsome return to the tenant, after payment of rent, labour, and all charges. We observed that two-thirds of the gross returns were the produce of dairy stock, sheep, and green crops, the remaining third arising from the sale of corn; and these proportions tally very closely with other instances of profitable farming which have fallen under our notice. The rent, rates, and taxes of this farm amount to 600*l.*, which, considering its proximity to an excellent market, seems moderate enough. It is situated at an elevation varying from 150 to 400 feet above the sea level.

The farms within a circle of some miles round Brighton are extensive, ranging from 400 to 2,000 acres and upwards. Labourers' wages vary from 9*s.* to 12*s.* a week, and there is no lack of employment. On some estates cottage rents are 1*s.* a week, but in villages, where cottages are run up cheaply by speculators, 2*s.*, 3*s.*, and as much as 4*s.* is exacted.

The "custom" of the county with regard to the payments by incoming to outgoing tenants, on the Weald, and generally in East Sussex, is very much the same as that we described in our

letter from Surrey. The "inventory" consists of manures, and
half-manures, rent, taxes, ploughings, and harrowings on land
fallowed for wheat, the expense of any green crop left for the
incoming tenant, the growth of underwood in the hedges, the
value of old lays, &c. The " manures " mean those from which no
crop has been raised; "half-manures" are those from which one
crop has been produced. Lime is calculated in the same way.
It is almost impossible to value with accuracy the half-manures,
either as to the quantity applied or the quality, the only evidence to be had being that of an interested party. Old lays
are such as have remained in grass over one year, and which
by custom the tenant might have ploughed and cropped with
wheat. A year's rent of such lays is the usual allowance if the
land is in fair condition, or less, according to its condition. The
allowance for underwood is for the value of the growth
to the stem; and where the fences are very wide this is a
considerable item. In the hop districts underwood land, properly managed, yields large returns for hop-poles, — in many
cases larger than the land under cultivation. The incoming
tenant, in paying for the different articles of this "inventory," must, on an extensive farm, sink a large amount of
capital, probably, on an average, not much under 2*l.* an acre,
and on hop farms considerably more. The effect of this has
been to limit the competition for farms, and to produce a moderate rate of rental. It also enables the tenant, if necessary, to
borrow money, which is readily lent to him on the security of
his "valuations;" and these, in fact, are very frequently mortgaged. But the system has serious drawbacks. It obliges an
incoming tenant to sink a large portion of his capital at the
commencement, and in that way cripples him of much that
would be required in carrying on the cultivation of his farm.
It encourages trickery and deceit, a man who has been taken
advantage of at his entry thinking himself quite justified in
retaliating on his successor. Indeed, some men are such adepts
at this that they find it profitable to change from farm to farm,

their profit arising from the difference they receive when they
go out above what they paid at their entering.   The subject is
of such importance, that we quote two cases in illustration,
supplied by Lord Liverpool, in a letter published in the *Sussex
Advertiser*, on the 15th of January last : —

"Some years ago," writes his Lordship, "I had a farm on my hands,
in East Sussex, for one or two years. The quality of the land not
being very good, I had some difficulty in procuring a tenant. At
length one appeared, but as his purse was not very full, I allowed the
inventory to lie ; that is, in other words, he did not pay it. He only
remained one year—namely, from Michaelmas to Michaelmas, used
the farm extremely ill, but contrived to swell his outgoing inventory
by every trick, in which he was an adept ; and upon his leaving the
farm received a difference upon the inventory of above 100*l*. In
another case of a tenant leaving upon this estate, the valuers met long
after Michaelmas ; the wheat seedings had taken place, and the manure
having been previously carried out, spread, and ploughed in, no trace
or record remained to guide the valuers as to the quantities of manure,
except what the outgoing tenant and his people chose to tell them.
The outgoing tenant was asked the quantities. He said he had put
100 loads of manure to the acre. This was such an astounding un-
truth, that the valuers looked aghast, but, as the outgoing tenant had
quarrelled with his waggoner, the latter showed the valuers a ' chalk'
account which had been made of cartloads taken out of the yard,
whereby the 100 loads an acre were reduced to 25."

While the system leads, in some cases, to such roguery, we
cannot say that it appears to produce good farming, the cultivation of the Weald, where it exists in greatest force, being far
inferior to that of West Sussex, where the payments are confined to acts of husbandry and the value of the hay and straw.
We draw no conclusion from this, however, at this stage of our
survey, against the principle of payment for unexhausted improvements though it shows the necessity for caution in dealing
with the subject.

LETTER XVII.

ESSEX.

VICINITY TO THE METROPOLIS HAS NOT HAD THE EFFECT OF MAKING ITS
AGRICULTURE PROSPEROUS — VARIETY OF SOILS — CHIEFLY CLAY — LAND-
LORDS HEAVILY MORTGAGED, AND THEIR ESTATES CONSEQUENTLY INJU-
DICIOUSLY MANAGED — DRAINAGE BY MOLE PLOUGH — AND WITH STUBBLE
— FARM BUILDINGS INFERIOR AND VERY COMBUSTIBLE — TENURE — RENT
— DEPRECIATION OF AN ESTATE IN VALUE — WAGES — COTTAGE RENTS —
"CUSTOM" — FARMING IN THE ROOTHINGS — BURNING OF SOIL WITH
STUBBLE — MR. HUTLEY'S FARM AT WITHAM — MANAGEMENT OF CROPS
AND STOCK — YARD FEEDING OF SHEEP — LARGE RETURNS FROM PIG
FEEDING — MR. MECHI'S FARM — MODE OF MANAGEMENT — PROXIMITY TO
THE METROPOLIS NOT TAKEN FULL ADVANTAGE OF BY FARMERS.

COLCHESTER, April, 1850.

THE position of the county of Essex, almost touching London,
and constantly traversed by the best farmers of Suffolk and
Norfolk on their way to the metropolis, has exposed it to much
agricultural criticism. With every facility which railways,
roads, and navigable rivers can supply for the disposal of produce and fetching back manure, this county might be expected
to be eminently well cultivated, the landlords wealthy, the
farmers prosperous, and the labourers fully employed. This is
far from being the case however, and there must be some
peculiar causes at work to produce results so different from
what might have been anticipated.

The soil is principally clay, varying from the coldest and most
stubborn quality, to a marly loam. There are also turnip soils
on the chalk, on the north-western side of the county, and on the
gravels which in different tracts are intermixed with the clays.
To the north-west of Chelmsford, is the district called the Roothings, which is a marly clay. On the banks of the Chelmer, the
Stour, and the Colne, are tracts of gravelly soil, chalky clay, sand,

K 3

and loam.   The Dengie Hundred on the east, is a specimen of the
heavy clays of Essex, that stubborn and unprofitable description
of soil which, from the expensive nature of its cultivation, leaves
a smaller free return to the farmer than any other kind of land
in England.

In a county where this heavy soil predominates, it must be
evident that great exertions are necessary to render its cultivation profitable.   The landlords of Essex generally, however, do
not co-operate with their tenants in carrying out permanent
improvements.   With few exceptions, they have shown complete indifference to agricultural enterprise, neither laying out
capital themselves, nor offering such security as would induce
their tenants to do so.   They impose restrictive and ill-con-
sidered covenants even on their most intelligent tenants, and
preserve their hedgerow timber with the utmost rigour.   The
" root ditches," by which the farmer in some parts of the county
cuts off the connection between the hedgerows and his fields, to
prevent them from robbing his corn crops of their nutriment, are
not allowed to be made on certain estates.   An explanation of
all this suicidal and unaccountable mismanagement, may be
found in the fact, that the landed property in the county is incumbered with mortgage debts and other liabilities to the extent
of half its value, while the proprietors are nevertheless extremely
tenacious of the influence which their position gives them over
their tenants, and are afraid to entrust them with such security
of tenure as might diminish that influence.   These mortgages
and embarrassments naturally throw the landlords into the
hands of solicitors, who, having themselves no practical knowledge of the subject, send down land valuers from London to fix
the amount of rent to be charged.   But that intelligent supervision, which the personal knowledge of either the proprietor or
a duly qualified resident agent should give, is in such cases
wholly wanting; and a tenantry who are encouraged neither by
sympathy nor example, and who are positively obstructed in
their voluntary efforts for improvement, soon lose the spirit of

enterprise by which alone the difficulties of clay-land cultivation can be overcome.

In the heavy clay district tile drainage is not approved of. The land is there laid into narrow stetches, with water furrows to carry off the surface water. On soils adapted for it the mole plough is used to a depth of 16 inches below the plough furrow of 6 inches, thus making a drain 22 inches in depth, with much advantage, at very moderate cost. On the lighter clays, drains are made by the tenant 22 inches in depth and 32 feet apart, where he thinks the outlay will repay itself in a few years. These drains are filled with "haulm" (stubble), and the subsoil, being a stiff clay, forms an arch which remains open after the haulm has decayed. This temporary mode of drainage is resorted to because in very few instances does the landlord contribute one farthing to the permanent improvement of his land. Such drains are therefore made on the more friable clays, as may be expected to last a short tenure; and on the heavy clays, where nothing but a thorough and more expensive system would produce any effect, the work is left undone altogether.

The farm buildings are usually of wood and thatch, old, and in warm weather as dry as tinder; the yards are plentifully littered with straw; the hay and corn stacks close at hand; and the whole pile has such a combustible appearance, that one cannot wonder at frequent cases of incendiarism. Lord Petre and some other proprietors have begun to erect more substantial and commodious accommodation for their tenants; but the common practice in the county is for the proprietor to give the wood, very frequently grown at his tenant's cost, and for the farmer to erect the buildings and provide the thatch at his own expense.

The holdings are generally from year to year, especially where the farms are small and the tenants a less enlightened class. Leases when granted are from 7 to 14 and sometimes 21 years. The main improvement on the stiff clays of the south-eastern part of the county is made by marling or chalking, the chalk being brought across the Thames from the Kentish coast, at an expense of 4*l.* to 5*l.* an acre. But this outlay cannot be safely incurred by a farmer holding on a yearly tenure, and if a lease is denied to him the due cultivation of the land is frustrated.

The rent of land in the north-eastern part of the county varies from 20*s.* to 30*s.* per acre ; in the Roothings 15*s.* to 20*s.*, poor rates being about 3*s.* 6*d.*, and tithe from 5*s.* to 6*s.* an acre. On the stiff clays the rent may be stated at from 10*s.* to 15*s.* an acre. A farm of 400 acres of good strong land, with good buildings, and residence, and within four miles of a railway station, one hour from London, a good corn-land farm, was lately let at 13*s.* an acre ; the rent charge, poor rates, &c., also payable by the tenant, being 6*s.* 6*d.* ; altogether 19*s.* 6*d.* an acre. This farm was bought a few years ago for 9,000*l.*, and the purchaser becoming embarrassed, it was again sold last year for 6,500*l.*

Labourers' wages are 8*s.* a week, and in some cases beer besides, which is valued at 1*s.* more. Cottages are scarce, and the rent is in consequence run up to 3*l.*, 4*l.*, and even 5*l.* a year.

The only payments customary between incoming and outgoing tenants are for acts of husbandry, which include the value of dung, the rent of naked fallows, and the cost of tillage for turnip sowing and hoeing.

Though the county, as already described, consists of various soils, the system of agriculture followed on the heavy and lighter lands respectively is pretty uniform. Corn farming is the distinguishing feature of the district, and long fallows and diligent hoeing keep the land very clean and free from weeds. The four-course system is generally adopted. Barley is sometimes sown after wheat, when the land is in a rich state, and excellent crops are got by this management. To illustrate the mode of cultivation we shall take a farm of 200 acres near the Roothings —a clay marl district, some miles west of Chelmsford. The soil is on a gentle slope, by no means strong clay, mixed with small stones and chalk, the fields large and in this case not incumbered with wood or wide hedgerows. The farm buildings are abundantly commodious for the stock at present kept by the farmer. They are erected by the landlord and kept in repair by the tenant, who farms on a lease of 14 years. Where the tenant thinks it necessary, he drains the land at his own expense, making the drains 22 inches deep and about 32 feet apart, and filling them with haulm (stubble). They are made in the division that is to be fallowed.

The fallow is ploughed and harrowed as often during the summer as the farmer thinks it necessary, never less than five or six, and occasionally as often as eight times. A portion of it is burnt annually, and that which is in the most foul condition is chosen for this operation ; indeed it is found advantageous to sow rye grass occasionally with the preceding crop, in order to get plenty of roots and organic matter to assist in the combustion of the clay. Early in May the land to be burnt is ploughed very light, well dragged about, and then gathered into heaps, a quantity of haulm having been previously placed in the centre. This is set on fire, and the earth packed round it, care being taken not to let the fire burn through without putting on more earth, while too heavy a quantity at a time must also be guarded against, as that would extinguish the fire. The fires are kept burning slowly night and day till the whole is reduced to ashes. These are spread over the ground at the rate of 100 or 120 yards an acre, and at a cost in labour of from 20s. to 25s. The effects of the burning are that, after it, the land dries sooner, can be worked and sown earlier in spring, and that both the quantity and quality of crops are improved, especially so of barley and clover. Experienced farmers say that the oftener it is burnt the more the soil is improved, and in many cases the process is repeated every sixth year. It is most necessary that the land should be well under-drained before being burnt.

Four or five acres of the division in fallow are sown with swedes and mangold wurzel well manured. Half of this division is sown in autumn with wheat (six pecks an acre), the other half early in spring with barley (four bushels an acre); the barley taking the place of the wheat in the next rotation.   The barley land is sown with 14 lb. of red clover to the acre, the greater portion of which is fed off and a small portion mown for hay.   The wheat is followed by beans, the land being dunged and the seed dibbled in.   The beans are not horse-hoed, but kept remarkably clean with the hand-hoe. They are hand-hoed by men at a cost of 3s. an acre for each hoeing, and that is repeated five times in a season if necessary, but never less than three times.   The clover and beans are both followed by wheat.

In the 200 acres there are thus annually — 45 acres in long fallow, 5 acres in roots, 75 in wheat, 25 in barley, 25 in clover, 25 in beans.   The "haulm" already mentioned is the stubble which is mown and stacked up in long heaps after harvest.   The stubble for this purpose is left about two feet long, the farmer arguing that the less bulky he can make his crop in harvest, when wages are high, the better.   In this way a much greater number of bushels of grain are carried home in the waggon, stored in the rick yard, and finally much more easily passed through the thrashing machine.   When the busy harvest period is over, the haulm or stubble is cut with the scythe, and carried to the field where the operations of burning and draining are to be effected next season.

Under this management the crops average 28 bushels of wheat, 40 of barley, and 32 bushels of beans.   The whole stock kept on this 200-acre farm, is 80 sheep in summer, 5 or 6 cows, and 12 or 14 straw-yard cattle.   Eight work horses do the horse work; three in a plough in winter, two in summer. The rent is 20s. an acre; tythe and rent charge, 6s.; poor rate, &c., 4s.; or about 30s. an acre altogether.   There is no hay or straw sold, and about a ton of guano is annually bought.

On the farm of Mr. W. Hutley, of Witham, we found a much more enriching system adopted.   By heavy applications of purchased manures, and the conversion of all his straw into excellent dung — by using his roots in conjunction with cake and corn, for feeding his cattle, he keeps his land in a high state of fertility. He drains his lighter land, at his own cost, with 2-inch pipes, laid 32 inches deep and 32 feet apart.   He thinks, with many others in this county, that heavy land receives no benefit from tile drainage.   His fences are kept very narrow, and the land ploughed close to their roots.   On land which is not too heavy for roots he thinks it advisable to have a long fallow, perhaps once in eight years, on the principle " that soil which is generous to him should be treated gratefully in return."   A crop of tares preceding a fallow " draws " the land, in his opinion, to the extent of 20 bushels of barley an acre; that is, he would expect 56 bushels an acre without a crop of tares, and only 36 bushels when a crop of tares had been previously taken. He manures highly for his mangold wurzel, the yellow globe variety, using 30 loads of dung, 4 cwt. rape dust, and 2 cwt. guano, per acre.   The result is a yield of 35 tons an acre over his whole crop, and that he is now selling at 15s. per ton on the spot, to be sent to London, which is equal to 26l. 5s. an acre. The green crop thus appears to be much more remunerative than the corn crop when it can be disposed of on such advantageous terms.   40 bushels of wheat per acre, and 56 bushels of barley, are reckoned equivalent crops when the soil is in equally favourable condition.   Wheat is sown broadcast, after clover and beans, at the rate of six pecks an acre; barley is drilled in after a long fallow in spring.   Mr. Hutley's system is to have one fourth of his farm in wheat, one fourth in fallow and roots, one fourth in barley, and one fourth in clover, trefoil, and beans. By changing the latter every rotation red clover is repeated only once in 12 years, and a plant seldom or never fails.

In the management of stock Mr. Hutley's practice is to turn his horses into a large open yard in front of the stable after they have had their bait of corn, and here they remain out night and day, when not in the yoke, summer and winter.   He is never troubled with grease or other ailments among his horses.

His sheep are fed partly in the field and partly in yards. The couples are fed in the field on roots. 300 teggs are kept in two adjoining yards, 150 in each — one provided with shelter sheds, the other quite open. Both yards are well littered with straw, and in these the sheep have been kept during the winter. They receive roots, cut chaff, and 200lb. oil cake daily among the 300. They are now being sold out at an increased price, between carcase and wool, of 18s. to 20s. for 30 weeks' keep; thus leaving the cost of the cake (about 9s. a head), a large quantity of rich manure, and 4d. a week for the roots and chaff.

The feeding of pigs is carried on to a great extent by Mr. Hutley. He breeds none, but buys pigs at about 18s., and feeds them five weeks, when they are ready for the London market. They are fed on meal of different kinds, and sometimes on boiled Indian corn. The money realised, including prime cost, from the pig stock for one year has reached more than 2,000l., and seldom falls below 1,200l. or 1,500l. As this sum goes to pay for the corn consumed by the pigs, it shows how much Mr. Hutley is every year adding to the fertility of his farms. Oxen are fed on meal and chaff; few are kept, as they are not, with this management, found to be a paying stock. Mr. Hutley attributes his success in farming to a liberal application of capital to the land, both by drainage, chalking, artificial manures, and, above all, by keeping a large stock and employing sufficient labour. To do this he has been encouraged by a moderate rent and entire confidence in his landlord.

At Tiptree-hall we examined the well known and much discussed farming operations of Mr. Mechi. The regularity and luxuriant appearance of the wheat crop on the fields next the public road led us to anticipate an instructive visit, and we were not disappointed. It is of course quite unnecessary to enter into any history of the farm, as Mr. Mechi himself has already made that public. We shall, therefore, confine ourselves to a short description of some interesting matters of detail. The farm is 170 acres in extent, principally a strong soil, with a very impervious clay subsoil. It adjoins Tiptree-heath, which is naturally very barren. By drainage, ample manuring, and liberal expenditure, the whole farm is kept in constant tillage, one-half of it every year under wheat, the other half in clover, Italian rye grass, tares, and roots. The wheat is drilled on stetches, about 7 feet wide; it is twice horse-hoed. Beans, pease, and tares are also sown by the drill. Red clover is sown on the same land once in eight years, and never misses a plant, 12 lb. seed being allowed to the acre. The Italian rye grass, though thin on the ground, is the most forward spring feed we have seen this season. Having walked over every field on Mr. Mechi's farm, we have no hesitation in saying that for clean cultivation and healthy appearance of wheat and other crops, it is equal to any, and superior to most, farms we have met with in this county.

In the management and accommodation for stock Mr. Mechi is yet experimenting. The stock regularly kept on the farm are 150 sheep, 200 pigs, young and old, 24 fatting bullocks and cows. Besides roots, 10 sacks of meal are used daily in feeding them. 700 to 1,000 quarters of corn are bought annually for this purpose. All the animals are kept on boards to economise the straw, and, with the exception of some of the pigs, they looked clean and comfortable. Mr. Mechi considers it proved that pork at 6d. a pound will pay for barley at 36s. a quarter, and at 4d. a pound for barley at 24s., over and above the manure. Two bullocks are placed in each box, 10 square feet of space being allowed to each. The boards on which they stand are 3 inches broad, with 2-inch interstices. For calves 1½-inch interstices, and for sheep 1¼, are found best. One man feeds 200 pigs, mixing and carrying them food. They are fed thrice a day. A 6-horse steam-engine is employed in thrashing the crop, cutting chaff, grinding meal, bruising linseed, hoisting sacks, &c. In short, no expenditure is spared by Mr. Mechi;

and whether it has on the whole been profitable to himself or not, there can be no doubt whatever that his example has in many points been instructive to the agricultural community.

In concluding our observations on Essex, we may point to the fact that hitherto the chief dependence of the farmer has been on his corn crops, cattle being kept for manure, but not generally as a source of profit. Considering the proximity of this county to London, and the consumption of milk which might ensue if the enormous population of the metropolis could obtain a supply of good milk on moderate terms, we are of opinion that farmers would find it very advantageous to turn their attention to this source of profit. Intersected as the county is by railways, there is nothing in the distance to prevent daily supplies being sent to London, and milk is a commodity in which we are not likely to have much foreign competition. To obtain a supply of this on the clay farms, we should expect a system of house-feeding on clover and tares in summer, and on mangold wurzel, cabbage, &c. in winter, the most economical, while it would also provide an advantageous means of converting into fertilizing matter the " haulm " which is now looked upon almost as a nuisance by many Essex farmers.

LETTER XVIII.

ESSEX.—SUFFOLK.

MR. FISHER HOBB'S BREED OF ESSEX PIGS — EXCELLENCE OF HIS OTHER STOCK
AND MANAGEMENT. —— SUFFOLK. —ARTHUR YOUNG — HIS SUGGESTIONS
THOUGH MUCH SNEERED AT BY " PRACTICAL " MEN OF HIS DAY HAVE
SINCE BEEN GENERALLY ADOPTED — SWEDES AND MANGOLD INTRODUCED
BY HIM — AGRICULTURE GREATLY INDEBTED TO HIM FOR ITS PROGRESS
— UNCONCERN OF LANDLORDS IN THE IMPROVEMENT OF THEIR ESTATES
— INJURIOUS PRESERVATION OF GAME — FARMERS WITH BORROWED
CAPITAL NOT EXPECTED TO KEEP THEIR POSITION — RELATIVE VALUE
OF WAGES AND FOOD — AGRICULTURAL IMPLEMENT MANUFACTORIES —
MESSRS. RANSOME'S — MESSRS. GARRETT'S — IMPLEMENTS SENT ABROAD.

SAXMUNDHAM, SUFFOLK, April.

FROM Colchester, in Essex, eastwards towards the sea, stretches a tract of well farmed land. We may especially notice that held by the Messrs. Ward, of Great Bentley, whose cleanly cultivated fields and admirable stock of every description are sure to claim the attention of the observant traveller.

From Colchester in the opposite direction, and nearly on the borders of Suffolk, is Boxtead-lodge, the residence of Mr. Fisher Hobbs, a distinguished stock breeder. His improved breed of Essex pigs is well known at all the great agricultural shows. They are perfectly black, rather small size, and of somewhat delicate appearance, peculiar for early maturity and fineness of flesh. They are superior to any other breed as "jointers" of 50lb. weight or so, for the London market, which weight they make at about three months old. When kept to a greater age they feed well, making, with good management, a score, or 20lb., weight a month; a ten months' pig usually weighing 10 score. The breed, being kept perfectly pure by Mr. Hobbs, is in much demand for crossing. With the Berkshire it makes an excellent cross, keeping its properties of early maturity with increased size. All the perfect animals are retained by Mr. Hobbs for

breeders. They are sent to many parts of the United Kingdom
and abroad, and are so much in demand that 10 guineas each is
commonly realised.   One litter of 12, which we saw, Mr. Hobbs
expects to sell before the end of the year for 100*l.*   Besides
pigs, Mr. Hobbs keeps a very pure breed of Leicester sheep,
Hereford cattle, and Suffolk horses.   His agricultural management is also deserving of notice.   Having lately succeeded to
the estate he now occupies, he is actively engaged in its improvement, levelling down unnecessary fences and abrupt eminences, clearing the land of superfluous timber without sacrificing
its natural beauty, grubbing out useless underwood, forming irrigated meadows, opening up better roads of access to the
different parts of his farms, and by a better style of cultivation
adding fertility to his fields, and affording constant employment
to his labourers.   The residence of such men as Mr. Hobbs, Mr.
Mechi, Mr. Hutley, Mr. Baker, and the Messrs. Ward, in different parts of the county, must give a great stimulus to the
development of the agricultural resources of Essex.

Crossing the river Stour we enter the county of SUFFOLK,
passing through a fine undulating country to Ipswich.   This
county possesses a peculiar interest to the agriculturist, as
having been for many years the residence of Arthur Young.
We had the good fortune to meet with Mr. Biddell, of Playford,
himself an extensive farmer, who was acquainted with Arthur
Young, and had frequently conversed with him on agricultural
subjects.   His ideas are represented to have been much in advance of the period in which he lived, and, though they were
ridiculed by the great body of the " practical " men of his day,
our informant has lived to see most of his recommendations
carried into practice, and considers the county indebted to him
for much of the progress that has been made in the cultivation
of its soil and the economical application of labour.   Swedish
turnips were early introduced by him, but not generally cultivated for 20 years afterwards.   When mangold wurzel was first
introduced by him into this county its value was so little under-

stood that for some years the leaves only were given to the stock, and the roots thrown away as worthless. It was at length noticed that the hogs seemed to eat with great relish the despised roots cast out on the dunghill, and this led to the general cultivation of a plant of inestimable value to the heavy lands of this county, and now indispensable to their profitable occupation. The accurate agricultural surveys of every county in the kingdom, set on foot by Young, as secretary of the Board of Agriculture, gave a great stimulus to improved practice throughout England by affording *data* for comparing the practices of different soils and districts, and in that improvement the county of Suffolk largely participated.

The relations subsisting between landlord and tenant are canvassed by the tenants of Suffolk with a degree of earnestness which we have not met with in any other county. They complain that, until within the last two or three years, their landlords gave themselves very little concern about the welfare of the tenants, or the management of their estates; that expenditure on drainage or farm buildings was hardly ever made by the proprietor; and that their intercourse with the tenants, and the supervision of their estates, was in many cases carried on through the medium of solicitors and others not practically acquainted with the management of landed property. There has been no leading man in the county, for a long series of years, to infuse into it a spirit of agricultural improvement, as the Earl of Leicester did for Norfolk, or the Yarborough family for Lincoln. The repeal of the corn laws and the fall in prices have at last compelled attention to a business which has been far too long neglected, and the landlords of Suffolk now begin to appreciate their true position. The larger proprietors in general are believed to be sufficiently unincumbered to undertake their share of the permanent improvements which must be made; and some of the smaller landlords, whose necessities have obliged them to provide for their families by such heavy incumbrances as to render them unable to make

L

the requisite outlay, will be obliged to part with the estates which
they can no longer hold with advantage.   Though farms are not
let by tender, the farmers complain that their landlords take advantage of every kind of competition, to increase the rent beyond
the valuation which may have been made by competent parties
for the owners' private information.   The injurious operation of
the present law of distress is also complained of, and likewise the
unnecessarily restrictive clauses as to the course of cropping
generally introduced into the leases.

The preservation of game on some estates is carried to an
enormously injurious extent.   In one parish the tenants have
subscribed among themselves 30*l.*, 40*l.*, and 50*l.* each, making
altogether a rent of 200*l.* a-year, to take the game from the
landlord on lease, and thus keep it within bounds.   Nor is the
mischief confined to the actual depredations of game; for
hedgerows are preserved to harbour it, and constant heartburn-
ings between landlord and tenant are the result.   The abundance
of game entices the labourer to become a poacher when employment is slack, and so leads to the demoralization of the lower
classes.

Under circumstances hitherto so discouraging, the position of
the Suffolk farmer has been gradually reduced; and in the
opinion of persons well qualified from local knowledge to judge,
it is thought that a continuance of low prices will bring ruin on
those who have been farming with borrowed capital.   And
there are many such; for it has been common for a young man
beginning business to increase the capital at his command by
borrowing from relatives, or from a neighbour of substance who
has confidence in the ability and integrity of the party to whom
he lends.   Though few failures have yet taken place, it must
not be concluded that the farmer's complaints are louder than
necessity warrants.   A farmer may manage, by curtailing his
expenses, diminishing his stock, and living very frugally, to
stave off actual failure for a considerable period; but when he
goes at last, his ruin is complete.   Unlike a tradesman, he has

no wholesale manufacturer to set him up on credit a second time.   The landlord's preferential claim secures the safety of the rent, and the whole loss falls on the friends and relatives, who have no inducement to run the same risk again.   To such parties it is of course unnecessary to talk of better farming and larger produce, as a compensation for reduced prices; and we can only regret that they were ever tempted to enter on a business for which they had not adequate capital.

With regard to the wages of labourers, they have in many instances been reduced 1s. to 2s. a-week; — 7s. and 8s. a-week are now the average wages of the county, and cottage rents being high — from 3l. to 5l. per annum — the balance left to the labourer is sufficiently small.   Yet the reduction in the price of what he has to buy is nearly equivalent to the fall in wages.   We were informed by a labourer, that for himself and his wife, without children, the following articles of weekly consumption cost respectively at present, and formerly when wages were higher and corn dearer, these several sums: —

			s.	d.			s.	d.
1 stone of flour	-	-	- 1	10	-	- 2	6
½lb. of butter	-	-	- 0	6	-	- 0	8
1lb. of cheese	-	-	- 0	7½	-	- 0	10½
1½oz. of tea.	-	-	- 0	4½	-	- 0	6
½lb. of sugar	-	-	- 0	2	-	- 0	2½
			3	6	-	- 4	9
Rent of cottage	-	-	- 2	0	-	- 2	0
			5	6	-	- 6	9
Weekly wages	-	-	- 8	0	-	- 10	0
Balance for sundries	-		- 2	6	-	- 3	3

If cottage rents had fallen in the same proportion as the other items, the present rate of wages would be fully equivalent to the former; but that has not been the case.   And it is but fair to the farmers to say that, though they are quite conscious in many cases that well-paid labour is the cheapest, and that it would be better for them to have fewer labourers, and those better fed, they are precluded from adopting this course by the necessity imposed on them of maintaining, either in the workhouse or out of it, all the able-bodied labourers of the parish. In most cases, as the land is at present managed, the supply of labour is redundant. During harvest and at task work considerably higher wages are earned.

Before entering on a description of the agriculture of Suffolk, which we reserve for our next letter, we think it may be interesting to give a brief description of that which forms so prominent a feature in Suffolk — the manufacture of agricultural implements. The names of Ransome, Garrett, and Smith of Peasenhall, are familiar to agriculturists throughout the kingdom; and the general adoption of their improved implements has given a character of neatness to the cultivation of the land, which it may be feared has in many cases been accepted as a substitute instead of a help to that enriching system of farming which is required by the lighter and more hungry soils of this country. Messrs. Ransome and May, of Ipswich, who unite implement-making with mechanical engineering, possess an establishment there which is one of the most extensive and well arranged of the kind in the kingdom, covering 10 acres of ground, and employing 800 to 1,000 men, nearly one-half of whom are constantly engaged in the department of agricultural implements. The highest engineering skill and abundant capital are applied to the perfecting and cheapening of our simplest farming machines; and the degree of finish which is given to every article, either in wood or iron, is equal to what is usually found only in the most costly descriptions of iron and wood work. It would be quite out of place to enter here into any detailed description of the different processes carried on in this extensive and most orderly establishment, but one or two points may be adverted to. Upwards of 300 distinct varieties of the plough are manufactured here, each of which is in greater or less demand. Each pattern has its distinguishing mark, and all the different parts of the plough can, by reference to that mark, be at any time supplied by the manufacturer. By the simple process of case-hardening the ploughshare on the under side, a sharp, thin edge is maintained by the wear of the upper and softer part in its work. Scarifiers, harrows, thrashing machines, clod crushers, horse rakes, iron rick stands, are a few of the great variety of implements made by Messrs. Ransome and May, who, besides, carry on an immense business in the manufacture of railway chairs and trenails. The latter manufacture involves an amount of ingenuity in the process by which it is prepared, little, if at all, inferior to the celebrated block machinery at Portsmouth.

The works of Messrs. Garrett are situated at Leiston, near Saxmundham, and give employment to between 300 and 400 men. This establishment is, in the strict sense of the word, an agricultural implement manufactory, and from it have been produced the description of drill and horse-hoe which are now held in such great estimation. Portable steam engines and thrashing machines are also manufactured largely by Messrs. Garrett. They pay great attention to the quality of the work turned out by them, and make it their study to adapt their implements to the requirements of agriculture as it exists around them. Guided by this principle, they have endeavoured to render the machinery for farm purposes portable, instead of being fixed and permanent. On large holdings their steam engine and thrashing machine can be readily put into use at any point which may be considered convenient; and, should the farmer wish it, they can be rendered stationary. The difference of price and efficiency between fixed and locomotive farm machinery is certainly in favour of the former.

Both Messrs. Ransome and Messrs. Garrett are at present extensively engaged in supplying orders for the other corn-growing countries of Europe, and they state that the implements sent out by them are less for use than to serve as models for the general introduction of similar implements. Had it not been for this foreign demand, their trade would have been considerably depressed, as it is at present not up to the average of more prosperous times, even with that addition, and with the fact which at both establishments we heard repeatedly stated, that farmers are every day becoming more and more alive to the facilities which mechanical science affords for economizing labour.   Messrs. Ransome are at present completing a very perfect thrashing machine for a Polish nobleman, an annual grower of 10,000 quarters of wheat, who, when recently in this country, assured those gentlemen that free trade with England would be of no use to him unless we could afford to pay 40s. a quarter for his wheat at Dantzic.

LETTER XIX.

SUFFOLK.

STOW-MARKET, April.

THERE are three distinct varieties of soil, which give to the agriculture of Suffolk its leading characteristics. Along the eastern
coast lies a narrow tract of sandy land mixed with shells and
other marine deposits, interspersed with salt marshes, in some
places so light as not to be worth 5s. an acre, and in others a useful
free soil fetching a rent of 28s. From this tract the coprolites
used in the manufacture of superphosphate of lime are chiefly
obtained. The central and south-western districts are occupied
by the prevailing soil of the county, a strong loam on a subsoil
of clay marl, and, towards the course of the rivers, a rich friable
loam. On the north-west is a light sand on a substratum, at
greater or less depth, of chalk, which, in the time of Arthur
Young, was rabbit warren, or very worthless sheep-walk, but
has since been to a great extent brought into cultivation.

The lighter soils on the eastern and western side of the county
are held, in conjunction with portions of fen and moor land,
by large farmers, generally men of capital, who have leases of

from 7 and 8 to 14 years.   On the heavy lands the farms seldom exceed 300 acres, and are sometimes not more than 100 or 50 acres in extent.   They are held chiefly from year to year by men of little capital, and who are suffering severely from the pressure of the times.   Farm buildings throughout Suffolk, being erected by the tenant, principally at his own expense, are made in a very unsubstantial manner, the side walls being of wood, the roof thatched, and the whole requiring constant repair, and being a fruitful source of inconvenience and waste.   The cattle sheds, the barns, and in fact all the premises, are deficient in economical arrangement.   Some of the more modern buildings are constructed of clay dried in blocks, but not burnt, which is found to make a very cheap and durable wall.

The chief characteristic of Suffolk agriculture is the success with which heavy land farming is carried on.   The strong land of the county, already referred to, is not a continuous tract, as it is everywhere interspersed with fields of a more friable texture, which are found very valuable when held along with a clay land farm.   The clay land forms not a flat, but a gently undulating country, affording ready means for drainage.   The soil contains generally a considerable admixture of gritty sand and some pebbles; while in the subsoil in many cases are found beds of chalky marl, which, after exposure to the air, are applied with much advantage to the surface.   Drainage is of course the primary improvement on this description of land, and it is effected in the cheapest manner, as in scarcely any instance has the landlord hitherto contributed any portion of the outlay.   Drains, two feet deep and 15 feet apart, filled with bushes, are giving place to drains three feet deep and 24 feet apart, still filled with the same material.   The cost of the operation in labourers' wages is under 30s. an acre, and the benefits are expected to last for a 14 years' lease.   At the beginning of a new lease the land is gone over again, the direction of the drains being now made to cross the old drains obliquely, and thus to *bleed* such as still remain open.   As a long fallow is regarded as a routine operation twice or thrice in the course of a short lease, so is draining looked upon as a matter of regular recurrence once every 14 or 16 years.   There can be little doubt that, if executed in a careful manner with tiles, the work would be made complete, and this constant draft on the tenant's capital rendered unnecessary.   For main drains, pipes are generally used at present ; and in all cases where it is found desirable that the work should be permanent, pipes or tiles are used throughout, and the drains cut from three to four feet in depth.

Next to under drainage, the great principle of the Suffolk heavy land farmer is to keep his surface soil dry and clean. For this purpose it is ploughed into " stetches " about 8 feet 2 inches in width, the furrows between each " stetch " acting in heavy rains like gutters, and being used in every operation, either in sowing the seed or after the seed is in the ground, as a trackway for the horses, which are thus never permitted to trample or injure the tender soil.   The different implements used are all constructed of a size to suit this width of " stetch " The drill covers the whole space, the horses walking in the two furrows on either side of the " stetch ; " the harrows are of the same width and drawn in the same way.   The roller is frequently made in two halves to adapt itself to the curvature of the " stetch," and the shafts are placed so that the horses walk in the furrows.   The gateways in the different fields are eight feet and a half in width, to suit the size of the implements. These " stetches " are found to answer well by giving a dry bed for the plant during the winter, and especially by preventing the injurious treading of horses on a soil so susceptible of injury. The space occupied by the furrows, or trackway, is little more than what is left betwixt each row of corn, and doubtless acts beneficially in admitting air more freely through the crop.

The course prescribed on all the heavy lands is the four-field. Beginning with the wheat crop, this is usually sown on one furrow after clover or roots.   The " stetches " are simply reversed, and after being harrowed, seed at the rate of five or six pecks an acre is drilled in *along* the "stetches," in rows of about nine inches apart, and 10 rows on the stetch. The seed is sown during the months of October and November. As early in spring as the land is in suitable condition, the crop is either horse or hand hoed, the expense of the latter being about 2*s*. 3*d*. an acre. Nothing can exceed the delicacy and precision of Garrett's horse-hoe, which is looked upon by the farmers of this county as an implement indispensable to clean cultivation. After being hoed, the roller is usually passed over the wheat, and in many instances both operations are repeated a second time. Should any root weed still exist, the whole ground is carefully gone over by children, and every weed picked off. By this management scarcely anything is left in the ground except the wheat. When the wheat is reaped, the ground is all gone over by men with forks, and any patches of twitch carefully taken out. It is afterwards worked about with the scarifier and harrows, the clods being gathered together and burnt. The land is then ploughed up in "stetches" for the winter. Early in spring, that portion on which roots are to be taken is ploughed into ridges 32 inches in width, or three on each stetch, and dung applied at the rate of 15 loads an acre. The ridge is then split over the dung, and mangold wurzel seed drilled in on the top of it. Some farmers prefer drilling it on the flat, and in this case the dung is usually applied in the autumn, and the seed drilled in spring, three or four rows on the stetch. In either way the land works well, as the soil which has been exposed to the ameliorating effects of winter is kept on the surface, and no time is lost in preparing the land in spring. For swedes the same system is followed. By the terms of agreement a portion must be wrought in naked fallow, and that receives during the summer the usual course of ploughings, is laid up in stetches for the winter, and sown with barley the following spring. The whole of this division, after roots and fallow, should, in strict terms of lease, be sown with barley; but wheat is very frequently substituted after the root crops. Barley is drilled in on the stetch in spring, and the ground is then sown with a peck of red clover seed per acre. In some cases this is covered slightly by having the ground hand-raked, which is done for about 7d. an acre. One half of this division is usually sown with clover; the other, after the corn crop being removed, is prepared for beans in the same manner as already described for roots. Every crop is repeatedly horse and hand hoed, and the soil kept remarkably clean. Thirty-two bushels an acre of wheat, 44 of barley, 36 of beans, may be reckoned average crops on the better description of heavy lands, where the details just mentioned are carefully pursued.

The management of stock is not attended with anything like the same success as the corn crops; and this department is felt by the farmers as not only barren of profit itself, but also as trenching heavily on the returns of the other. The land being chiefly under the plough, the stock of cattle kept is usually purchased in autumn, to be fed during the winter, and sold off in spring. They are put into large yards supplied abundantly with straw, and with 14 lb. to 18 lb. a day of corn and cake each, and one to two bushels of mangold wurzel. As few of the farmers breed their own stock, they usually comprise many varieties — Polled Galloway Scots, Shorthorns, Irish cattle, &c.; and as the breeders of the best description of cattle in their native districts are now, by the extension of green crops and facilities of rapid communication, becoming the feeders also, it follows that the worst specimens of each breed now find their way to the feeding counties. The quality of the polled Scots now sent to Norwich, the great market of the Eastern counties, is quite inferior to what it used to be, and nothing pays worse than a bad animal of this breed. Beginning with bad animals of their several kinds, the Suffolk farmers grudge no expense in trying to make them fat. Each bullock costs for its food not less than 40s. a month, and as the returns for the last two seasons have not probably exceeded 20s. in the increased value of the animal, there is an apparent loss of 6l. on each animal for the winter's keep. The farmer looks to the manure to make good his loss; but a little consideration will show that this is a most expensive mode of making manure. On a farm of 350 acres we shall suppose a stock of 40 cattle to be fed during six months of winter. At present prices, and with the usual mode of feeding, these lay a charge of 240*l.* on the manure. These 40 cattle make about 1000 yards of manure; but at least one fourth of this must be deducted for the value and bulk of straw. We have thus 750 yards of dung, costing 240*l.*, or nearly 6*s.* 6*d.* a yard. This applied at the rate of 15 yards to an acre will manure 50 acres of land. But if the same sum were expended on guano, superphosphate, and rape cake, at the present prices of these articles, 120 acres of land could be annually manured at the rate of 2 cwt. per acre of each of these substances, or 6 cwt. altogether, with the certainty, in our opinion, of a much heavier crop from each acre than would be yielded by the application of 15 yards of manure. By the adoption of this plan it would be unnecessary to tread down the straw in yards merely for the purpose of getting it converted into dung, as the greater amount of green crops would admit of a double stock of cattle, and would afford to these a supply of much less expensive food. This, however, leads to a large question, which, as it involves a change in the stock-farming altogether, we have not at present space to enter upon.

On the farms of Mr. Capon, of Dennington, who holds upwards of 5,000 acres of land, 2,000 of which are of the heavy land already described, we found some variations in the management adopted. On one farm, which is his own property, he is trying whether he cannot on the heavy land grow crops every year, without any naked fallows or root crops, which are at present found so unremunerative. The course he follows here is to have (1) a bean crop, followed by (2) wheat and (3) barley, which is sown with (4) clover, and followed by (5) wheat and (6) barley. Every crop is most carefully horse and hand hoed, all being drilled, and the land is kept quite clean. The practice has not been long enough followed to be accepted as proof of its

correctness ; but this much has been ascertained, that, by carefully horse and hand hoeing every crop, weeds are extirpated,
and the yield of each of the grain crops has been quite equal,
and of barley generally superior, to what is got under the regular
four-course.   For the work of his different farms Mr. Capon
keeps a stock of about 120 horses.   Two-horse ploughs are
universal on both light and heavy land, and the land is ploughed,
when necessary, with a deep strong furrow. The whole management of this extensive holding is conducted with great neatness
and skill.   The bill for oil-cake, &c., for feeding, sometimes exceeds 1200*l.* in a year, and an equal sum is expended on artificial
manures.

Much of the heavy land has been broken up from pasture
within recent years.   The native vigour of the soil in such cases
is very great, and it is usual to take several crops of wheat in
succession without any manure.   The mode adopted in breaking
up the land at first is to pare and burn it, at a cost of about 25*s.*
an acre.   If this can be done early enough in the season a crop
of oats is taken.   If too late for oats, the land is sown with rape
and fed off.   Wheat is then taken in succession — four or five
times — and great crops are reaped.   On the farm of Mr. Bond,
of Earl Soham (which is very neatly and well managed), the
second wheat crop yielded 40 bushels an acre, weighing 69 lb.
per bushel.   On this farm, besides the usual careful management
of all the corn crops, the clovers are gone over by children early
in spring, and every weed picked out by hand.

In the cultivation of the lighter soils of the county, the usual
details of the four-course system are followed out with a peculiarity in the preparation of fallows for green crops, which has
arisen from an absurd clause in the leases and the mode of payments between incoming and outgoing tenants.   Farmers are
required to plough their winter fallows five times, no matter
how light the land may be, even though it should be a blowing
sand ; and, where covenants are strictly enforced, this unnecessary expense must be incurred.   But the hardship is peculiarly

great to an incoming tenant, who must pay his predecessor for each of these operations, though they are in most cases rather injurious than otherwise. The year's rent and rates are likewise charged, so that the entry to a large light soil farm is a very expensive matter, the whole amounting to a charge of about 5*l.* an acre on the fallow division. Now, if this were an absolutely necessary expense, we should have less to say against it; but on the farm of Mr. Bond at Wickham-Market, we found that his light land is only ploughed once in preparation for roots; which ploughing is delayed till spring. The land turns up finely pulverised; it is perfectly wrought, and produces crops which contrast favourably with those of any farm managed according to the usual prescription. We had an opportunity of ascertaining the opinions on this point of several of the most intelligent light land farmers in the county, and found them unanimous in their condemnation of the compulsory five-furrow system.

Manure is used rather sparingly on the light land farms, and very moderate green crops are grown. So little is the turnip crop valued, that in many places it is sold for consumption by sheep at 1*l.* to 2*l.* an acre, and sometimes even less. On farms where breeding stocks of sheep are kept, it is thought that turnips which have been manured with guano or other forcing manure are injurious to the ewes, and accordingly, to guard against this danger, the turnip crop on such farms is sown without any manure. It is not easy to understand the *rationale* of this, but it may be, perhaps, owing to the more succulent nature of the root and its stronger growth in spring, purging poorly fed stock, and causing them to " warp " their lambs. On a farm managed on this plan, we found the tenant complaining of meagre crops; but it is difficult to see how they could be otherwise. A light sandy soil, turnips with no manure, eaten off by a breeding flock of ewes, are not the most favourable preparations for remunerative corn crops.

In the arrangement of farm buildings the occupier of a large farm prefers having several barns and feeding sheds at different points of his farm to having them all placed in one centra position near his own house, and more immediately under his eye.   He argues that a great saving of cartage ensues from this practice, as the corn crops are stacked at a barn near where they have been reaped, the roots are carried to a yard at no great distance, and the manure from both is returned to the land without heavy cartage.   To suit this arrangement of buildings, portable thrashing machines, whether of steam or horse power, are preferred to fixed machines.

The mode of conducting harvest work is somewhat peculiar. It is usually done by task work.   All the labourers are joined in the engagement, and the earnings divided among them. 7s. an acre for cutting and securing the crop of wheat, barley, and beans, is paid by some farmers.   This usually includes, also, the hoeing of the late turnip crop twice, which is done in the mornings or in weather not suitable for corn harvest.   For this sum the people cut the crop, pitch it into the carts, and build it in the rick yard.   The carters are paid separately, and the thatching is done separately by task work. At this rate of payment the labourers occasionally earn 1l. a week, but work hard for it.

Soot is used pretty extensively as manure, especially on farms much overrun with game.   It is believed to protect the crop in a considerable degree, by rendering it distasteful to the game.   On asking a farmer the present price, he said it was rather scarce this season in his neighbourhood, on account of a large demand from his landlord, who requires it for application to a farm overrun by game, which has been lately thrown on his hands.

The farm carts seem very cumbrous and heavy, and might with much advantage have some of the skill, which is so conspicuous in the drills and horse-hoes, applied to their construction.   Ploughs are generally of wood, with the wearing parts of iron, and almost universally with only one handle, and a cross pin by which to guide.

Near Euston Park, a few miles south of Thetford, we visited

an extensive light land farm on the Duke of Grafton's estate, which may be taken as an instance of management common to both Suffolk and Norfolk. It is about 1700 acres in extent, one half being under the plough, and the other half unbroken heath and pasture. In many respects it resembles the Down farms of the southern counties, saintfoin being grown regularly as a hay crop, and the system of folding the sheep on the cultivated land at night, and driving them a mile or two to the pastures during the day, being steadily carried out. 800 acres are managed on the four-course system. For turnips the ground is dressed with artificial manures, the dung being reserved for the wheat crop. $6\frac{1}{2}$ cwt. of rape dust, or 2 cwt. of guano, or 16 bushels of bones, are reckoned equivalents, and one or other of these substances is applied to the turnip crop. It is sown in ridges 27 inches apart. The greater portion is consumed on the ground by sheep, and part carried home to the farm-yard. To secure the roots from frost, four rows from each side are thrown into a central row, and covered with earth by the plough. They are easily picked out when required. Two classes of sheep stock are fed with turnips, — first, a lot of 500 fattening sheep, which receive cake and beans daily, besides having their turnips cut and placed in boxes; and, second, the breeding stock of the farm, 900 ewes in number, which are folded on the turnips at night, and driven to the heath during the day. When the turnips are consumed, the land is sown with barley and seeds. A small portion is sown annually with saintfoin at the rate of $3\frac{1}{2}$ bushels an acre. The seeds are partly mown and partly folded over and fed. They are then manured with a compost of dung and earth. The earth, which is turf cut from the heath, full of vegetable mould, is carted on to the seeds during the winter, when the teams are not otherwise occupied, and laid in heaps conveniently placed for future distribution. Dung is afterwards mixed with it in equal bulk, and the compost is laid on the land as a manure for the wheat at the rate of 20 loads an acre. The layers are thus dunged as far as the compost goes, and what remains is folded by the sheep, fed on cake also.

The quantity of stock kept on this farm is 900 ewes constantly, besides 500 fattening sheep in winter, and 40 bullocks in the yards.   The bullocks are fed loose in open courts with warm sheds.   They are each receiving daily, at present, 7lb. of cake, three quarters of a peck of meal, one bushel of cut mangold-wurzel, and saintfoin hay.   The horses are worked eight hours a-day in summer, in one yoking, leaving the stables at 6 o'clock in the morning, and returning home about half-past 2.   In winter they start as soon as there is light and stop at half-past 2.   When they return to the stable, one man takes charge of his team of six horses, cleaning and feeding them, which is reckoned his sufficient occupation for the rest of the day ; and, as more than two horses are seldom used in a plough, the other two men, now disengaged, are employed, after they have dined, at any other work about the farm till 6 o'clock.   Such is the routine management of a well-conducted light land farm on the borders of Suffolk and Norfolk.

On the Duke of Grafton's estate the rent of labourers' cottages is charged very moderately, — 1s. a week, down to 20s. a year, being the rule in the villages which are the property of the Duke.   In others it rises to 3l. or 4l. a year.   Labourers have in many cases to go a long way to their work, on account of the scarcity of cottages on the large farms.   On the farm just mentioned there are two labourers regularly employed, who walk every day from Thetford and back, 9 miles, — or 54 miles a week.

LETTER XX.

NORFOLK.

NATIVE BREEDS OF STOCK SUPERSEDED — VARIETY OF SOILS AND APPEAR-
ANCE OF COUNTRY — MR. COKE'S IMPROVEMENTS — ENCOURAGED TENANTS
OF CAPITAL BY GIVING LEASES, AND GOOD RESIDENCES AND BUILDINGS —
ESTABLISHED ANNUAL MEETINGS AT HOLKHAM FOR DISCUSSING AND
EXAMINING AGRICULTURAL OPERATIONS — IMMENSELY INCREASED PRO-
DUCTIVENESS THAT ENSUED — FARMING AT HOLKHAM PARK — BENEFIT
OF APPLYING NITRATE OF SODA AND SALT TO WHEAT — MR. HUDSON'S
FARM AT CASTLE ACRE — MANAGEMENT DETAILED — INCREASE OF STOCK
AND CORN — SAVING OF WASTE BY THE USE OF RAILWAYS IN TRANS-
PORTING FAT STOCK TO MARKET — FARMS OF MR. OVERMAN AT WEASEN-
HAM, AND AT BURNHAM SUTTON — SPRING HOEING OF CORN CROPS
RECKONED INJURIOUS — FOUR-COURSE ROTATION EXPANDING, AND CROPS
INCREASING IN PRODUCTIVENESS — MR. BLYTH'S FARM — CHICORY CULTI-
VATION — EASTERN DIVISION — GREAT EXTENT OF MARSHES STILL TO BE
IMPROVED — MESSRS. HEATHS' GRAZING FARMS — CONDITION OF LABOURERS
— ADVANTAGE OF ENLARGING SETTLEMENT PROVED.

NORWICH, April, 1850.

THE high agricultural reputation of Norfolk does not arise
from the superior quality of its native breeds of stock, but from
its early development of the system of alternate husbandry.
The native breeds, indeed, are nearly superseded by superior
breeds from other counties, the restless Norfolk sheep having
been supplanted by the Southdown, the " Norfolks " or home-
bred cattle by Short Horns and Polled Scots, and the abo-
riginal breed of pigs by the Berkshire, or improved Essex. The
trotting hackneys, for which the county was celebrated fifty
years ago, are almost extinct, and the dairy system which then
prevailed has nearly disappeared. But the graziers of Norfolk
are justly celebrated for bringing to perfection the best breeds
of other counties; while the systematic management and improved cultivation, which have been produced by the capital
and enterprise of the large farmers of West Norfolk, are not
surpassed in any district of England.

The geographical position of the county, as the nearest land to Holland and Belgium, and hence at one time the point of intercourse between these countries and England, is said to have given it the first start in agricultural improvement. But this was at an early period; and the present pre-eminence of the county in improved husbandry is due alone to the celebrated Coke of Norfolk, the late Earl of Leicester. It will be instructive to trace the progress of his enlightened system of improvement; and in order to do so, we must shortly describe the physical character of the county.

The southern boundary is traversed by the rivers Brandon and Waveney, which may be seen near Lophamford, flowing slowly to the right and left of the road—the one westward till it gains the Ouse, with which it falls into the Wash, at Lynn; the other eastward to the sea at Yarmouth. The course of these rivers is almost entirely through low peat or fenny marshes, which thus separate the dry land of Norfolk from the adjoining counties. The Yare, the Bure, and their tributaries, in their sluggish course towards the sea, form extensive marshes in the flat country, some of which are very valuable grazing grounds, and some unwholesome flats. In the south-west part of the county is an extensive tract of the same description. Between this and Norwich there is much light drifting sand. To the north and north-east of Norwich is a fertile sandy loam, with a pleasant undulating and wooded country; and to the south of the city, there is stiff cold clay, wet, and difficult to drain. The district chiefly celebrated for its agricultural improvement lies on the western side of the county, on the chalk, between Swaffham and Holkham. The general appearance of the country is flat, and unpicturesque to the eye of the tourist, though the experienced agriculturist will find much to admire in the large, open, well-cultivated fields, divided from each other by straight lines of closely-trimmed thorn hedges, and tilled with garden-like precision and cleanliness. This, however, is not the universal characteristic of Norfolk husbandry; for perhaps no other county presents greater

contrasts between the best and the worst farming.   The climate
is dry, and cold east winds prevail during part of the winter
and spring.

In 1776, Mr. Coke came into possession of his estate.   He was
then a young man, passionately fond of field sports, and taking
no peculiar interest in agricultural improvements.   The tenants
of part of the land which now forms Holkham Park declined
to continue at 5s. an acre, which was then the average rent
of the whole estate.   It fell into Mr. Coke's hands, and made
him a farmer.   Entering into this new business with the same
energy which distinguished him as a sportsman, he soon discovered latent properties in his land, which amply repaid his attention.   But he also found that to develope these thoroughly
required the expenditure of considerable capital, and a degree of
personal supervision which the owner of a large estate could not
possibly bestow.   No man with considerable capital would be
willing to lay it out freely, unless he had the prospect both of
good returns, and the security of such lengthened tenure as
would fully reimburse him for his adventure.   Knowing this
perfectly, Mr. Coke determined to get men, if he could, to take
into their own hands that improvement of his farms which he
could not accomplish himself; for it is through an enlightened
tenantry, that large estates can be permanently and most profitably improved.   As opportunities arose, he offered his farms to
men of capital and intelligence, under long leases, on liberal terms,
content to part with the control of his land for a time, if he could
thereby secure its improvement.   He expended large sums in
erecting substantial farm buildings, and suitable residences for a
superior class of tenantry.   Feeling the importance of having a
competent steward to superintend his estate, and to assist him
in its improvement, he secured from one of the best farmed districts of Scotland the services of Mr. Blakie, whose name is still
held in affectionate remembrance.   But it was not enough that
he should bring his tenantry to the standard of the day — they
must be prepared to advance with the general progress; and this
they were enabled to do by the annual meeting at Holkham,

where leading practical men from the surrounding country, and
from distant counties, were invited by Mr. Coke to inspect his
improvements, and to discuss their merits with him and his
tenants. The intercourse thus produced among leading agriculturists, at the Holkham sheep-shearing, was highly beneficial to
the estate. The old mode of cropping as long as the land would
yield anything without outlay, was soon changed into a system
by which stock was kept more extensively, and a more equal
proportion observed between corn and stock. Light lands
thus became fitted for wheat growing. To give firmness to the
light soil, it was top-dressed with clay marl, a stratum of which
was fortunately found, at various depths beneath the surface,
throughout nearly the whole district. Nature having done her
part, art was called to the assistance of the farmer; — rape cake,
as an artificial manure, having been successfully applied for the
wheat crops. Clover and artificial grasses were introduced, and,
with the increased means of feeding, an improved stock of cattle
and sheep were maintained. Thus the Devon cattle and Southdown sheep were brought into the district, the latter of which
has now generally superseded all others, and the former partially
so. The four-course, or Norfolk system of husbandry, was then
founded, and the advantage of keeping a large stock, to enrich the
land for corn crops, was established. — After fifty years of un-
deviating attention to his duties, as the landlord of a great estate,
Mr. Coke might truly boast that he had converted West Norfolk
from a rye to a wheat growing district. From 5s. an acre his
rents rose to 20s. and 25s.; and his tenants became prosperous
and wealthy. In that period as much as 400,000l. is said to have
been expended by him in farm buildings and other permanent
improvements; and this liberality drew from his tenants an
equal spirit of enterprise. It is calculated that they expended
for artificial food and manures, in the same time, not less than
half a million, to their own great advantage, as well as that of
the estate. The influence of these improvements rapidly extended to other districts, leases for 21 years being now common

over Norfolk, and the leading principles of the cultivation adopted on the Holkham estate having spread into all the best farmed counties of England. The wisdom of the course taken by Mr. Coke is shown not only in such results, but in the continued progress of improvement; for here there is no standing still, no belief that perfection in farming has been attained, though probably few farmers in England have more right to rest content with the point they have reached, than the best of Lord Leicester's tenants, as will be subsequently seen. And ample though the expenditure in erection of farm buildings was, the accommodation for stock is now, with still heavier crops, found far from sufficient; and the outlay by the present earl, in repairs and additions, on the estate was not much under 10,000*l*. for last year.

To convey an idea of the present state of farming in West Norfolk, where it is carried out in the best style, we shall give a brief account, of some of the farms we examined.

Holkham Park, the seat of Lord Leicester, is situated within a mile or two of the sea, on the north coast of the county. The home farm, which is now managed by Mr. Keary, is within the park, and extends to 1,800 acres, 1,300 of which are under tillage, and 500 in pasture. The farm buildings are in three different situations, for the convenience of carting the corn in, and the manure out. They are commodious and well arranged for the object in view—viz., the storing in large barns of a considerable portion of the crop, and the accommodation of cattle, in yards and sheds, littered with abundance of straw. A herd of pure Devons are kept on the farm, many of which are very fine specimens of the breed. Their aptitude to fatten on a somewhat scanty supply of food is very remarkable. A stock of 40 cows is kept, and their produce reared and fattened. They do not yield much milk, but it is of rich quality. The usual stock of cattle on the farm is 250 head, inclusive of 20 working bullocks. The sheep stock numbers about 2,500—700 of which are pure Southdown ewes. About 150 pigs are also kept—a cross between the Neapolitan and the Suffolk, combining the

fineness of flesh and fattening properties of the former, with the
hardiness of constitution and fecundity of the latter. 32 farm
horses and 20 working bullocks are the working stock of the
farm, the bullocks working in pairs, but changed at each yoking,
a fresh pair being taken in the afternoon ; four bullocks thus do
the work of two horses, in either plough or harrow.

The farm is managed in the usual four-course rotation.
Mangold wurzel, of which the red globe variety is preferred, is
cultivated on 27-inch ridges, in which ten 3-horse loads of dung,
and 20s. worth of artificial manures per acre have been deposited. For swedes 7 to 8 loads of dung, and 15s. worth of
artificial manure, are applied. White turnips are sown, on the
flat, in rows 18 inches apart, 30s. worth of artificial manure being
drilled in beneath them. The manures chiefly used are guano
and Lawe's superphosphate. In the other operations of the
course the usual details of good farming are practised. The
whole of the young wheats, 280 acres in extent, have a top-
dressing applied to them, in spring, of 6 stones of nitrate of soda,
mixed with 16 stones of salt, to the acre. This quantity is
applied, in equal moieties, at intervals of three weeks or a month,
beginning early in March, and ending about the 20th of April.
It has been found in practice better to apply it so, than to
lay it all on at once. The cost at present is 15s. an acre; and
the increase of crop not less than 6 bushels an acre.* Spalding's
red wheat is the only variety sown. The rent charged against
the farm is 20s. an acre, the expenditure in artificial manures
and food is as much more, and the labour amounts to 32s. an
acre on the land under tillage. Labourers are paid 9s. a-week,
but earn 10s. 6d. on task work.

The farm of Mr. Hudson, of Castleacre, on Lord Leicester's
estate, is between 1,400 and 1,500 acres in extent, held in two
adjoining occupations of nearly equal size. About 1,200 acres
are regularly under crop, and somewhat over 200 acres in pasture.

---

* See Note at end of volume, for description of experiment by Mr. Pusey
to test the practice here recommended.

The four-course rotation is followed throughout, there being
annually 300 acres in wheat, 300 acres in barley, 3C0 acres in
turnips, &c., and 300 acres in clover, and trefoil and white
clover alternately. No rye grass is sown with the clovers, as,
being a cereal, it is considered injurious to the following wheat
crop.

The principle adopted here is to manure for *every* crop. Thus,
for the wheat, eight 3-horse loads of dung are laid on the
clover after it is mown. This promotes a rapid growth of clover,
which is ploughed in for the wheat crop. Salt is sown over it
in spring to strengthen the straw. Turnips are manured partly
with dung and partly with artificial manures, 25s. an acre being
expended for this purpose in guano and Lawe's superphosphate.
A large proportion of the turnips are consumed on the ground
by sheep, which are also cake-fed, and the soil is thus prepared
for barley. The land is covered with clay marl once in a lease
of 21 years. In the feeding of stock 10 tons a-week of oil cake
are consumed during the winter, in addition to the green crops
and herbage of the farm ; Mr. Hudson never using less than 200
tons of cake in a year, and sometimes considerably more. Each
bullock gets 10lb. of cake a day, besides roots ; and each fatting
sheep on an average ¾lb., beginning with ½lb. and ending with
1lb. daily. The cattle are fed in open courts with sheds, well
littered with straw.

Thirty-six work horses, and 16 working bullocks, are required
for the operations of the farm. The bullocks work in pairs,
two in a plough, the same as the horses, and walk quite as
quickly, and in either plough or harrow, get over as much
ground as the horses.· Each ploughman drives his own pair.
The farm horses are fed on sprouted barley, 12lb. each daily,
besides fodder. The barley is steeped 24 hours, then placed in
a heap on a floor, where it is turned once a day for five days, by
which time it is ready for use. A new heap is steeped daily,
and one consumed, the first being laid at one end of the house,
and, being daily turned, arrives in due course at the other end,
where it is taken out. With this feeding, on this light land,

where the furrow seldom reaches six inches in depth, the horses are maintained in good healthy condition. A small dairy of Devon cows is kept, the produce of which is reared and fed. The ewe flock is Southdown, the fat hoggets now (April) going off to the London market. The sheep are at present being fed on rye, sown after wheat for spring feed, and now affording a full bite. Mr. Hudson does not sow tares, as he considers the tare very exhausting to the land. The whole harvest work of the farm is done by 32 mowers, each of whom is attended by one woman to gather, and one to bind up the sheaf. In carrying the crop, the stacks are always made in the field where they grew, and carted home to the barn, as required, during winter and spring. The only kind of wheat sown is Spalding's Red, which is found by much the most prolific hitherto tried in West Norfolk.

Twenty-seven years ago the stock annually kept on this farm was 400 sheep and 30 bullocks; it now averages 2,500 sheep and 150 bullocks. The wheat and barley crops then did not exceed $22\frac{1}{2}$ bushels an acre; that average is now nearly doubled. Every crop is drilled, and the land kept perfectly clean. The roads and fences are all maintained in the best order; and the beauty and regularity of all the crops now growing on the farm, sufficiently attest the enterprise and skill of the farmer.

As exemplifying the saving of waste, in the transportation of fat stock to the London market, by the introduction of railways, Mr. Hudson mentioned to us a fact which may be interesting. Formerly, when several days were occupied in driving to London, a sheep was found on the average to have lost 7lb. weight and 3lb. inside fat, and a bullock 28lb. These weights were ascertained by a series of trials, average animals being killed and weighed on the farm and compared with the weights of similar animals when slaughtered in London. This difference of weight was waste, entirely lost to everybody. On the quantity of stock annually sent out by Mr. Hudson, this loss was equivalent in value to upwards of 600*l.* a year; nearly the whole amount of which now finds its way to the market, as the stock are put into the trucks in the morning, and reach London in the afternoon,

without fatigue.   When it is considered over how great a quantity of stock throughout the country a similar saving has been
effected, there can be no doubt that the increased weight so
saved has had a perceptible influence in increasing the general
supply of the market.

A few miles further north lies the farm of Mr. H. Overman,
at Weasenham.   It is managed generally like other well cultivated farms in the district, but is peculiar in having on it a
stock of Ayrshire dairy cows, the produce of which, in butter,
is sent to the London market.   During the winter the cows are
fed on swedes and a little hay, and those which are giving milk
receive also three pounds of oil cake daily, till the grass is forward enough for turning them out to pasture.   Mangold wurzel
is not given to the cows, as, though productive of milk, it is not
found to yield butter.   The taste of the turnips is taken from
the butter by scalding the cream in a pan placed in a boiler of
hot water, and when the cream is hot, saltpetre (at the rate of
half an ounce to a gallon of cream) is dissolved in it, which is
found completely to neutralise the peculiar taste of the turnip.
Excellent butter is made by this management and feeding.

At Mr. Overman's, of Burnham Sutton, another of Lord
Leicester's tenants, the same system of high cultivation has
long been successfully followed.   In top-dressing his wheat,
Mr. Overman uses 1 cwt. nitrate of soda mixed with 2 cwt. of
common salt, applied in two dressings, as already described.
He has long given up horse or hand hoeing his wheat crops in
spring, from the conviction, founded on experience, that it is
injurious on light lands, by increasing the proportion of tail or
inferior wheat.   This opinion we find is common among the
best light land farmers, and hoeing the wheat crop is nearly
discontinued in this part of the county.   We were sorry that
we had not an opportunity of visiting Mr. Bloomfield, of Ware-
ham, and others of Lord Leicester's tenants, equally noted for
the excellence of their farming.

The question of how far the four-course system of husbandry

continues applicable to the present circumstances of agriculture, we heard frequently discussed by the most intelligent and experienced farmers in West Norfolk. We found but one opinion on the point, and that was, that covenants are necessary to bind bad farmers, not good ones; — and that, with our facilities for obtaining artificial manures and food, a strict adherence to the four-course system is no longer necessary or expedient. Accordingly, in every case we met with, where a man farmed his own property, or was not restricted by rigorous covenants, the four-course was departed from. On this point we were favoured with some valuable information by Mr. Blyth, of Sussex farm, Burnham, whose style of cultivation and the high condition of whose land may challenge comparison with any in the county. For a number of years back he has been growing more and more wheat, diminishing his barley, and taking an additional corn crop in the course. His system now is as follows : — (1) Clover, trefoil, or peas, (2) wheat, (3) oats, (4) turnips, (5) wheat or barley. He manures for every crop where he thinks it is required. On land where the turnips are fed off before Christmas, the crop of barley was never found so good, by six or eight bushels an acre, as on that which is fed off later, and on that portion he now sows wheat.

To show the gradually increasing produce which follows a constantly improving system of agriculture, we give the following figures from the farm books of an equally competent authority : —

				Bushels an acre.
The average wheat crop for 7 years ending	1839	-	25
Ditto	ditto	„	1846	-	29
Ditto	for 2 years	„	1848	-	36
The average oat crop for 7 years ending	1839	-	54
Ditto	ditto	„	1846	-	57
Ditto	for 2 years	„	1848	-	68
The average barley crop for 7 years ending	1839	-	31
Ditto	ditto	„	1846	-	33
Ditto	for 3 years	„	1849	-	45

It will be observed, that in each case the increase of average has been greatest since 1846. This may be in part accounted for by the use of artificial manures, as a direct application to every crop, having since that year become matter of system; and also by the very rapid strides which our agriculture has made within the last four years in the hands of all who had the means, and who were not hampered by unwise restrictions from following the obvious course of improvement. It is further important to find that these increasing averages of wheat crops have taken place simultaneously with more frequent calls on the land, there having been on this farm, for the first period of seven years, an annual average of 214 acres in wheat; for the second seven years, 268 acres in wheat; for the last two years, 340 acres in wheat.

On Sussex Farm, Mr. Blyth follows the same system of top-dressing his wheat with nitrate of soda and salt, as has been already described. He does not hand or horse hoe his wheat crops in spring, but carefully rolls them; having found, by frequent experiments, that where rolling is omitted, there will be a large deficiency in harvest. He breeds and fattens his own stock of cattle, and keeps a large flock of Southdown ewes, crossed with the Hampshire down, which produces a much more profitable sheep than the pure Southdown, but not one suitable for prize gaining. Twenty-five lambs to the score is the average yield, though that is frequently exceeded. Millers' offal is given to the lambs along with their green food. During winter, they receive daily ¾ lb. to 1 lb. each of oats and peas crushed, while feeding on turnips; and are now being sent off fat to London, out of the wool, at 27s. to 28s. each. The wool is worth 7s. more. On account of the low price of corn, Mr. Blyth has been using it exclusively in the feeding of his stock this season.

Chicory is grown on a large scale on this farm, there having been 50 acres under this crop last year. It is taken after wheat instead of an oat crop, drilled in May with artificial manure, and treated exactly like carrots. It is taken up in autumn in quantities as it can be manufactured; for it is not injured by frost, and receives no damage by being left in the ground. After being washed, the roots are cut with a small guillotine machine, and then kiln-dried; 12 to 15 tons, the produce of an acre, drying to about one ton, in which state it is sent to London. The price this year has ranged from 15*l.* to 23*l.* a ton. Last year it was not more than 10*l.*, and did not pay, as the manual labour in the cultivation of the crop, and the expense of drying, &c., amount to 5*l.* a ton. At the present price it pays well.

In the eastern division of the county, between Norwich and the sea, lies a great extent of fine land. Along the hollows the waters of the Bure and Yare find their way to the sea at Yarmouth, passing many thousand acres of marsh land, of which an immense extent is still unreclaimed. Windmills are chiefly used for pumping out the water from those marshes which have been embanked. After a marsh has been embanked and pumped dry, the soft spongy soil consolidates, and subsides several feet below the level of the river. Ditches are formed through it, which convey the water to one point, where, by means either of a steam-engine or a windmill, it is pumped over the bank into the river. The whole cost of embanking, ditching, and erecting a steam-engine, is said not to exceed 10*l.* an acre in situations at all favourable for the operation; and as at such an outlay a worthless and unwholesome marsh may be converted into rich grazing ground, it is surprising that so much of this yet remains to be done, in a county so celebrated for its agricultural progress as Norfolk. If the new Government drainage loan can be applied to this purpose, probably it could not be laid out in a more profitable manner.

At Horning, in this district, we visited the extensive grazing and arable farms of the Messrs. Heath. On their marsh farm, 700 acres in extent, they keep during the summer 400 bullocks and 700 sheep. Having large arable farms adjoining the marsh, the stock is wintered on them and turned out to the marsh as soon as the grass is ready, where part is fattened and sent direct

to Smithfield from the grass, and part is brought on to be finished
in the yards as prime fat for the Christmas show.   The best
polled Galloway Scots are grazed here, as they have been by
the Messrs. Heath for a long period of years, but the quality
now sent to Norwich is reckoned very inferior to what it used
to be some years ago.   They are, therefore, now going more
into Herefords and Welsh runts, both of which, in the order
just mentioned, they reckon superior for their purpose to any
other breed.   Having land of very rich fattening quality, they
purchase animals which are nearly fat, to finish them; and their
great aim is to get stock of the best quality of its breed,
and to " dwell on it," as such is sure to pay, and to pay most for
the last month of its keep.   In the yards the cattle are not
treated so profusely to cake as in some other parts of the county,
a larger proportion of roots being given to them.   Those, however, which are being prepared for the Baker-street show are
fed without stint.   They are beautiful specimens of their several
breeds, Hereford and Galloway.*

On their arable lands, which are of fine quality, dry and
friable, the Messrs. Heath do not restrict themselves to the
four-course, or even to an alternation of white and green crops.
They frequently take wheat after wheat, or oats after wheat, or
barley after wheat; and find that by liberal treatment of the soil
they can do so without injury to it, and with manifest advantage
to themselves.   In the Blofield Hundred, we found the same
practice followed by Mr. Tuck on his own property; and were
informed that in all cases where the land is farmed by its
owner, or where a rigid adherence to rotations is not enforced,
the four-course is scarcely ever adhered to.   On the farms just
mentioned, the appearance of the soil and the crops indicated
clean cultivation and a high state of fertility.

The marsh lands of East Norfolk are of very various fertility.
For the first 8 or 10 weeks cattle thrive and swell out greatly

* Both specimens subsequently gained the highest prize of their class at
the Smithfield Club Show in 1850; the Hereford ox being reckoned the finest
fat ox ever exhibited.

upon them, but after that they do not continue to progress in
the same way ; in fact, they seem to require a change of food.
This may, perhaps, arise from the nearly uniform character of
the natural marsh herbage, which in that respect differs from
the natural herbage of a rich meadow containing a great
variety of grasses, early and late, the one springing up as the
other begins to fade.

The average number of labourers employed at present on a
light land farm of 1,000 acres has been ascertained, by comparison of 10 different farms in West Norfolk, to be, —

Men able to do harvest work.	Men of inferior ability.	Lads above 16 Years.	Boys under 16 Years.	Women.
23.1	4.5	7.8	11	6.25

The wages of labourers in Norfolk are at present 8s. a week ; in
some places a reduction to 7s. is spoken of. A great proportion
of the work on farms, however, is done by task work or contract,
and the rate of wages, therefore, does not afford any correct estimate of the condition of the peasantry. Task work will generally bring larger pay to the labourer, but this is more doubtful
where the farmer resorts to contract   Hand hoeing, and other
light operations of husbandry, which can be carried on by
children, are sometimes paid for by contract, — a man engaging to do what is required for so much, and employing all
the children he can collect, in gangs, to get through with it. The
evils of such a state of things are obvious.   The boys and girls,
thus brought together from considerable distances, frequently do
not return home at night, and sleep in stackyards or barns, or
wherever they can find shelter.   Another point connected with
farm labour in Norfolk is the employment of women in the
fields, which some of the most intelligent agriculturists here
strongly condemn.   They contend that it has a most demoralising effect, causing women thus employed to lose all feeling of
self-respect, rendering them bad housewives when married, and
unfit, from want of experience, to exercise that strict economy
in expenditure, and to provide those small fireside comforts

which are so necessary in a labourer's wife. It is further said
to be very questionable whether, even with the low wages paid
to them, they are employed remuneratively to the farmer, as
they are generally slow and indifferent workers. These objections apply to the regular employment of women in field
labour, not to their assistance in harvest time.

An association had been formed at Docking, which distributes
prizes, to the amount of 120*l.* annually, to successful competitors
in ploughing, stacking, mowing, and other operations of husbandry, for the purpose of encouraging expertness in the industrial
processes of the labourer. Prizes are also awarded for good conduct, for knitting done by the labourer's wife and children, and
for various other objects calculated to stimulate the mind of the
peasant, and to elevate the character of his employment.
Though the association has been only a few years in existence,
it has already been productive of much benefit, and of an inter-
change of good feeling between employer and employed. An
attempt was made some time ago at Burnham to associate the
farmers and principal people of the locality with the labourers in
the formation of a benefit club, for the support of the latter in
sickness, and to defray their funeral expenses when they die.
The plan, however, failed, the labourers themselves refusing to
co-operate in such an object.

The law of settlement is felt by many of the Norfolk farmers
to press unfairly upon some parishes, both as respects the comfort of the poor and the inequality of the rates. To obviate
this, the Docking Union have, by private arrangement, extended
the settlement of their labourers to the whole Union, with very
beneficial results. The disputes between parishes have thus been
terminated, and the labourers of the Union are fully employed,
and at a rate of wages rather higher than those of the adjoining
districts. The listless indifference of the labourer, who trusted
to his parish to give him either labour or support, is being overcome, as he is now at liberty to go to any parish in the Union
where his labour may be most in request.

LETTER XXI.

THE FEN COUNTRY.

EXTENT OF FEN COUNTRY — EARLY AND PROGRESSIVE IMPROVEMENTS — TWO KINDS OF FEN LAND — " BLACK " THE MOST FERTILE AND REMUNERA- TIVE AND THE LOWEST RENTED — MANAGEMENT OF SHEEP — AND CATTLE — DETAILS OF FEN FARMING — ADVANTAGE AND EXPENSE OF CLAYING BLACK LAND — COURSE OF HUSBANDRY ON DUKE OF BEDFORD'S ESTATE AT THORNEY — AVERAGE CROPS — RENT AND RATES — CONDITION OF LABOUR- ERS — DETAILS OF FARMING IN THE CLAY DISTRICT OF THE FENS — DRAINAGE DEFECTIVE — AVERAGE CROPS — USE OF CAKE AND GUANO — RESULT OF AN APPLICATION FOR A REDUCTION OF RENT.

BOSTON, LINCOLNSHIRE, April 1850.

THIS extensive district, which drains into the Wash, comprises portions of six counties, Norfolk, Suffolk, Cambridge, Hunt- ingdon, Northampton, and Lincoln. It is estimated to contain 680,000 acres, forming a continuous plain 70 miles long by 20 to 40 in breadth. From the background of these counties, five large rivers find their way to the sea through this plain, — the Ouse, the Nene, the Cam, the Witham, and the Welland. Great works have been formed to keep the rivers to their channels, that they may not again overflow and convert to a marshy wilderness the fertile region rescued from them by the enterprise and ingenuity of man. The works begun by the Romans were after many centuries continued by the monks; the elevated spots, still called the " islands," or " highlands," which appeared above the waters when this great plain was submerged by the winter floods, having been occupied and cultivated by them. In 1630 the reclamation of the fens became a matter of systematic enterprise; Francis, Earl of Bedford, with 13 gentle- men adventurers, having then undertaken to drain the Bedford Level, " on condition that they should have 95,000 acres for their

N

satisfaction." During the Commonwealth the work was continued with such success, that in 1652 " about 4000 acres were sown with coleseed, wheat, and other winter grain, besides feeding innumerable quantities of sheep, cattle, and other stock, where never had been any before." This success caused the work of improvement to be vigorously prosecuted till the reign of Charles II., when a corporation was established for the regulation and continuance of the drainage works in the Bedford Level. This system of general superintendence spread to other districts of the vast plain, and in later times the skill of eminent engineers was enlisted to aid the efforts of individual enterprise. The floods from the surrounding country were cut off from overflowing the fens, windmills were employed to pump out the accumulated waters, and, last of all, powerful steam engines have been erected at certain points to which the drainage of a district is conveyed, and where it is pumped in vast quantities into the channel of the great main drains or rivers. Mr. Clarke, to whose description of the fens in the Journal of the Royal Agricultural Society we are indebted for much information, estimated, in 1847, that 250 windmills and 40 to 50 steam engines were constantly at work in pumping out the drainage. By improvements in the outfall, it is found that many districts can be drained naturally, in which wind or steam-power was formerly employed. The mud deposited in the Wash, with its drifting banks, was continually closing up the outfall channels of the rivers; but that is now receiving attention with much advantage to the general drainage of the fens. It is believed that, by all these means, 680,000 acres have been brought into cultivation from being, as Dugdale describes it, —

" A region of wild and swampy country, partly cultivated and " partly overflowed, by which overflowings in the winter time, " when the ice is strong enough to hinder the passage of boats, " and yet not able to bear a man, the inhabitants upon the hards " and the banks within the fens can have no help of food nor " comfort for body or soul, nor supply of any necessity, save what

" those poor desolate places do afford.   And what expectation of
" health can there be to the bodies of men where there is no
" element good? the air being for the most part cloudy, gross,
" and full of rotten harrs; the water putrid and muddy, yea, full
" of loathsome vermin; the earth spongy and boggy, and the fire
" noisome by the stink of smoking hassocks."

The whole district is now traversed by excellent roads and
railways, and being mostly freed from the overflow of floods, the
further progress of improvement has no insurmountable obstacle
to contend with.

The fen land is of very various fertility.   It may be divided
into that portion which is more inland, and consequently a fresh
water deposit, and that which lies nearer the sea and has in the
course of ages been formed by the reflux of the tide.   The first
has usually a surface of black vegetable mould, varying from a
few inches to a foot or more in thickness, lying on a bed of
clay; the last is a strong silty loam of more or less thickness,
also incumbent on clay.   The first, from its porous character,
does not generally require under drainage; the last cannot be
profitably cultivated without it.   The first produces every
description of crop with the greatest luxuriance, being friable
and easy to till, and readily cleared of root weeds.   The other
is generally too stiff for the profitable growth of green crops, is
easily injured by the trampling of horses, or the use of the
plough in wet weather, is much more dependent on the seasons
for the quality of its produce, and much more difficult to keep
clean and free from root weeds.   Strangely enough, with all these
advantages it seems to bear the higher rent of the two, so that
it is not to be wondered that the farmers of the clay fens are
not so prosperous as their more fortunate neighbours on the
black land.

On the dry lying lands of the fens, — that is, those more
elevated portions called " Highlands " or Islands, — the land is
principally in grass, and that of very rich quality.   It is grazed
by short-horn bullocks and Lincoln sheep.   The bullocks are

bought in lean at the autumn fairs, wintered in the straw yards with cake, and turned out to these rich pastures in summer. The Lincoln sheep is a mixture of the Leicester with the old Lincoln, the produce being of larger size with less fat meat, and somewhat coarser wool than the Leicester.   They are fattened at 20 months old, or may in some cases be kept a month or two more, when, after yielding a second fleece, they are sold fat and have attained great weight.   The fen farmers buy their stock of sheep at the spring fairs, and are paying for hoggets in the wool, at present (April 1850) 30*s*. each — an unusually low price. These hoggets are put partly on the " seeds," partly on the old pastures.   In autumn they are folded on cole or rape, receiving also cake; and as soon as this is consumed, generally before Christmas, the sheep are sent fat to market.

In the winter feeding of bullocks some farmers are substituting cut straw, with meal, and boiled linseed poured over it, for oil-cake; and in one case, when 900*l*. used to be expended in a large holding in the purchase of cake, a better effect is said to be produced by 600*l*. expended on linseed and meal.   At Thorney, Mr. Whiting adopts an ingenious plan for preventing yard-fed bullocks from annoying each other when feeding.   The food is all placed for them in a long manger under the shed, and when each animal goes forward and puts his head into the manger between upright spars, the feeder pulls a rope attached to a simple contrivance, by which the whole of them are confined to their places, and released again when they have consumed their food.

The principal thing to be attended to in the management of a fen farm on the black land, is to keep the outfalls and ditches throughout the farm open.   Some think that all the water should be drained off from the ditches; others believe that in dry summer weather the crops are much benefited by the damp stratum which is maintained beneath the working soil, by the presence of water.   Practically, the water is kept in the ditches at about two feet from the surface, thus serving as a fence betwixt the fields, and affording water to the stock. The land is so very level, that by arranging the sluices at the main outlets of the farm, each man may keep such depth of water as he chooses. The next process in fen farming is to give solidity to the black vegetable surface mould, and that is done by digging trenches into the clay, and throwing it over the surface. A trench two feet deep and two feet wide is made along the field, and the clay which is taken out of it is laid four yards over the surface on either side of the trench. The same process is repeated throughout the field, a new trench being opened eight yards apart from the last. The cost of this operation is about 35s. an acre, but it is a permanent improvement, not requiring to be repeated during a lease. It is usually done in the division to be fallowed, as the clay has then time to work down among the soil before a crop is sown on it.

The course of husbandry pursued is not very definite, most farmers being permitted to farm as they think best, or at all events according to the custom of the country; and as that custom is by no means certain, the farmers have sufficient latitude. The following is the system we found pursued on a well managed farm on the Duke of Bedford's Estate at Thorney. The farm contains 600 acres, 200 of which are "highlands" in permanent pasture. The rest is good fen land, cultivated as follows: — (1) fallow with roots or coleseed, (2) oats, (3) wheat, (4) seeds, (5) wheat, (6) beans, (7) wheat. The fallow is well wrought, pulverised and cleaned. It is then manured with 8 or 10 loads of dung an acre, and a portion planted with potatoes, carrots, mangold, and swedes. The larger portion is reserved for coleseed, which, after the dung has been applied and ploughed into the land, is drilled, with 16 bushels of bones mixed with ashes to the acre. This is usually sown in June, and is ready for the sheep in autumn, one acre affording keep for 8 or 10 large sheep, 12 weeks. The other root crops having been taken up when ready, the whole

of the land is ploughed, and sown with oats early in spring,
as it is then in so rich a state that a wheat crop would run too
much to straw and be spoiled.   The oats yield from 80 to 100
bushels an acre, and as much as 120 bushels have been got of
a black coarse variety.   After the oat crop is removed, the
ground is ploughed, and sown with wheat, which is drilled and
carefully hoed in spring, wherever weeds make their appearance.
Clover seeds are sown among the wheat, in equal quantities
of red, white, and trefoil, for sheep pasture, and of red alone
where it is to be cut for hay.   Next year the "seeds" are
pastured with hoggets, of which they carry from 7 to 8 an acre,
and a few beasts on the field besides.   They are ploughed
up in October, and the land is again sown with wheat.
After the wheat is removed, a slight dressing of dung is
ploughed in, and the following spring the land is drilled with
beans.   The beans are kept as clean as possible by horse and
hand hoeing; but being a dirty crop, it is necessary to plough
twice, and work the land well after them, in preparation for
the last crop of the course — wheat.   There is not much
difference in the yield of the wheat in the various places it
takes in the rotation, 44 bushels an acre being reckoned a good
crop.   The grass land, as already mentioned, is old pasture
land of rich fattening quality.   The labour bill amounts to 30s.
an acre for the land under tillage.   The farm is let tithe free,
—rent 30s. an acre, and poor rate 2s. 6d. a pound.

Labourers' wages here are 9s. to 10s. a week, and their cottage rents 3l. to 4l. a year.   The picturesque village of Thorney
(standing in the centre of the Duke of Bedford's estate in this
quarter, 18,000 acres in extent, of rich and valuable land,) is
ornamented by a new street of labourers' cottages, recently
completed by the Duke.   They are very handsome, with many
conveniences, though the apartments are complained of as too
small.   With good gardens attached to each, they are let at
very moderate rents.   Not many of his Grace's tenants here
hold on lease, though some families have been on their farms

for generations; but the most perfect confidence is felt by them in the high character of the House of Bedford. Their rents, previous to the late fall in prices, may be reckoned moderate.

As an example of the second description of fen land, we shall take the Wildmoor Fen above Boston, which is a stiff clay loam, incumbent on clay. On this soil it is necessary to drain with pipes or tiles, and the outfall now is sufficient to admit of this being done to a depth of three or four feet. At this depth the drains are usually made 11 yards apart, a very common arrangement between landlord and tenant being for the former to supply the tiles, and the latter to put them in. A vast amount of this kind of drainage yet remains to be done, the difference in condition of a drained and undrained farm being very perceptible to the traveller. Where draining has not been done, the farmers are rapidly losing money; and even where it has, this land is so entirely dependent on the prices of grain, that the present depression is telling seriously on them.

The following four-course shift on a farm of 176 acres may be taken as a specimen of the routine of one of the better managed drained farms in the clay district: — (1) fallow with roots and coleseed, (2) wheat or oats, (3) clover and beans, (4) wheat. More than half of the fallow is sown with coleseed, to be eaten off, with cake, by sheep; the rest is sown partly with mangold, partly with swedes and white turnips. Night soil, got from Boston, and mixed with earth, is drilled in with the seed of these different crops, at the rate of 10 loads an acre. All the ground that can be early enough cleared of roots is sown with wheat, and the rest in spring with oats, the clover seed being sown at the same time on the half of the division. The other half, after the crop is removed, is dressed with dung and ploughed for beans, which are drilled early in spring. After the clover has been eaten off about the end of August, the land is ploughed and harrowed, and ploughed again, receiving, as it were, a second fallow, when it is got into a fine friable

state, and sown with wheat. Dung is either laid on previous to
the last ploughing, or is carted on to the ground during frost,
and laid over the young wheat, which is considered an excellent
plan. Some only plough the land once, but harrow it five times
(the horses walking in the furrows which divide the "lands"),
and then drill in the wheat: 30 bushels of wheat and 50 bushels
of oats are reckoned average crops. The best farmers give cake
in their yards to the cattle, 5 tons per 100 acres being considered
a very liberal expenditure. Guano or other artificial manures
are but little used. Rents vary according to circumstances,
from 27s. to 40s. an acre.

The wheat crop on this description of land yielded very badly
last year. Unlike the dry light lands of the Eastern Counties,
which produced more than an average crop, these heavy soils
fell short of their usual growth nearly 8 bushels an acre. The
low price tells therefore very strongly here; and many of the
farmers are permitting necessary operations (such as the cleaning out of the main drains — so essential in this flat district)
to remain unperformed. No reduction of rent of any importance has yet taken place; and in one instance, where a representation by a body of tenants was made to their landlord on this
subject, the only reply they received was a notice to quit to the
man whose name stood first in the list. Much of the land in
the clay fens is held by small proprietors, many of whom
are understood to have their estates heavily mortgaged.

LETTER XXII.

LINCOLN.

GREAT VARIETY OF SOILS — DESCRIPTION OF THE COUNTRY — OPENED UP BY NETWORK OF RAILWAYS — WANT OF DRAINAGE — GREAT IMPROVEMENT EFFECTED BY THE APPLICATION OF CLAY MARL TO SAND — MANAGEMENT OF CLAY SOILS INFERIOR TO ESSEX AND SUFFOLK — RENT DEPENDENT ON THE LANDLORD — DETAILS OF FARM MANAGEMENT ON THE WOLDS — INFERIOR TO THAT OF WEST NORFOLK — CHALK A CURE FOR ANBURY IN TURNIPS — DETAILS OF FARMING ON LINCOLN HEATH — ADVANTAGE OF BRINGING UP BROKEN CALCAREOUS ROCK BY DEEP PLOUGHING — RENT — PROCESS, AND EXPENSE OF "WARPING" ON THE HUMBER AND TRENT.

LINCOLN, April 1850.

THE county of Lincoln presents many features of interest to the agriculturist. It embraces a great variety of soils and modes of cultivation, varying from the richest pastures to the most sterile sands, and exhibiting on its various soils the treatment which experience has taught in the management of stiff clays, fens, warp-land, sands, wolds, and heath. In this county, too, has chiefly risen into prominence that system of compensation to the outgoing tenant for unexhausted improvements, which is believed by many to have been the foundation of the agricultural progress of Lincolnshire.

Entering the county from the south, an extensive district of fen land is traversed, reaching up to the city of Lincoln, where, on the summit of the hill, rise the towers of the stately cathedral. At this higher level, some 150 feet above the vale, stretches a tract of dry turnip land running north and south of the city about 40 miles, and still known as Lincoln Heath. Nearly parallel with this, but separated from it by the great central vale of the county, lies the district called the Wolds; and between that and the sea extends a tract of richer and heavier land. On the north-western boundary of the county, on both banks of the Trent, is that low-lying tract of land on which the peculiar process of warping is carried on. The fen lands we have already described, and purpose now to enter into some details of the present state of agriculture in the Wold and Heath portions of the county, and then to give a very brief description of the process of warping and its effects.

The system of husbandry general throughout the district is the four or five course. This will no doubt come to be modified by the change which railway accommodation will afford; this county, which was formerly somewhat remote, being now covered with a network of railways, giving ready access on the one side to the seaports, and on the other to the great central markets for supplying the dense population of the manufacturing districts with agricultural produce.

Taking the line of country traversed by the railway from Lincoln northwards, by Market Raisen and Caistor, the land is generally very insufficiently drained, and by no means well managed. On the stiffer clays little seems to have been done for removing the stagnant water; the grazing lands being much covered with rushes, and the fallows, on the land in cultivation, being foul and out of condition. Approaching Caistor, a large extent of weak sandy soil is passed, on part of which a very great improvement has been effected by claying. We were favoured with some interesting information by the gentleman who commenced the practice, Mr. Dixon, of Holton. Twenty-two years ago, when he came into possession of the estate, there were 500 acres of rabbit-warren, which the tenant refused to hold at a rent of 50*l.* Mr. Dixon took it into his own hands, and by covering it all with clay, and by under-draining, he now considers the same land worth 16*s.* an acre, or 400*l.* a-year. The expense of claying varied, according to the distance from which the clay had to be carted. In some cases it was got in pits in the field where it was applied; but in general it had to be brought from spots at a higher level than the sandy tract.

There, beneath the surface, beds of whitish marly clay, mixed with chalk and flints, are found four or five feet in thickness. These are opened out like a quarry, and the face is wrought down by pick and shovel. It is then filled into carts, and conveyed to the sand-land, where it is laid on at the rate of 60 yards an acre. It is usually laid on land in preparation for green crops, where, after being exposed some time to the action of the weather, it breaks down and mixes with the sand. The whole field is at the same time under-drained with tiles, the cost of both operations averaging from 12*l.* to 14*l.* an acre. After the field has been green cropped, and when the subsequent barley crop has been removed, 30 yards more of clay are spread over the seeds; the whole dose being thus 90 yards an acre, not applied all at once, but in two separate applications. The effect of this mixing of soils has been to convert a weak sand, unfit for the production of any valuable crop, into light land of fair quality, on which, by good farming, clover, wheat, turnips, and barley, may be taken in regular succession. The quality of the clay should not be overlooked, as it is highly calcareous, and is probably not often to be found in proximity to tracts of sand. A different description of clay, more unctuous, and of a black ferruginous character, was tried by Mr. Dixon, but instead of proving beneficial, it was very injurious. Beyond Mr. Dixon's estates, this tract of sandy land seems, with few exceptions, to be badly managed.

The strong clays are by no means so well managed in Lincoln as in Essex and Suffolk. In executing drainage it is common for the landlord to supply tiles, and the tenant to put them in; but a great deal yet remains to be done by both. The old form of raised ridges is still maintained, from the difficulty experienced on this kind of land of levelling them down without injury to the surface. The best soil on the top of the ridge, when levelled into the hollow, exposes a barren subsoil, which requires many years' exposure to the air and to the influence of manure before it becomes fertile. Where the land is well

drained and carefully tilled, it yields large crops. The usual
succession is clover, or seeds, wheat, turnips, barley, fallow,
wheat or oats sown down with seeds. It being common to
hold a quantity of grass land along with a clay land farm, the
farmer is enabled to keep a considerable stock, which, when
wintered in his yards on cake, converts his straw into valuable
manure. The rent of this description of soil, like that of all
others in this county, varies more according to the character of
the landlord than its intrinsic qualities. On Lord Yarborough's
estate the clay farms, which are close to a line of railway, and
where the landlord gives tiles for drainage without any other
charge than that the tenant must put them into his land, the
rent varies from 18s. to 22s. an acre, tithe free, and the poor-
rate is very moderate.

Ascending from the clay lands to the Wolds, we enter on that
tract of country which, with the Heath, to be subsequently described, has given celebrity to the farming of Lincolnshire. It
is situated on the chalk formation at about the same elevation
and of much the same character as the land round Holkham, in
Norfolk, but lower and less exposed than the Down farms
of the southern counties. This tract, extending to more than
200,000 acres, varies considerably in quality, being best where
it dips to the lower ground, and very light and sandy towards
its highest points of elevation. The rent of the best land rises
as high as 32s. an acre, though the average is 20s., tithe free, and
with very low poor-rates. The mode of culture is the four or
five course, the " seeds " remaining one or two years down.

On one of the better-managed farms on the Wolds the following is the routine of cultivation : — Beginning with the
turnip crop, the land, after being properly cleaned and wrought,
is sown with yellow and white turnips in succession, 1 cwt. of
guano and 4 bushels of dissolved bones per acre, mixed with
ashes, being drilled in before the seed. The crop is consumed
on the ground by sheep, which get no cake unless the turnip
crop proves very inferior. If the crop is good, part of it is

drawn for consumption in the yards by cattle. It is succeeded by barley, a small proportion being in some cases sown with wheat. The barley is followed by clover and seeds, on which the sheep are turned very early in spring, there being no other provision made for them; and, *as swedes are seldom or never grown,* the yellow and white turnips do not stand late in the season. It is not usual to give cake to any of the sheep on the " seeds," except ewes with twin lambs. The seeds are ploughed in autumn, after being dunged with yard manure, which many farmers prefer applying the previous winter, that the seeds may be benefited as well as the wheat. The wheat is then sown in the usual manner. The strawyard cattle are seldom fed fat, being principally stores, kept through the winter in fair condition, on 4 lb. per day of oil-cake, and sent down to the richer grazing lands, which most Wold farmers hold in the low country, to be fed during the summer. The sheep stock is Leicester or improved Lincoln, a breeding flock being kept, and the produce sold at two years old, seldom earlier. The result of this management is an average crop of 26 to 28 bushels of wheat per acre, and not much more of barley.

The reader who has perused our description of West Norfolk, must be struck with the difference of management in the two districts. The alleged impossibility of growing swedes advantageously here, will be at once referred to the very scanty supply of manure applied to the turnip crop. We should not expect a good crop of swedes, in any other district, with no more enriching application than one cwt. of guano and four to eight bushels of dissolved bones per acre. One farmer grew last year a quantity of swedes for the first time (four or five acres in upwards of 100 acres of soft turnips), and, having manured the land with dung, in addition to the above quantities of artificial manure, he succeeded in getting a good crop. This year he intends to grow 20 acres. But one is surprised to hear that in Lincolnshire this discovery has been now made for the first time. Nor is the cultivation of the land attended to with anything like the same neatness and care which distinguish the best farmers in West Norfolk. It is permitted to get foul, and the same minute attention in the extirpation of weeds from the corn and grass fields is not here observed. The land is chalked as often as it appears to require it, perhaps once in 20 years; the want of it being shown by the appearance of the disease in turnips called " fingers and toes," for which chalking is a perfect cure. Four horses are often used in a plough ; but it is a two-furrow plough, turning over at this season of the year three acres a-day, and usually managed by one man without a driver.

Leaving the Wolds, and recrossing the central plain, we arrive at the Heath farming. This is part of the great tract of land on the lower oolite formation which commences at Bridport, in Dorsetshire, and runs in an unbroken chain through Gloucestershire, Oxford, Northampton, Rutland, and into Lincolnshire. It is a dry reddish turnip soil, varying from a few inches to a foot or more in depth, lying on a porous calcareous rock. Here the style of farming very much resembles that of the Wolds, except that the crops are somewhat more generously treated. In feeding turnips or seeds, however, cake is very seldom given to any of the sheep stock, except ewes with twin lambs. Top dressings for the corn crops are quite uncommon. A well managed farm of 500 acres will winter 1,000 sheep on turnips. Swedes are successfully grown here. The favourite breed of sheep is the improved Lincoln, which clips from 7 lb. to 8 lb. of wool, and weighs at two years old 30 lb. a quarter. This is a strong-necked sheep, differing in that respect from the pure Leicester, and producing more lean meat than that breed. In the yards the cattle receive cake during the winter, some farmers feeding them fat, others keeping them as stores. Half a ton of oilcake per head is the largest quantity expended by the best farmers. In sowing wheat after clover, the land after being ploughed is first well harrowed ; the seed is drilled, then covered by the harrows. The land is then rolled and harrowed again after the roller. On the better land wheat is usually sown after turnips up to the beginning or middle of March, " Hunter's white wheat" being the variety found most productive. Hornsby's drop drill is much used for sowing the turnip seed and manure together. Great benefit has been found by ploughing deep in the heath land with a subsoil plough, thus bringing up broken calcareous rock, and at once deepening and manuring the soil. The wheat is built in lofty oblong stacks, containing 60 or 70 quarters; and a great deal of labour is expended in carrying these up very high, and in thatching, and finishing them off neatly. The rent is about 20s. an acre, no tithes, and rates low. The better class of land near Lincoln lets at 30s. an acre; but rent is scarcely thought a criterion of value, as some landlords let better land at 20s. than others at 40s.

The farm-buildings on the better class of farms in Lincolnshire are superior to most we have met with in the more southerly counties. On the estate of Mr. Chaplin, of Blankney, many very substantial and complete ranges of buildings have been and are being erected.

A good idea of the process of warping may be got by sailing up the Trent from the Humber to Gainsborough. The banks of the river were constructed centuries ago, to protect the land within them from the encroachments of the tide, or rather to exclude the tide from the land, which was left dry at low water. A great tract of country was thus laid comparatively dry, the tide rising every day within the embankments several feet higher than the cultivated land. The wisdom of one age thus succeeded in restricting within bounds the muddy tidal waters of the river. It was left to the greater wisdom of a succeeding age to improve upon this arrangement, by admitting these muddy waters to lay a fresh coat of rich silt on the exhausted soils, and so to restore them to their original fertility. The process began nearly a century ago, but has become more of a system in recent times. Large sluices of stone, with strong doors to be shut when it is wished to exclude the tide, may be seen on both banks of the river, and, from these, great drains are carried miles inward through the low country to the point previously prepared by embankment, over which the muddy waters are allowed to spread themselves. These main drains being very costly, are constructed for the warping of large adjoining districts, and openings are made at such points as are then undergoing the operation. The mud is deposited, and the waters return with the falling tide to the bed of the river. Spring tides are preferred; and so great is the quantity of mud, that from 10 to 15 acres have been known to be covered with silt from 1 to 3 feet in thickness during one spring of 10 or 12 tides. Peat moss of the most sterile character has been by this process covered with soil of the greatest fertility; and swamps which, in the memory of our informant, were resorted to for leeches, are now, by the effects of warping, converted into firm and fertile fields. Near the mouth of the river the water is muddiest, and the process can there be more easily accomplished, but sluices are seen for nearly 30 miles up the Trent, so that even at that distance from the Humber the water has not entirely lost its fertilizing particles of mud. The expense of warping varies from 15l. to 21l. an acre. After the new land has been left for a year or two (in clover and seeds) it produces great crops of wheat and potatoes.

LETTER XXIII.

LINCOLNSHIRE — *continued.*

RAPIDITY OF IMPROVEMENT IN LINCOLNSHIRE — SIMULTANEOUS INCREASE OF RENT AND TENANT'S PROFIT — LIBERAL MINDED LANDLORDS — COMPENSATION FOR UNEXHAUSTED IMPROVEMENTS DESCRIBED — EXPENSE OF ARBITRATION, AND EVIDENCE OF CLAIMS REQUISITE — CLAIMS INCREASING IN EXTENT — LANDLORDS FIND IT NECESSARY TO LIMIT AND DEFINE THEM — FARMERS NOT MORE LIBERAL IN OUTLAY OR MORE HOPEFUL THAN IN DISTRICTS WHERE NO TENANT RIGHT EXISTS — LORD YARBOROUGH'S ESTATES — MR. CHAPLIN AS A RESIDENT LANDLORD — LABOURERS' WAGES. — COTTAGE RENTS HIGH — GREAT DISTANCES WHICH LABOURERS ARE COMPELLED TO WALK OR RIDE ON DONKEYS TO THEIR WORK FROM SCARCITY OF COTTAGE ACCOMMODATION.

GAINSBOROUGH, April, 1850.

THE agricultural reputation of Lincolnshire is due more to the stride it has made in a given time, than to any real pre-eminence above the best farmed counties. A hundred years ago it was almost a *terra incognita*, its land boundaries impassable fens, desolate heaths, and broad rivers, with no important sea-port, and lying out of the track of the traveller. Till the reign of George III. the county remained in a neglected state, the fee simple of the now cultivated wolds and heaths worth little more than their present annual rent: the fen districts an unwholesome reedy waste, prolific of ague and aquatic birds. Till even a more recent period the improvement was slow. In the parish of Limber, 60 years ago, four tenants renting 4000 acres of land at 125*l.* each, or 2*s.* 6*d.* an acre, became bankrupts. The same land is now yielding its owner upwards of 4000*l.* a year, paid by prosperous tenants. Lincoln Heath, whose improvement had begun in Arthur Young's time, excited his astonishment that farmers in prosperous circumstances could afford to pay 10*s.* an acre for land which, a few years before, had yielded nothing, or next to nothing, to its owner. For the

o

same land they now pay double; and at Blankney several
housand acres were let as rabbit warrens in his time, at 2s. to
3s. 6d. an acre, for which Mr. Chaplin now receives 20s. ; the increased rent being accompanied in both cases with the increasing
wealth of the tenants. The transition has therefore been very
rapid and striking, perhaps more so than in any other county
in England.

It was very fortunate that when the time for this transition
arrived, the leading landlords were liberal and enlightened men.
Among these may be named the late Earl of Yarborough and
Mr. Chaplin of Blankney. They saw the advantage of encouraging tenants to embark their capital freely ; and as leases
were not the fashion of the county, they gave them that security
for their invested capital, which is termed " tenant right," or
compensation for unexhausted improvements.

Though this tenant-right may not be a strictly legal claim, it
is universally admitted in Lincolnshire, the landlord paying it
when a farm falls into his own hands, and refusing to accept a
tenant who declines to comply with the custom. It varies,
however, considerably in different parts of the county; and
appears to have enlarged in its obligations with the greater
development of agricultural improvement. In North Lincolnshire, the usual allowances claimed by the outgoing from the
incoming tenant, include draining, marling, chalking, claying,
lime, bones, guano, rape dust, and oil-cake. The following is
the scale on which these allowances are usually made : —

" When the landlord has found tiles, and the tenant has done the
labour, if done within twelve months before the end of the tenancy,
and no crop has been taken from land after the draining thereof is
completed, the whole cost is allowed. If one crop has been taken
from such land, three-fourths of the cost is allowed, and so on, diminishing the allowance by one-fourth for each crop taken; but this
allowance is made only when the work is well and properly done by
the tenant, to the satisfaction of the landlord or his agent, expressed
in writing.—For marling or chalking, if done within twelve months
before the end of the tena cy, the whole cost is allowed ; for that

done in the previous year, seven-eighths of the cost are allowed ; and so on, diminishing the allowance by one-eighth for each year that shall have elapsed since the marling or chalking. — For lime used within twelve months before the end of the tenancy, if no crop has been taken from the land limed in that year, the whole cost, including labour, is allowed ; if one crop has been taken from such land, four-fifths of the cost are allowed ; and so on, diminishing the allowance by one-fifth for each crop taken from such land. — For claying on light land a similar allowance to that for lime. — For bones used within twelve months before the end of the tenancy two-thirds of the cost are allowed, and for those used in the previous year one-third of the cost. — For guano and rape dust used within twelve months before the end of tenancy, for turnips or other green crop, two-thirds of the cost are allowed. — For oil-cake, given to cattle and sheep, one-third of the cost price of that so used within twelve months before the end of tenancy, and one-sixth of the cost price of that so used in the previous year is allowed."

If the tenant is entitled to a waygoing crop, it is of course mentioned in the agreement, as are also payments for acts of husbandry, such as the carting out of manure, or other labour, for the sole benefit of the incoming tenant. The amount of these allowances is settled by arbitration, the award being made in the gross without particulars, that there may be no room for cavil. The arbiters and their umpire are generally farmers who are paid 2*l.* each per day, the whole expense of an arbitration being from 30*l.* to 50*l.* The evidence of claims for manure purchased, and for cake, &c., consist of the dealers' receipts for these articles, which are sometimes fraudulent, especially when the outgoing tenant has another farm in his possession. On the whole, however, the system is believed to have worked well; though the landlords are beginning to find the claims of the outgoing tenants so serious, that, in order to check their increase, they prefer embodying the claims in a special agreement to trusting to the indefinite "custom" of the county.

The Lincolnshire system, as at present in operation, has not led to the frauds practised in Surrey and Sussex ; partly, perhaps, because it has not been so long a period in use. But an indefinite custom of this kind is liable to great abuse, and it must possess advantages of no common kind to compensate this risk. In a large district of country, where it is most liberally observed, we did not find the farmers one whit less desponding than in other places where they had no such security; and they were limiting their outlays and complaining of their landlords quite as much as in Essex or Suffolk. The best farmers we had an opportunity of visiting are still behind the agricultural proficiency of the leading men of West Norfolk, whose capital is protected by a 12 years' lease and a liberal landlord. And there is a vast extent of land in Lincolnshire in a very backward state, and where much has yet to be done by both landlord and tenant. Nor do the farmers themselves attribute so much benefit to the system of tenant-right; as compensation for unexhausted improvements, however valuable in itself, is in their opinion of less importance to the progress of agriculture than moderate rents, and the existence of perfect confidence between good landlords and good tenants.

On the Earl of Yarborough's estates farms are held by the same families for generations. Besides continuing the liberal treatment of his tenantry which distinguished his predecessors, the present Earl has engaged in vast undertakings for developing the resources of Lincolnshire, by means of railways and docks.

Mr. Chaplin of Blankney is generally regarded as the *beau ideal* of a resident landlord. He spends his time and his income on his estate, and devotes himself to its improvement and to the welfare of all who reside on it. Though the owner of a large estate, his practical knowledge of details enables him to dispense with the interposition of an agent between himself and his tenants. His farms are not only moderately let, but his personal acquaintance with the wants of his tenantry gains his acquiescence in all permanent improvements. They dine with him on the rent day; and as he always lives among them, they communicate freely with him on matters of mutual interest. The labourers on his estate share in his solicitude, their comfortable cottages and gardens being let to them at moderate rents.   In every quality of a landlord, a magistrate, and a neighbour, the influence of his example among the other landlords of the county must be of the greatest advantage, while it has produced on his own estate the most perfect feeling of mutual confidence and attachment.

Labourers' wages in Lincolnshire are at present 10s. a week. In some localities they pay very high rents for their cottages, being swept out of close parishes which are under the control of one or two large proprietors, and obliged to compete with each other for the possession of the limited number of cottages which speculators, naturally taking advantage of their necessities, run up in open parishes for their accommodation.   They are thus in many cases compelled to live at a great distance from their work, to which it is quite common for them to ride on donkeys a distance of six or seven miles.   The farmers, to save the exhaustion of the men, willingly give the donkeys accommodation.   But this abuse of the rights of property is now giving way; and landlords, feeling the impropriety of driving off the labourers required for the cultivation of their estates, are beginning to build good cottages, to be let at moderate rents to well-conducted men.   The system of boarding farm servants in the farmer's house is again coming more into practice, and is likely to continue to do so if provisions are moderate in price. Some board the servants with their bailiffs, but this plan is said not to work well.

LETTER XXIV.

NOTTINGHAMSHIRE.

CHANGE FROM AGRICULTURAL TO MANUFACTURING DISTRICTS — THE "DUKERY."— COMPETITION FOR FARMS NOT UNDULY ENCOURAGED — DUKE OF PORTLAND'S ARRANGEMENT WITH HIS TENANTS — ONE OF THE EARLIEST PROMOTERS OF AGRICULTURAL IMPROVEMENT — FARM BUILDINGS — TENANT RIGHT — NECESSITY FOR LIMITING IT — PARTICULARS NOT SPECIFIED IN BILL OF VALUATION — PROPORTION OF ENTERING TENANTS' CAPITAL AB-SORBED BY THE PAYMENT OF VALUATIONS. — SIZE OF FARMS. — RENT. — CONDITION OF LABOURERS — DIMINUTION OF COTTAGES.

MANSFIELD, May, 1850.

LEAVING the purely agricultural counties of the eastern coast, we enter the midland districts, beginning with the county of Nottingham. On the western verge of the county the coal formation makes its appearance; and the busy scenes and sounds of mining and manufacturing industry are now intermingled with the older and less bustling processes of agricultural employment. In the counties we have already described, the occupations connected with the soil are superior in importance to all others; but as we enter the manufacturing districts, the interests of the country become more immediately subservient to those of the town, and the producers of food find themselves best remunerated when they adapt their management to the varied wants of the great communities which are growing up with such rapidity amongst them.

In the northern division of the county this influence is not directly felt, on account of the unusual number of the nobility who have made that picturesque part of the county their residence. Near Worksop, the district called "the Dukery" is occupied by the seats of the Dukes of Newcastle and Port-land, and the Earl Manvers and Earl of Scarborough; Clumber

Park, Welbeck Abbey, Thoresby Park, and Rufford Hall, being all within the compass of a few miles. The land in the county is chiefly in the hands of large proprietors, who possess great influence, letting their farms from year to year at moderate rents.

Some of the proprietors are very wealthy, while others are heavily embarrassed. On the estates of the latter the rents are usually higher than on those which are free, and drainage and other permanent improvements proceed very slowly. Still, even where the pressure on the landlord is understood to have been very great, no unfair competition to raise the farmers' rents has been encouraged. It is thought that if no deduction of rent can be made at present, the farmers must fall into arrear, and that this will compel the sale of such encumbered properties.

The Duke of Portland has lately made an arrangement with his tenants, to meet the change of circumstances caused by free trade. For this purpose his Grace's tenants were divided into three classes, the high-rented, who are tenants of newly purchased estates, the fairly-rented, and the under-rented, who occupy the old hereditary property on the easiest terms. The whole estates were to be revalued, and one half of the rents to be commuted into a corn rent, taking 56s. a quarter as the basis. This half is to fall with the average price of wheat, however low, but is limited in its rise to 64s. per quarter. The effect of the valuation was to lower the high-rented land 10 per cent, after which the corn rent comes into play. The low-rented tenants prefer to remain as they are without a new valuation. To the first two classes the modified rent this year is equal to a reduction of 14 per cent, giving a benefit to the second class to that amount only, and to the first class a total reduction from their original rents of 24 per cent. The Duke's rents are payable six months after they become due. The tenants are said to be well satisfied with these arrangements, and continue to farm with confidence.

But the Duke of Portland, besides treating his tenants in this way, has for many years back been carrying on vast improve-

ments on his various estates.   While he is one of the largest, he
is also one of the best landlords in the kingdom.   Long before
Mr. Parkes, before even Mr. Smith of Deanston, he was an
energetic tile drainer, having the entire work done systematically,
and at his own cost, and then charging a moderate per centage
on the outlay.

Few of the other great proprietors of Nottinghamshire have
had the same means at their disposal, or the same taste for
agricultural improvement.   Where they have assisted their
tenants in drainage, it has been by the landlord finding the
tiles, and the tenant the labour.   But those who are encumbered, excuse themselves from making any outlay, on the ground
that their lands are low-rented, and as they ask no increased
rent, their tenants reap the sole benefit if they choose to make
the outlay.   For the same reason farm buildings are imperfect
and incommodious; though in North Nottinghamshire they
are on the whole more substantial than is common in the eastern
counties.

In the northern division of the county there is a system of
tenant-right similar to that of Lincolnshire.   It has been longer
in operation, however, and embraces a greater variety of allowances; and yet there is no uniformity established, different
estates, and sometimes even adjoining farms, having different
allowances.   The custom is determined by the award of
allowances paid for at the preceding entry, for evidence of
which, the tenants carefully preserve the written awards between them and their predecessors.   Where such award cannot be had, the arbiters fix the allowances with as much equity
as they can from the custom of the estate or the neighbourhood.
Being themselves farmers, they examine each field, and weigh
such evidence as is laid before them in reference to its management, and form their conclusions accordingly.   The award is
settled by two payments at intervals of some months, so that
any accidental mistakes or deception may be rectified.   The
new tenant having come into possession some months before

the last payment is due, may become aware of any unfairness on the part of his predecessor, if such there were, and speedily reports it to his arbiter, who takes care to have the matter adjusted before the final settlement. Formerly it was the custom for the outgoing and incoming tenant to adjust these matters without the interference of the landlord; but on account of the great increase in the use of artificial manures and food, the allowances have been every year increasing, and for his own protection the landlord finds it necessary to appoint a representative to guard him against excessive liabilities. The awards are made, as in Lincolnshire, without specifying any of the items of charge, the whole amount being set down in one sum at the end. Some experienced valuers attempted to introduce the particulars, but these were so much discussed, that it was found better, or at all events more convenient, to revert to the old plan. In North Nottinghamshire, buildings erected by the tenant, if not claimed by the landlord as fixtures, are included in the valuation.

It is calculated that about one half the capital requisite for the occupation of a farm, is paid by the incoming to the outgoing tenant for these allowances, 3*l.* to 5*l.* an acre being the variable amount of valuations of tenant-right in North Nottinghamshire. In this district, farms are from 300 to 500 acres in extent, good turnip and stock land, chiefly on the red sandstone formation.

In the south-eastern parts of the county, strong loams and tenacious clays prevail, and farms do not exceed 300 acres in extent, the best tracts being the Trent bank land and the Vale of Belvoir. Between Newark and Nottinghamshire, on the rich lands, the rents rise as high as 65*s.* an acre, ranging from 35*s.* upwards. On the colder and stiffer lands the tenants suffer most from the low prices of corn. These lands require drainage, farm buildings, and a greater application of capital and energy ; and as these are plants of slow growth in such localities, the prospect to both landlord and tenant is far from hopeful.

The position of the labourer is comparatively good.   Wages
are about 10s. a week, and cottage rents from 2l. 10s. to 5l. a
year.   A higher rate is paid by some of the great landlords to
the numerous people employed on their parks, woods, and farms,
2s. to 4s. a week being given above the common rate of wages.
But the men have to walk considerable distances to their work.
Both the Duke of Newcastle and the Duke of Portland have
established garden allotments, near towns and villages, for the
accommodation of the inhabitants.   They appear to be cultivated
with great care, and are much appreciated by the artizans and
tradesmen, as well as the labourers.   The Duke of Newcastle
has about 2000 such allotments on his estates.

The rule among the large proprietors throughout the county
has been to diminish cottages, and to drive the labouring population into villages and towns ; thus obliging them to walk unnecessary distances to their work, and exposing them to the
temptations of the beer-houses, and the greater expense of living
in towns, while the rate-payers of such towns are unjustly burdened with the support of persons who have no claim upon them.

Game is not preserved to any very injurious extent, and does
not form matter of complaint by the farmers.

LETTER XXV.

NOTTINGHAMSHIRE — *continued.*

DIFFERENT KINDS OF LAND IN THE COUNTY — DETAILS OF MANAGEMENT OF
A LIGHT LAND FARM NEAR WORKSOP' — PECULIAR AND PROFITABLE MODE
OF GROWING POTATOES — RENT — CLUMBER — WELBECK — CLIPSTONE FARM-
ING AND FAMOUS WATER MEADOWS DESCRIBED — SHERWOOD FOREST —
DEFECTIVE MANAGEMENT OF CLAY SOILS — EXCEPTION TO THIS — MR.
PARKINSON'S FARM AT LEYFIELDS — RENT AND PRODUCE OF LAND NEAR
NOTTINGHAM — MR. PAGET'S FARM — USE OF " SHODDY " AS MANURE —
INCREASING WHEAT CROPS — LARGE RETURNS FROM DAIRY.

NOTTINGHAM, May, 1850.

THE farming of the northern division of Nottinghamshire is in
many respects superior to that in the southern part of the county.
The land is of a lighter character, less expensive to till, and
better adapted for green crops and stock ; while it happens to
be pitched at a considerably lower scale of rent.   Much of the
land being of a light sandy nature, two corn crops in succession can seldom be taken with advantage ; and to prevent this
the chief restriction is imposed on the farmer by his landlord,
who does not generally insist on the exact observance of a four
or five course, which are the common rotations followed in
the district.

On a well-managed farm, 500 acres in extent, within a short
distance of Worksop, the following details there practised may
be taken as an example : — The whole manure is applied to the
turnip crop, which comes every fourth year on the best, and
every fifth on the lighter part of the farm.   Five cwt. rape-dust,
3 cwt. guano, and 10 loads of well rotted dung are all laid in
open drills, 28 inches wide, which are then covered by the
plough, and the seed sown.   Swedes are sown in the middle of

May, and grown in the proportion to yellow and white turnips of 70 acres of the former to 40 of the latter. The crop seldom falls much short of 30 tons an acre. White turnips are sown on all the inferior sandy land, on which the sheep are first placed; and the turnips being thus consumed early, the land is prepared and sown with wheat. As many bushels an acre of wheat, of good quality, can be got on this description of land as of barley, for which it is not good enough. On the lightest blowing sand rye is taken. A breeding flock of 400 Leicester ewes is kept on this farm, the produce of which is fattened off on it at a year old, generally before going to grass the second spring. They each get half a pound of oilcake daily, along with turnips, which are all taken up before the fold and given to them cut, in boxes. Besides the sheep, 30 cattle are fattened every winter, the heifers tied up in stalls, the bullocks fed loose in courts. They receive 4 lb. of oilcake daily, and swedes. The other crops are cultivated in the usual manner, except that in breaking up the light inferior land which has been two years in grass, instead of sowing it with oats, a different plan has been adopted with success for the last year or two. It is broken up in August, and well knocked about, and then sown with white turnips, which are eaten off by sheep, and the land planted in spring with potatoes, without manure. Good sound crops are got, worth three times as much as the oat crop for which they are substituted. The land is then fallowed for turnips in its usual place in the course. Much of this farm is a light poor sand, requiring a large expenditure in manures and cake to keep it in a productive state. It is let at 30s. an acre, inclusive of tithe, and with present prices it is said not to pay. The average rent of similar land in the neighbourhood does not exceed 20s. an acre.

Passing through the Duke of Newcastle's park at Clumber, where his grace holds a very large and neatly-managed farm, we proceeded through a finely-cultivated tract to Welbeck, the residence of the Duke of Portland. Going on some miles

farther through an undulating light land country, the great
feature of which is the extensive woods planted by the present
Duke, we reached Clipstone Park, remarkable for its water-
meadows, the most gigantic improvement of its kind in England.
These, extending to 400 acres, are held in conjunction with an
arable farm of upwards of 2,000 acres, and comprise together
one of the home farms of the Duke of Portland.   The arable
land is chiefly a light sandy tract, formerly part of Sherwood
Forest, which could only be kept in cultivation by a large outlay
in manures, or an equivalent, such as is afforded by the produce
of the water-meadows.   It is cultivated in a seven course, lying
four years in pasture, though during the two latter years the
pasture greatly deteriorates.   Nearly 300 acres are each year
in turnips; and as a large stock of cattle and horses are kept
constantly in the yards, summer and winter, chiefly on the
produce of the meadows, sufficient manure is made to admit of
an application of 30 tons to each acre.   No artificial manure is
purchased, but with this dressing of good dung, great crops of
turnips are grown, 40 tons an acre of swedes being reckoned
not uncommon.   On some heavier land, where beans are
cultivated, we may mention that three rows are sown pretty
close to each other, with a wider interval at every third row to
admit the horse hoe.   After the beans have flowered, men are
sent along these wider intervals with hooks, with which, taking
the three rows at a blow, they very speedily and cheaply shear
off the tops of the beans.   This prevents the *aphis* from effecting
a lodgment; and is found to protect the crop from the total
destruction which, in some seasons, has overtaken it through the
devastating attacks of this insect.

The water-meadows extend about seven miles in length along
the sloping bank of a valley, through the bed of which runs the
little river Mann, its opposite bank for a considerable part of
the way rising abruptly from the stream, covered with woods,
some of recent, some of older date.   The road winds along the
valley, and nothing can be more refreshing to the eye than the

constant succession of green meadows glistening with the trick-
ling water, or covered with flocks of ewes and lambs browsing
on the luxuriant herbage.   Formerly this rich and beautiful
tract was a succession of barren hill sides, covered with gorse
and heath; the bottoms a swamp of rushes, the haunt of the
snipe and the wild duck.   By catching the stream as it leaves
the town of Mansfield, charged with the whole sewerage of
the place, and confining its waters within a new bed at a higher
level along the hill sides, the means of irrigating this extensive
tract were obtained.   The ground was then thoroughly under-
drained, cleared of inequalities, and laid out in the most perfect
manner for letting on and taking off the enriching waters.
These, after flowing over the surface of one side of the valley,
are received into a brook, from which, some miles farther down,
they are passed over meadows on the opposite side of the valley.
At all seasons of the year the waters are laid on the meadows
with the best effect.   The flock of South-down ewes, beginning
to lamb in October, is immediately placed on the rich grass; at
Christmas the lambs are ready for the market, and continue to
be sent off during the early months of spring at the period when
they fetch the highest price.   As one meadow is eaten bare, the
flock is transferred to another, the water being then laid on to
the first.   When the clovers and pasture of the adjoining farm
are ready, the flock is removed to them, and the meadows shut
up for hay, or mown in succession for forage to the horses and
cattle which are kept in the farm-yards during the summer.
Two cuttings are yielded in the season, besides what remains to
be pastured with sheep and cattle in the autumn.   The annual
value of the produce is estimated at from 10*l.* to 12*l.* an acre,
and the whole expenditure from first to last has exceeded
40,000*l.*   This great and expensive agricultural improvement is
justly regarded as the pride of Nottinghamshire, unrivalled
as a work of art in irrigation, and in its cost worthy of the liberality of a wise and patriotic nobleman.

    Crossing the county from the town of Mansfield towards

Southwell we pass for some miles through an open heath, part of the ancient forest of Sherwood, still unenclosed. It is a succession of undulating eminences, covered with short heath and shorter grass, here and there some furze bushes, and occasionally a stunted oak. Beyond this a tract of very light sandy land has been enclosed from the forest, divided into small fields by very thriving thorn hedges.

A few miles farther on we reach a district of clay soils, the system on much of which has hitherto been two crops and a fallow. The land, which is imperfectly drained, undergoes a naked summer fallow, on which wheat is sown, in the autumn. This is followed by beans, sometimes drilled, sometimes broad cast, but very frequently, as we were told, so foul at harvest, that it is difficult to say whether the beans or the weeds are the strongest. Very little stock is kept; and of course very little manure is made, and that of inferior quality. Even here, however, we found an instance of well-managed clay land, some details of which we give, as they exhibit a rational method of dealing with this difficult kind of soil, which may possibly be instructive.

The farm of Mr. Parkinson, of Leyfields, on the estate of the Earl of Scarborough, contains about 300 acres of clay land, 70 of which are in meadow and permanent pasture. The rest of it is divided into 16 fields, as nearly of equal size as possible, which are managed in two six courses and a four, thus,—(1) turnips, (2) barley, (3) clover, (4) wheat; then (1) turnips, (2) barley, (3) Italian rye grass and white clover pastured, (4) ditto, (5) oats, (6) wheat; then (1) turnips, (2) barley, (3) Italian rye, grass, and white clover, cut for hay and feeding and then dunged, (4) ditto pastured, (5) oats, (6) wheat. Of the 16 fields there are thus, annually, three in turnips, five in clover and pasture, three in wheat, three in barley, and two in oats. The farm was first thoroughly drained with tiles, and then divided by hedgerows into square fields of nearly equal size. Immediately after the wheat is reaped, the ground is slightly stirred to encourage the vegetation of annual weeds, and in that state it is left till the wheat sowing is completed. The dung of the previous winter, which has been all kept for the purpose carefully covered with a layer of earth, is then laid on the ground at the rate of 20 loads an acre, and ploughed in. In this state the land remains untouched till the spring seed time is completed, when the first favourable weather is seized for cross-ploughing. It must be remarked that the soil is of that stiff clayey nature that the utmost caution is necessary to prevent it being "poached," or becoming cloddy in spring, the whole success of the turnip crop depending on this. It is, therefore, ploughed with two horses in length, to prevent the tender surface soil being trodden. After lying exposed for some time, it is again ploughed in a contrary direction, the horses still walking in line; and if the weather is favourable, and when the ground has become dry enough to bear them, it is well harrowed, and, if necessary, rolled. But great care is taken not to put the horses on the land to harrow it until it has become dry below; as, however fine they might make the surface with the harrow, they would do great injury by *fastening* the ground beneath. This preparation is usually sufficient, though occasionally a fourth furrow is given. When this is accomplished, as early in May as possible, the land is drawn into ridges 27 inches apart, on the top of which the turnip seed is sown with a drill, which at the same time deposits beneath it a layer of ashes, soaked in liquid, at the rate of 20 quarters an acre. The seed falling upon this moist bed springs at once; and the chief difficulty in clay land, that of getting the small seed of the turnip to vegetate, is thus completely obviated. This is followed by a vigorous growth, the thread-like roots penetrating through the now loose and tender land, and finding in every part of it the nourishment which was laid on in the shape of dung during the previous autumn. In the autumn the turnips are taken up and removed for consumption in the yards, dry weather or frost being chosen for taking them off the ground. The barley is sown in spring, the same care in ploughing dry being exercised.   Clover and other seeds are sown in the usual manner, red clover coming only once in 16 years, and never missing a plant.   The red clover is mown once or twice, as may be necessary.   One field of Italian rye grass and white clover is pastured both first and second year, the other is cut for soiling and hay the first year, and, after getting 10 loads an acre of dung, is pastured the second year.   Wheat is sown after red clover, the land being ploughed deep by three horses, with a plough provided with a broad skim, which shears off the entire surface, turning it into the bottom of the previous furrow, where it is covered up by the advancing plough.   The ground is then harrowed and the wheat seed sown by the drill.   Where the land has been down two years in grass it is ploughed in spring with the same three-horse plough, and sown with oats.   In autumn it is again ploughed, and turns up very mellow, the roots of the grasses now decayed being brought up again to afford nourishment to the wheat, which is then drilled in the usual manner.   Under this mode of management the land is kept very clean, and all the heavy operations are performed at the season when least injury can be done to this tender soil.   No artificial manure whatever is purchased, and the crops average 20 to 30 tons of swedes, 40 bushels of wheat, 40 bushels of barley, and 64 bushels of oats per acre.   No turnips but swedes are sown except on the headlands, and a small portion on the ground where tares for the horses had been grown.

The burnt ashes are prepared by digging up the corners of fields, and close to the roots of the hedges where the plough cannot work — the sods, full of vegetable matter, taken from which, after being dried in the sun, are burnt in large heaps with the trimmings of hedges to keep them on fire.   By close attention these heaps are burnt completely through, and the ashes are then drenched with liquid from the manure tanks.   In this manner no waste ground is lost, and the hedge trimmings are well worth the cost of keeping the hedges in order.

A flock of 200 Leicester ewes is kept on the farm, the produce

P

of which is sent off to another farm as soon as they are weaned. About 50 cattle are fed fat during the winter, each receiving 4lb. of oil-cake daily, and swedes. A few cows and eight to ten work horses are also kept on the farm. The whole liquid from the feeding yards and houses is carefully caught and preserved in tanks. Watering places for the stock are provided in every field by a simple arrangement of the main drains, and several of the meadows are irrigated in winter by a small stream which runs along the bottom of the farm. The whole of these improvements have been executed by the tenant at his own cost, though he has no lease, and holds only from year to year; yet he would rather be so than under a lease, his rent being fairly and moderately charged,—such is the mutual confidence here subsisting between landlord and tenant.

In the neighbourhood of Nottingham much of the land is in pasture for supplying the town with dairy produce; and the advantage of water carriage for manure is enjoyed by the farmers on both banks of the Trent. The best land here lets high, from 2l. up to 4l. an acre. Beyond the immediate influence of the town the rents vary from 35s. to 45s. an acre, inclusive of tithe. The land, being naturally rich, yields large crops; but the management generally cannot be commended either for neatness or industry. On many estates there are no restrictions as to cropping, and we had great difficulty in ascertaining what the usual course of husbandry was. On one large farm which we visited we found it common to take turnips, barley, seeds, wheat, oats, wheat (manured). The wheat was said to average 40 bushels an acre, oats often 80 bushels, and barley when grown between the two wheat crops, which it often is instead of oats, as much as 65 bushels an acre. The whole of the work horses, 20 in number, are under the charge of one man, the waggoner, who feeds them all himself, and has a number of boys at 8d. a day to clean them and work them in the fields in plough or harrows, under his eye, he taking one pair of the horses, and obliging the boys to go the same pace as himself. This is certainly the most short-sighted economy we have yet met with.

It is pleasant to turn from such management to that of Mr. Paget, of Ruddington Grange, four miles south of Nottingham, who here farms about 300 acres of his own property. This farm was all closely tiledrained many years ago, and has been for a considerable period in a high state of cultivation. A large dairy stock is kept for supplying Nottingham with milk, 50 acres of old grass surrounding the house affording them pasture in summer, and several water meadows at the lower part of the farm yielding hay for them in winter. The rest of the land is kept in a constant succession of crops; a green and white crop always alternating in the following manner:—turnips; wheat and barley; clover and Italian rye grass (10lb. red, 4lb. white, and 1 peck Italian); wheat; mangold wurzel: wheat; beans; wheat. Every green crop except the clover is manured heavily—the

Turnips receiving 20 loads of dung and 15 cwt. shoddy per acre.
Mangold   „   24   „   „   20   „   „
Beans   „   10   „   „   6   „   „

The swedes average 24 tons an acre, the mangold 30 to 33, and better crops of wheat are always got after the latter than the former, especially if the mangold leaves are ploughed in for manure. The average crops of corn are 46 bushels of wheat, 65 to 68 bushels of barley, and 42 bushels of beans. Shoddy, the refuse woollen rags of the shoddy cloth manufacture, costs about 40s. a ton. The corn crops are all drilled and repeatedly hoed by Garret's horse hoe, which Mr. Paget holds in high estimation. Nor has he ever observed that spring hoeing has been injurious to his wheat crop, as we found it complained of in some parts of Norfolk and Lincoln. When he first began to take wheat so repeatedly off his land, he was told by experienced neighbours that it could not last long, and that his crops would year by year become less productive. The very contrary has been the case, the crops becoming more luxuriant from the continued supply of manure; but now corrected in their tendency to produce too much straw by being top-dressed with 2 cwt. of salt, which is sown broadcast over the wheat in spring. In 15 years one field has borne eight crops of wheat, none of which has been less than $5\frac{1}{2}$ quarters an acre, and one year the average was 7 quarters. Neither the quantity nor the quality of the produce has in the slightest degree fallen off.

Fifty cattle are annually fattened on this farm, and 150 barren ewes, bought in August, are sent off fat in March. The principal stock, however, is from 48 to 50 cows, that number being kept always giving milk, by fattening those which are dry, and substituting others fit for the dairy. In winter they each receive 4lb. of oil-cake daily, besides roots, swedes early in the season, and mangold in spring. As soon as the grass is ready they are turned out to pasture, and then receive no other food. The whole produce in milk and butter is sent daily to Nottingham, and realises 1,100*l.* a year.

LETTER XXVI.

LEICESTERSHIRE.

LEICESTER, May.

THE county of Leicester is nearly equally divided by the lias
and sandstone formations, the former occupying the greater
portion of the eastern, the latter the western side of the
county. The coal formation exists to a considerable extent on
the west, and the clay slate on Charnwood Forest. Nearly
two-thirds of the county are in permanent pasture, the greater
proportion of which is on the lias formation, the corn lands being
chiefly on the sandstone. The quality of the pasture varies considerably, being richest on the low grounds along the banks of
the rivers, and there devoted to the fattening of stock, or for
meadows; while on the higher and colder land it is better
adapted to the dairy and the production of cheese.

From a variety of circumstances this county has long been one
of the highest rented in the kingdom. Its proximity to large
manufacturing towns, the facilities of communication it has long
possessed, first by canals and good roads, and more lately by
railways in all directions, and the number of men of fortune attracted to it during the hunting season, have contributed to en-

P 3

hance its value. The mode of farming and the quality of the land have had their influence, for where land is good enough to maintain its quality in old grass, there can be no doubt that under that system the largest share of the produce goes to the landlord. Labour costs little, the risks of season are less, the competition for land generally greater, and, though the gross produce may be diminished, the net return to the landlord is relatively high. We have accordingly found the pressure of rent more complained of in this county than in any we have yet visited: 30s. an acre, exclusive of tithes and rates, may be reckoned as an average rent for medium soils of the county, and we have found farms let at 35s. an acre, inclusive of all rates, certainly not superior in quality to many we have visited in counties nearer the metropolis and let at 25s. an acre. The neighbourhood of towns and manufacturing villages greatly enhances the value and rent of land; 4l. and 5l. an acre being quite common for good pasture land in such situations. And the fact that these are most numerous on the western side of the county has contributed to bring up the rent of that, though inferior in quality, to a par with the eastern, which is naturally more fertile.

The appearance of the county is picturesque, in so far as green fields, small enclosures, numerous hedgerows, and a succession of gentle eminences can make it. It possesses no bold outline of hills, and the streams which flow through it are more like navigable canals than rivers. On the eastern side, from Melton Mowbray to Lutterworth, you may ride for many miles through grass lands, a great proportion of which are laid up in the old high-backed, crooked ridges which our forefathers adopted for drainage. On the western side the pasture is chiefly applied to dairying, and most farms have a portion under crop, so that arable here divides the country more with pasture. In various parts of the county tracts of fertile light turnip land are met with, on which the style of farming is generally superior.

The grazing farms of the best class are frequently held in

conjunction with arable farms, on which the stock are wintered, and then turned out to be fattened on the grass lands in summer. The stronger land, which is under the plough, is difficult to manage, and, being high rented, the farmer has little encouragement to farm it with spirit.   But where well managed, the land is first tile-drained, a drain being made in every furrow in the hollow of the old high-backed ridges.   These are scarcely ever altered in form, as the labour of lifting off the surface soil, levelling down the subsoil, and replacing the surface, is too costly an operation for a tenant, and it is here thought that by no other means can the levelling of these ridges be safely accomplished. In many instances, where it has been attempted by the plough, the farmer has found it necessary to return to the old form. These ridges are of two breadths, the one being exactly double the other, the narrowest usually in the worst land, so that, practically, no great mistake is made in running the drains up every furrow.

Commencing with a fallow (for which on this stiff land five horses are not unfrequently used in ploughing the winter furrow), the land, after being dunged in spring, and wrought as well as circumstances admit, is partly ridged up and sown with swedes, and part sown with tares for the horses. When these are removed in autumn, the land is ploughed and left to the influence of the frost of winter.   In spring it is drilled with barley, which, after being hoed, is sown out with clover and grass.   The following year, it is pastured or cut for forage and hay.   About the end of September it is ploughed with a plough provided with a skim-coulter, which cuts the surface clean off, and throws it into the bottom of the furrow.   The day before each piece of land is ploughed, newly slaked lime is sown over it, at the rate of 5 cwt. an acre, to kill the slug, for which this dressing is found a specific cure.   Three horses draw each plough, and two ploughs are followed by a presser drawn by one horse, which consolidates the furrow, and prevents that hollowness which is often injurious to the wheat plant.   The field is then left for a few weeks to mellow with the weather. About the end of October the sharp edges of the furrows are harrowed down, and the wheat-seed drilled in across the land, and covered by the harrows. Under this management the stiff ground works down fine, more like a fallow than a clover lea. The wheat is hand-hoed in spring. After the crop is reaped the land is ploughed, and the following spring it is sown with beans, which are dibbled and hand-hoed during the summer. As soon as these are removed from the ground it is ploughed, and again sown with wheat. The land is manured only once in the course, on the fallow division, partly with farm dung, and partly with manure from the towns or manufacturing villages. No guano or other artificial manure is used. The crops are very moderate, — three to three quarters and a half of wheat, and from two to four quarters of beans per acre. On some of the stiff inferior dairy farms in the western part of the county the wheat crop after a bare fallow frequently yields no more than 10 to 15 bushels an acre.

On a farm of the better description of arable land, suitable for all green crops, and sound enough to feed them off with sheep, we found the following mode of husbandry practised, the farmer being a large holder of land, of great experience, and believed to have been successful. Beginning with swedes, and other turnips, the land is prepared for them by the application of 15 tons of dung, either farm-yard or good dung from the stables at Melton Mowbray, for which 5s. to 6s. a ton is paid, though it has to be afterwards carted several miles. The turnips are partly eaten on the ground, and partly drawn for consumption in the yards. The land is then sown with barley and grass, and clover seeds. These are pastured, and the following spring the ground is ploughed, and sown either with oats or barley, or dibbled with beans, according to the quality of the land and the taste of the farmer. Whatever the crop, as soon as it is removed the ground is ploughed and sown with wheat. There are thus three corn crops and two green crops every five years, the wheat always following a corn crop. The land is dunged only once, and bones, guano, or any other artificial manure is unknown, the farmer boasting that he never used an ounce of such in his life! His wheat crops average four quarters, and his barley five quarters an acre.   He pays 2*l.* an acre of rent, no tithe, and the poor-rates are moderate.   On both the farms just described about half of the land is in old grass.   A few cows are kept for dairying, the produce of which is reared and sold off the grass land in the second or third summer.   A flock of Leicester ewes is also kept, the produce of which are fed in winter on turnips and oil-cake, and made fat at a year old.

A great portion of the inferior grass land of the county is devoted to cheese-making.   Dairy farms vary in size from 100 to 500 acres in extent, some having a large portion under the plough, and some little more than affords a few turnips for the stock in winter, and a few acres of wheat to help in making up the landlord's rent.   A considerable outlay has, in many instances, been made in drainage, but from want of supervision on the part of the landlords or their agents, that outlay has been very inefficiently made.

Where landlords give their tenants tiles for drainage without exercising any supervision as to the mode in which the work is executed, the tenant, often from ignorance of the true principles of drainage, puts them in too shallow.   His immediate interest prompts him to do so, as, having to pay the expense of cutting the drains, he naturally does it in the least costly manner.   Inefficiently done at the first, the drains fail of their proper effect; and instances were mentioned to us in this county where a tenant, getting the tiles gratis, laid a new drain immediately over the top of the first, the defects of which he thus expected to remedy!   The original defect was want of depth, and this was no cure.   In despair the man resigned his farm, and his successor, on commencing operations, discovered first one row of tiles, and then a second beneath it.   By going considerably deeper than the lowest he has drained the field.   But how much disappointment and waste of capital might have been saved by the exercise of an intelligent supervision by the landlord, at the first, in the outlay of his own money.

Stilton cheese is produced in some parts of the county, and it varies extremely in quality according to the skill with which it is manufactured. But the kind chiefly made is Leicester, a full milk cheese, somewhat flat in shape, and varying from 30lb. to 50lb. in weight. The average yield of a dairy on moderate land may be reckoned at 3 cwt. to $3\frac{1}{2}$ cwt., and, on better land, up to $4\frac{1}{2}$ cwt. and even 5 cwt. of cheese per cow. The cows under ordinary management are wintered very poorly, receiving straw, sometimes a few turnips, and being turned out to poach and trample the rough pastures. Sufficient accommodation for housing them comfortably and for saving the manure is seldom provided by the landlord. The tenants themselves differ very much in their estimate of the advantages of manure as applied to grass land for dairying. One farmer assured us that by top dressing his old pastures with good dung he for three years rendered his cheese nearly unsaleable. Others again, believing that any injurious effects may be rectified by attention in the process of cheese-making, are adopting the system of house feeding on artificial grasses in summer as well as constant housing during the winter. By this plan they can keep from double to treble the stock on the same ground, the increased quantity of produce greatly overbalancing any slight deterioration of quality. Accurate trials show that warmth and care in feeding exercise a most important influence in the secretion of milk. A herd of cows to which water is usually supplied by pipes and troughs in the cowhouse were, from an obstruction in the pipes, turned out twice a day to be watered. Their milk instantly decreased, and in three days the falling off became very considerable. The pipes were mended, the cows received water in the cowhouse without being exposed to cold, and the flow of milk returned. In another case the person who had the principal charge of the herd was obliged to leave home for a couple of days; the cows were placed under the care of a youth, with strict charge

as to their feeding. This he neglected; the yield of milk immediately declined, and during the rest of the season it never
could be restored to its previous quantity. Where cows are
housed and abundantly fed they should be milked three times
a-day; the milking in the middle of the day being found to
increase the secretion of milk very materially.

The land in Leicestershire is more subdivided than in Notts.
There are considerable estates, as those of the Duke of Rutland,
Lord Stamford, Earl Howe, Lord Maynard, and Mr. Packe.
But the majority are small landholders or yeomen with 50
to 500 acres, which they generally occupy themselves. On the
heavy soils the farms extend from 50 to 300 acres, and from
150 to 500 on the lighter. The tenants almost invariably hold
from year to year; and from the frequent changes of property,
in a county where it is so much subdivided, they have not great
confidence in the permanence of such a tenure. The land is, in
consequence, imperfectly cultivated, and improvements make
slow progress. Some run the risk of uncertain tenure rather
than the certain loss attending bad farming in times of low
prices. Others are compelled, it is said by high rents, to exert
themselves; but many take merely what the land with ordinary
cultivation will produce, having no motive to increase its productiveness, as they have no assurance that they will reap the
fruits of their improvements.

It is not uncommon on large estates to find the proprietor
letting his land on apparently moderate terms, with the distinct
understanding that the rents are so fixed, in order that the
landlord may be exempted from all outlay for improvements.
But a tenant, with no better security than a yearly tenure, will
not make a permanent improvement, from which he may never
be benefited; and the consequence is that estates where this principle is adopted are negligently farmed, undrained, and without
adequate farm-buildings. Though let nominally low, they are
really high rented, and neither landlord nor tenant thrives by
them. On such estates the tenant has often three rents to pay;
one in money, another in feeding the game, and a third for the

support of the hedgerow timber which is allowed to overspread his fields.

It is to be regretted that hitherto the large proportion of landlords in Liecestershire have given little attention to the improvement of their estates. The fashion seems to have been, that all intercourse with their tenants should be through the medium of agents. This has arisen in part from an indisposition to business, and in part from pecuniary embarrassments, which compel them to turn a deaf ear to all claims for outlay on improvements. If the agents are competent men, with adequate powers, there may be no great injury done; but if their qualifications are judged of chiefly by their power of screwing out the rents, and shutting the door against all appeals for necessary outlays, there can be no progress under such management. This isolation is naturally productive of distrust on the part of the tenant; and there is certainly less sympathy between the two classes of Landlord and Tenant in Leicestershire than in any county we have yet visited. This may be in some degree attributed to the spread of opinion from the adjoining manufacturing towns. The farmers discuss the law of distraint, as a means of bolstering up the landlord's rent unfairly by inducing undue competition, and look for relief and encouragement to a statutory act on the subject of compensation for unexhausted improvements, by which they may lay out their capital with confidence and security.

The best landlords in the county are said to be capitalists from the towns, who, having purchased estates, manage them with the same attention to principles and details as gained them success in business. They drain their land thoroughly, remove useless and injurious timber, erect suitable farm buildings, and then let to good tenants on equitable terms. Nominally these rents are high; but farms provided with every facility for good cultivation can far better afford to pay a good rent, than can a dilapidated estate any rent, however apparently moderate.

LETTER XXVII.

WARWICKSHIRE.

LETTING VALUE OF VARIOUS SOILS — REMOVAL OF HEDGEROW TIMBER —
TENURE — ADVANTAGE OF RESIDENT LANDLORDS — CONDITION OF LABOURER
— IMPROVING FARMERS — CERTAIN LOSS IN FARMING STRONG LAND WITH
INADEQUATE CAPITAL — OLD-FASHIONED FARMER WHO PLOUGHS LITTLE,
MORE SUCCESSFUL — COURSE OF CROPS, MANAGEMENT, AND PRODUCE OF
STRONG LAND — DEFECTIVE DRAINAGE — MANAGEMENT OF LIGHTER SOILS
— LIBERAL APPLICATION OF MANURES — MANAGEMENT AND PRODUCE OF
STOCK — MILKMEN IN BIRMINGHAM — REMISSNESS OF FARMERS OF SUR-
ROUNDING COUNTRY IN NOT TAKING UP THIS BUSINESS — WASTE OF MANURE.

BIRMINGHAM.

FROM Leicester we enter Warwickshire, with its pleasant
undulating surface of hill and dale, its old castles of ancient
renown, its classic streams, its birthplace of Shakspeare, and its
modern hives of industry.

The red, deep, sandy loam in the centre of the county,
especially from Stratford by Wallesbourne to Warwick and
Coventry, is the most valuable tract of soil, as it is equally
adapted to turnip and bean culture. It lets at from 35s. to
45s. an acre. A second description, of a more sandy character,
on a subsoil of limestone, marl, or sandstone, brings from 25s. to
35s. The stiff clays, of which there is a considerable breadth
in Warwickshire, are let in some cases as low as 15s. Though
these figures mark the value at which experienced land-valuers
rate the different classes of soils in the county, the real rent
is regulated more by the views of individual landlords than by
the intrinsic quality of the soil. The valuations of a preceding
generation are often the basis on which the rent is fixed; so
that the stiff wheat soils which were then most valuable continue to bear a rent quite disproportioned to their comparative
value now. Other soils, again, which were then little esteemed,

remain at a rent extremely moderate, now that their convertible qualities have been brought fully into play.

The fields are, in many parts of the county, too small, and much encumbered with hedgerow timber. So anxious are the farmers to have this injury abated, that on one estate, where no one could be got to buy the trees, the tenant purchased them himself, taking the risk of loss in selling them afterwards, in order that he might be relieved from the injury caused to his fields. Landlords are believed to be aware of the evil, and disposed to remove it ; but they have great difficulty in getting purchasers of trustworthy character, and are, besides, reluctant to believe that their estates can be improved by the removal of objects which add so much to their picturesque beauty.

Farms are generally let from year to year ; tenants, on the large estates especially, being seldom disturbed in their holdings ; and a good understanding exists between the owner and occupier of the soil. The tenants do not appear to desire leases : some even say that they would have nothing to do with a landlord from whom such a security was necessary. They regard compensation for unexhausted improvements as a matter in which the legislature might justly interpose. But men of great experience and intelligence — such as Mr. Chapman, Lord Leigh's agent — do not expect that any measure generally beneficial could be enacted, and believe that such matters ought to be settled by private agreement between landlord and tenant. Farm buildings are, for the most part, inadequate ; and on estates which are heavily encumbered or neglected by their owners, they have been suffered to fall into such decay that they cannot be repaired.

The presence of a large body of resident landlords secures an outward neatness and order in the appearance of the roads, fences, gates, and buildings, which is not to be found in any district where the landlords are chiefly absentees. Though the tenants have given up the idea of a return to Protection, and are bent on a reduction of rent as the only compensation, in their opinion, adequate to the change of prices, the landlords are slow to believe this. Few reductions of rent have been made, or appear to be contemplated. The complaints of injury from game are not numerous.

The condition of the labourer in Warwickshire is tolerably good. In the south the rate of wages is 8s. a week, or 7s. with beer. Northward wages, rise to 9s. Cottage rents are very moderate; and on some estates the landlords pay much attention to the well-being of the labourers. Lady Leigh takes the cottages under her immediate care. In some parishes the law of settlement is taken advantage of to drive out labourers, who are thus obliged to reside two or three miles from their work.

The state of agriculture in Warwickshire is said to have undergone a very great improvement within the last 30 years. Men are still living who remember to have seen potatoes sown broadcast over the ground like corn, and then ploughed in; but such management as that has long gone out of fashion. The larger farmers are now, on many points, on a par with their brethren in the best cultivated counties, though on others they seem to a stranger to adhere with undue pertinacity to antiquated customs. Even on the best farms no great exertion has been made to get rid of the inconvenience of the old-fashioned high crooked ridges; while four or five horses in line, in plough, or harrow, are in too many instances still considered indispensable.

A large proportion of the county, however, is held by men of a different stamp. Here there is little change and not much progress. Where the land is drained it is not done deep enough; too much of it is kept under the plough in comparison with the capital employed by the farmer in its cultivation; and, though there appears on the whole a sufficient number of horses for its management, they are applied with such a waste of power that the work falls behind, the best seed time is lost, or, if taken advantage of, the crops are sown before the soil has been duly

prepared for their reception. The consequence is, that they are
more expensive to hoe, that operation is imperfectly done, and
every succeeding crop makes the operation more difficult. As
the crops diminish in strength the weeds increase, till at last the
produce bears but a small ratio to the expense of production.
Each step in the downward progress accelerates the evil, and increases the difficulty of restoring the land to fertility, especially
on the stronger red marl soils of this county. In fallowing, a
portion is intended for turnips ; there is not time to get it sufficiently cleaned, so the turnips are sown broadcast; they cannot
be hoed and wrought to advantage, and yield a scanty crop; the
stock to be fed by them are half-starved, and their dung is of
little value. After the turnips, the ground is sown with barley
and grass seeds, in foul condition. The barley is a failing crop,
and the seeds, growing up among twitch, are neither plentiful
nor nutritious. At every point the farmer finds himself beaten ;
year by year his land becomes more expensive and less pro
ductive ; to meet his landlord he contracts his labour ; his stock
gradually follows, and by-and-by his ruin is complete. Such is
too often the current of events with the man who occupies stiff
land with too little capital, and too much under the plough.
The careful old-fashioned farmer ploughs less, runs fewer risks,
and employing little labour, and paying a low rent, manages to
get on even in bad times. But such farming as his keeps no
pace with the increase of population, nor his intelligence and enterprise with the progress of events around him. On a farm of
this description we found the turnip crop managed thus : — A
little manure was scattered over the surface, the turnip-seed was
sown broadcast over the ground, which was then ploughed up
into raised ridges 27 inches apart, the farmer arguing that in this
form he had more surface on which to grow his crop, as you could
stick more pins into a round pincushion than a flat one !

The soils throughout the county are very much interspersed,
almost every farm containing a variety. There are no uniform
tracts of clay or of turnip and barley land, as in the eastern

counties. Even in one field you pass from the finest turnip and barley land into the stiffest red marl. Farms differ, however, in the proportions they contain of heavy and light soils, some having so large a proportion of the former as to be called strong or heavy land farms, and others being principally turnip and barley farms.

On the strong soils the course of crops taken by the best farmers is as follows : — (1) wheat, (2) beans, manured, (3) wheat, (4) fallow, with a few turnips, (5) wheat or barley, (6) seeds. The corn crops are all drilled from seven to nine inches apart. They are hoed and kept clean. The beans are usually dibbled in each furrow, the land having been previously dunged. They come up in rows, eight inches or so apart, and are generally twice hand-hoed. The dibbling by hand costs 5s. an acre, and each hoeing about 2s. 6d. Winter beans are now sown to a considerable extent, though they have not been long introduced into this county. After the bean crop is removed, the land is ploughed, and, if possible, well cleaned before the following crop of wheat is drilled in. After the wheat crop the ground is fallowed, many of the best farmers believing that on these strong soils an occasional naked fallow is indispensable. Part of this division is sown with swedes, and in some cases with mangold wurzel, that plant, though of recent introduction, being much approved for such soils. The fallow and turnips are followed by wheat or barley, with which the land is sown down with clover and seeds. On land of this description, well managed, and rented at 35s. an acre, 30 bushels of wheat, 32 to 40 bushels of barley, and 30 bushels of beans are reckoned fair average crops. Thirty-six bushels of beans are considered a very heavy crop.

Much of this land, as already mentioned, still lies in high crooked ridges or lands, in draining which the mode generally followed is to place a tile drain in every furrow, at about 18 inches beneath the surface of the furrow, and that is believed to be about 3 feet below the uniform surface, if the ground were

Q

all levelled down. The effect of this kind of drainage is, that
the water falling on the adjoining steep-sided lands runs rapidly
over the surface to the furrow, where it is at once carried off
by the shallow drains. But it is evident that much of the
benefit of drainage is thereby lost. If the ground were level,
and the drains not less than 3 or 4 feet in depth, the water
falling upon it, instead of washing over the surface, would sink
into and permeate the subsoil, assisting, by its action, and that
of the air which follows it, in the processes of decomposition
and nutrition, and then slowly passing off by underground
channels to the drain.

On the lighter soils the course of crops is, (1) wheat, (2)
peas, followed by white turnips, (3) barley or spring wheat, (4)
turnips, (5) barley, (6) seeds. The corn crops are drilled and
hoed, peas being treated in the same way as a white crop.
White turnips are usually sown after the peas, which ripen
early; they are manured with guano, and eaten on the ground
by sheep. They are followed by spring wheat or barley, principally the former. The land is again cleaned and wrought for
turnips, farm manure and guano together being applied by the
best farmers, and the turnips drilled on the flat. Part of the
crop is drawn for consumption in the yards, and part eaten on
the ground, with cake, by sheep. The field is then sown with
barley, and clover and grass seeds.

Much difficulty is sometimes experienced in getting a plant
of turnips in the hot scorching weather of June, and this has
been in great measure obviated by the mode of preparation
adopted by one of the best farmers in this county, who dungs
his stubbles and then ploughs them in autumn. The ground is
never again ploughed, but is worked in spring by grubber,
scarifier, roller, and harrow; and, as the land is kept quite clean
in all crops by careful hoeing, there is not found to be any
necessity for spring ploughing. The moisture thus remains in
the land, and the seed, when sown, comes up at once. The
general practice is to sow swedes about the middle of June,

from a fear of mildew if sown earlier; yet, on inquiry, we found that the heaviest crops had been got by sowing early in May.   Many farmers manure their turnip crops liberally, using three cwt. of guano in addition to 15 or 20 tons of good dung per acre, and four or five cwt. of guano for white turnips, without dung.   Two cwt. of guano is also occasionally applied to the wheat crop in spring, as a top dressing, though in this form it has not been found so productive of grain as of straw.   Our experience would lead us to expect better results by harrowing the guano in with the seed at the time of sowing, in autumn or spring.

The management and products of cattle and sheep form a large part of the Warwickshire farmer's business.   One-half of the county at least is under permanent pasture, and that on the river sides, and towards the east, being the richest, is used for fattening; westwards from Kenilworth it is principally under dairy management, cheese being the chief product.   Near the towns the farmers make butter, but, even in the vicinity of one of the large manufacturing towns, we found the grass lands stocked with young cattle and sheep, just as if such a market for dairy produce had been 100 miles distant.   The arable farms have also a considerable proportion of grass land, and most farmers keep a mixed stock of dairy cows, young cattle, and Leicester sheep.   A calf is reared from each cow, and kept till turned out fat at three years old, or, if a cow calf, till fit for the dairy, the rest of the produce being made into cheese.   On most farms a stock of Leicester ewes is kept, the produce of which is sold fat at a year or 15 months old.   Cheese has declined very considerably in price; but we found, on inquiry at one of the principal butter merchants in Birmingham, that his price for butter to the farmers in the neighbourhood, by whom he is supplied, is precisely the same as it was last year, being from $10\frac{1}{2}d$. to $1s$. $0\frac{1}{2}d$. per lb., according to quality.   The same dairy varies in quality from $1d$. to $1\frac{1}{2}d$. in a week.

The chief supply of milk for the manufacturing towns is

provided by the cow-feeders within the towns. Birmingham, for instance, with a population of 200,000 inhabitants, contains within it, as we were informed, about 1,000 cows, and upwards of 200 milkmen, some of whom keep from four to seven cows each, and others from fifteen to twenty. Nine-tenths of the milkmen are of the former class : having little means they keep their cows wretchedly, and make a very sorry living by the trade. The others find it their interest to feed well, which they do on hay, grains, bran, and swedes in winter, and turn their cows out in summer to pasture on the aftermath, in hay fields near the town. Milk sells at 3d. a quart, the largest consumption being in winter, the supply of vegetables in summer being found to limit the demand for milk at that season. Within a circle of two or three miles round the town the farmers send in milk for sale ; but we were surprised to learn that scarcely any milk comes in by the railway from a greater distance, as it does for thirty miles round Liverpool and Manchester. Surely there is some great remissness here in the farmers of Warwickshire and the adjoining counties. The supply of so large a population with milk and other dairy produce, and with early potatoes and other vegetables suited for field culture, might, we should imagine, furnish very profitable occupation for the farmers along the different lines of railway leading into Birmingham.

The care with which the cow-feeders preserve the solid manure from their cow-houses, to sell to the farmers, forms a striking contrast to the waste of the far more valuable substance which is daily carried away in the sewage of the town, and comparatively lost. For it is a somewhat singular consequence of improved sanitary arrangements, that the most valuable properties of nightsoil are now washed away, and the ashes collected in the dust-carts, thus deprived of these, have lost their chief value as a manure.

LETTER XXVIII.

STAFFORDSHIRE.

EXTENT AND DIVISION OF SOILS — MARLING IN SOME PARTS DISCONTINUED — MINERAL AND MANUFACTURING WEALTH — PROPRIETORS AND TENANTS — ARRANGEMENTS MADE, OR CONTEMPLATED, TO MEET REDUCTION IN PRICES — TENURE — TENANTS DO NOT DESIRE LEASES — SMALL FARMERS ON COLD CLAYS VERY POOR — BUILDINGS GENERALLY GOOD — CONDITION OF THE LABOURERS — LORD HATHERTON'S EXPERIMENT.

STAFFORD, May 1850.

THE county of Stafford contains an area of 780,000 acres, of which about 150,000 are said to be occupied by roads, woods, and wastes; of 630,000 remaining, the larger portion is under the plough. Gravelly and sandy soils, varying in strength, but generally well adapted for green crop husbandry, are most prevalent; but there is a considerable quantity of heavy land in different parts of the county, and the proportion between the two is estimated, by persons qualified to give an opinion, as being two-thirds of the former to one-third of the latter.

The practice of marling, which 60 or 70 years ago was carried to a great extent, is now almost abandoned. Arthur Young makes repeated mention of it, in his tour through the county, as a great improvement to light soils; but it appears to have been carried to injurious excess, rendering light useful soils over tenacious, and less valuable in the modern system of agriculture.

The mineral wealth and resources of Staffordshire are of national importance. Birmingham, Wolverhampton, Walsall, Newcastle-under-Lyne, and many other towns of importance contain large populations, which draw their supplies from the

adjoining district.   The Potteries yield employment to a population of more than 80,000, and the annual produce of the
manufactures there is estimated by Mr. M'Culloch at 1,700,000*l.*
Canals and railways give communication to both seas, and to all
parts of the country.   The county is watered by broad streams
and rivers; the Trent, after traversing its centre, with its
tributary the Dove, dividing it from Derbyshire on the east,
while the Severn touches its south-western boundary.   The
banks of all the rivers abound in rich pastures and irrigated
meadows, those in the Trent valley being liable to summer as
well as winter floods.   The climate is rather cold and damp, the
high grounds arresting the moist vapours from the Irish Sea.

Much of the land in the county requires drainage to render
it fruitful, and the benefits of this improvement are fully appreciated.   The work is generally well done; the experience
acquired in the mining operations of the district having proved
very valuable in this respect.   The landlord usually supplies
tiles, and the tenant puts them in; and where a landlord is so
fortunate as to have a good class of tenants, he may do so in
this county with safety; as both tenants and labourers are very
skilful in dealing with water under ground.

The small size of fields, and prevalence of hedgerow timber,
is still much complained of in many parts of the county, especially in the neighbourhood of Trentham.   But this evil is
expected to be gradually abated, as the more intelligent landlords and farmers are fully alive to the loss they occasion.

A considerable proportion of the soil is held by large proprietors; the Earl of Lichfield, Lord Willoughby, Earl Talbot,
the Earl of Harrowby, Lord Bagot, Lord Hatherton, Sir Robert
Peel, the Marquis of Anglesea, and Lord Stafford being the
chief.   There are a considerable number of the class of yeomen
proprietors also, cultivating their own land.   Several of the
principal landlords have done much for the advancement of
agriculture, Lord Hatherton and the late Earl Talbot having
been enthusiastic farmers.   There has been no encouragement

given by the proprietors to excessive competition for land; and the tenantry, sensible of the fair and liberal treatment which they have received, speak of them with every feeling of respect. Very many, however, are heavily embarrassed; and the owners of cold clays especially, where there is most necessity for a reduction of rents, will be very hard pressed.

As yet few reductions have been made, though several are now promised. Earl Talbot proposes to commute his rents, one-half into a corn rent calculated on the basis of 60s. a quarter for wheat; the other, which is to be a fixed money rent, being reduced 12½ per cent.; and the whole reduction being estimated by the current rate of prices at about 25 per cent. The Duke of Sutherland's tenants have long had corn rents determined by the average prices for the three preceding years; and though their farms are small, and principally of cold difficult soil, they are making less complaint than others in the county. Lord Hatherton's estate is peculiarly circumstanced, and he has announced no intention to make a reduction. In 1813 his estate was valued, and let at 5 per cent. below the valuation. In 1816, 15 per cent. was taken off the rents. In 1821, 10 per cent. more was taken off, and in 1835 a further reduction of 10 per cent. was made; making altogether a reduction of 35 per cent. below the letting, and 40 per cent. below the high valuation of 1813. But rents are not exorbitant by any means, the present average of arable land throughout the county being from 26s. to 28s. an acre. Tithe, where it exists, is seldom more than 5s. an acre, and the extent of titheable land is not great. Poor rates also are moderate, the labourers being fully employed.

The tenantry hold chiefly on yearly tenures, and prefer to do so. Several of the large landlords are ready to give leases, but the tenants do not desire them. They know that at each renewal of a lease their farms would be revalued, and the rent increased; a course which is very seldom taken with yearly tenants. Great care is used by the best landlords in their choice of tenants, men being looked out for who will carry out improvements vigorously.   One landlord encourages his tenants to travel to the best farmed districts for information, sometimes paying their expenses himself to induce them to go.   He finds himself amply repaid in the improved cultivation which is introduced on his estate, and in the good understanding which exists between himself and his tenantry.   There is, however, a large class of tenants, with small holdings and little capital, chiefly located on the stiff clay soils, who are said to be worse off at present than the agricultural labourers.

Farm buildings are erected by the landlord, and many estates are very well provided with substantial and suitable accommodation of this kind.   They are generally superior to any we have yet met with in other parts of the country.

The labourers' wages are from 9s. to 10s. a week, and at task-work, which is much resorted to, they earn more.   Cottage rents are about 3l. 10s., with a good-sized garden.   For many years there has not been an able-bodied pauper in the Penkridge union.   Lord Hatherton is carrying out an interesting experiment at Teddesley, for improving the industrial training of the agricultural labourers, and superseding the employment of females in the field, except during harvest time.   He has a gang of about 30 boys, between the ages of 10 and 14, under the charge of a steady labourer, who works with them, and teaches them all the different kinds of light work on a farm, — hand-weeding, hoeing corn and roots, haymaking, picking couch or stones.   They are paid 6d. a day each, and some qualify themselves by this systematic training for taking service with the neighbouring farmers, who are glad to have such disciplined lads.   The 3s. carried home at the week's end by each boy is a very useful help to the labourer's family; while the boy is kept from idleness, and brought up to industrious habits.   If any plan could be devised by which his education in other branches could at the same time be attended to, the system would be followed with advantages of still greater importance.

LETTER XXIX.

STAFFORDSHIRE, *continued*.

CLIMATE AND ELEVATION OF COUNTRY — MR. HARTSHORNE'S FARM — GRAZING OF PIGS — MANUFACTURE OF CONCENTRATED MILK — MR. LEWIS'S FARM — MANUFACTURE OF SUPERPHOSPHATE OF LIME FOR MANURE — LORD HATHERTON'S IMPROVEMENTS AT TEDDESLEY — SKILFUL APPLICATION OF DRAINAGE WATER — EKPENSE AND PROFIT OF THIS IMPROVEMENT — DETAILS OF MANAGEMENT OF LARGE ARABLE FARMS — INCREASING CROPS — EFFECTIVE AGRICULTURAL IMPLEMENTS — CANNOCK CHASE, ITS CAPABILITIES UNDEVELOPED.

STAFFORD, May 1850.

THE state of agriculture in Staffordshire is influenced by such a variety of circumstances, that examples of every system pursued in England may be found in this county. The soil, as already mentioned, varies from a light blowing sand to the stiffest and most obdurate clay; the altitude comprises the level low lying lands along the beds of the principal rivers, and ascends by degrees to an elevation of 700 feet at the highest point of Cannock Chase, while the climate of course changes with the altitude and exposure. The dense population in the " black " or iron country around Wolverhampton on the one side, with that of the Potteries on the other, offers a constant home market for every kind of agricultural produce. We were therefore prepared to find the farmers generally an active and intelligent class, and, judging from those we met, we certainly were not disappointed.

On Earl Talbot's estate, after a brief glance at that fine old English mansion, Ingestre Hall, we visited the farm of Mr. Hartshorne, of Brancott. This gentleman, who holds largely under Lord Talbot, has completely removed all straggling fences and laid his arable farm into four fields of about 100 acres each. The land is a fine turnip and barley soil, and all has been drained that required it. His system is the four-course, applying all the dung of the farm, except what is needed for his mangold wurzel, to the seeds, and using artificial manures for his turnip crop. He has extensive farm buildings, which include feeding houses, with a railway in front for wheeling in the food, a steam engine for thrashing, &c., and extensive yards for feeding swine. A tank is being constructed to collect the liquid manure, and a field adjoining the tank is in course of preparation for being sown to grass, for the purpose of being dressed with the liquid, as each cutting is removed, for consumption, to the farm buildings. A stock of 200 pigs is kept on this farm. They are cheaply fed on roots in winter and on clover in summer, receiving little or no meal until they are finally put up to fatten. They are driven out every morning, and folded on the clover like sheep, returning in the afternoon to the farm yard, where they remain during the night well littered with straw. They eat the clover very bare, and in the following wheat crop the benefits of the pig fold are readily recognised by its superior luxuriance.

On the next farm to this we examined a process which has not yet spread among the agricultural community — the manufacture of concentrated milk. Mr. Moore, who has a license from the patentee, has fitted up an apparatus by which he manufactures annually the produce of about 30 cows. The milk, as it is brought from the dairy, is placed on a long shallow copper pan heated beneath by steam to a temperature of about 110 degrees. A proportion of sugar is mixed with the milk, which is kept in constant motion by persons who walk slowly round the pan, stirring its contents with a flat piece of wood. This is continued for about four hours, during which the milk is reduced to one-fourth its original bulk, the other three-fourths having been carried off by evaporation. In this state of consistency it is put into small tin cases, the covers of which are then soldered on, and the cases and their contents are placed in a frame which is lowered into boiling water. In this they remain a certain time, and after being taken out, hermetically sealed, and duly labelled, the process is complete. The milk thus prepared keeps for a lengthened period. It supplies fresh milk every morning on board ship, and may be sent all over the world in this portable form. The process seems very simple, and might probably be adopted with advantage in districts remote from markets. The expense and profits we had not an opportunity of learning.

The farm of Mr. Lewis, of Groundslow, on the estate of the Duke of Sutherland, near Trentham, offers a good example of the four-course system of husbandry. Mr Lewis for some time cultivated it under a five-course—that is, two years in grass, but has been much more successful under the four-field. He plants every year from 10 to 20 acres of potatoes, which he finds a very paying crop. He sows mixed rye grass and clovers with Italian rye grass, and never misses a plant. In sowing his wheat he always harrows in with the seed two tons an acre of newly slaked lime, which he finds completely to prevent the attack of grub or wireworm, besides materially benefiting the crop itself. Much of the land is of a light sandy nature, but has been greatly improved by marling, many thousand loads of marl having been carted to the farm by Mr. Lewis. His thorn hedges are a perfect fence, and yet, being kept with much care and neatness, occupy very little ground. His farm is divided into two sets of four fields each, one of the upper and lower divisions being every year in the same kind of crop. His farm buildings are very commodious and handsome, comprising a barn with water power and thrashing machinery, feeding houses with passage in front, convenient root houses, steaming house, stables, yards, &c. Part of Mr. Lewis's farm, now under excellent crops, was uninclosed common when he commenced operations; and on the other side of the fence there still remains a tract of 500 acres of land of precisely the same character, entirely waste and unproductive.

Mr. Wood, a tenant of Lord Hatherton's, near Penkridge, conjoins with his business as a farmer the manufacture of superphosphate of lime. He purchases bones from collectors in Wolverhampton and Lichfield. These he boils by placing them in a perforated cylinder, which is then lowered into a great cauldron of boiling water. After remaining there a sufficient time they are hoisted up, the water draining off through the holes in the cylinder, and laid out to dry. The fat which has been boiled out of the bones is then skimmed off the surface of the boiling water, and placed in barrels, in which it is sent to the soap boilers. The bones yield about a twentieth of their weight in fat, and that sells at about 25*l.* a ton. After being boiled, the bones are easily ground by the crushing mill; when crushed they are carried to a large wooden trough, into which 10 bushels at a time are thrown. Upon these, previously moistened with water, a carboy of sulphuric acid is emptied; the mass is then well stirred about, and in 15 or 20 minutes removed from the trough and thrown into a heap, where the dissolving process slowly goes on. After a day or two the mixture is passed through a set of small crushing rollers, by which it is reduced to powder, and is then ready for sale. For the convenience of the farmers, mixtures of guano and superphosphate are made up here in such proportions as may be desired: 2 cwt. of guano and 2 cwt. of superphosphate, mixed together, are found an excellent application for the turnip crop.

Lord Hatherton's estate at Teddesley, near Penkridge, affords many examples of enlightened improvement, of which a brief description cannot fail to be both interesting and instructive to the agriculturist, whether landlord or tenant. That part of the estate which is farmed by his Lordship embraces about 1,700 acres, a large proportion of which was originally part of Cannock Chase. It extends from the river Penk over the wooded heights which bound the view eastward from the Penkridge station on the London and North-Western Railway. Thirty years ago, the whole of this tract was in a most neglected state, great part of it a worthless waste, without roads, undrained, and open and exposed to the wintry blasts which sweep over the elevated grounds of the midland counties.   It is now a rich and fertile domain, carrying luxuriant crops of wheat and barley, the upper parts ornamented with sheltering woods, the pastures folded over with flocks of Southdown sheep, the extensive farm buildings filled with cattle, while the lower slopes are covered with verdure produced by irrigation.   The skill which has been shown in turning to advantage every provision of nature in each step of progressive improvement is what is chiefly instructive here.   The water which soaked through the bogs and elevated swamps, rendering them barren, is drained and collected into a reservoir.   From this it is conveyed to the farm buildings, which it supplies with water for the stock ; it turns the machinery which manufactures the products of the farm, and then glides off to enrich by irrigation a tract of meadows, 111 acres in extent, the produce of which has been doubled by the process, at an annual cost of 4s. 6d. an acre for attendance in laying on the water.

The ease with which a constant supply of water for driving machinery may be obtained, is well illustrated here.   A bog, 30 acres in extent, left unplanted in the middle of a plantation, having been considered irreclaimable, was thoroughly drained. Besides the surface water, some strong land springs were tapped, and the whole conveyed by main drains to a reservoir a few acres in extent, whence the water is carried underground about half a mile to the farm buildings.   The drainage of this swamp, and that of 140 or 150 acres more adjoining it, gives an ample supply of water for working machinery of 12-horse power every day throughout the year ; and, before the lands were drained, this water was not only lost as a motive power, but did immense injury by stagnating beneath the surface, and extending its chilling effects to every portion of ground through which it slowly oozed from its source.   At the farm buildings to which the stream is conveyed, a mill-wheel, 38 feet in diameter, is sunk into the solid

sandstone rock to such a depth, that the water discharges itself
into it " overshot." The tail water is taken from the bottom of
the wheel by a tunnel driven through the solid rock for nearly
500 yards, whence it is conducted into channels for irrigation.
When the mill is stopped, the water between the reservoir and
the wheel, which would otherwise run to waste, is conveyed by
pipes to the different yards and buildings for the use of the
stock, from which any surplus finds its way to the meadows.
The purposes to which the water power is applied are these : —
It turns two pair of stones (one, as we saw it, grinding wheat,
the other pease); it grinds malt, works a circular saw, a lathe, a
chaff-cutter, and a thrashing machine. The whole of these can
be worked at the same time, though in practice that is seldom
necessary. It has been in operation for several years, working
every day and all day, summer and winter. Independent altogether of the improvement of the land by drainage, and the
subsequent use of the water in irrigation, its direct value, as a
motive power, is estimated to exceed 500*l.* a year ; and that was
obtained by a total expenditure of about 1,700*l.* In a mul-
titude of cases a similar power to this could be as easily got,
which at present is suffered to stagnate in the ground, or, if collected in drains, then heedlessly allowed to run to waste ; for
there were no unusual facilities on this estate for obtaining a
supply of water. All that is required is procured from the
drainage of about 200 acres of land. It is carried in earthen
pipes along a gentle declivity, and with very little leakage,
about 600 yards from the reservoir to the mill, and is then discharged through a tunnel ; the whole distance, from the reservoir to the outfall, being 1,200 yards, and the total fall
being about 50 feet. On this point we take the opinion of
Mr. Williams, civil engineer, page 26 of his Pamphlet, published
by Ridgway, *On the Application of Drainage Water to Mill
Power* : —

" There is surely nothing here described which would lead a casual
observer to imagine that water power equal to 12 horses was to be

found in such a situation. The estimate usually formed of the requisites for the use of water as a motive power, is a large stream or
brook of water, having a considerable fall in its course; and the term
' millstream ' which is given to such currents, invariably conveys the
impression of a large body of water, flowing very rapidly down a
channel having considerable inclination; the only other idea which
seems to be associated with the driving of waterwheels is the nearly
vertical descent of a comparatively small body of water down the face
of some sharp declivity, and in cases of this kind large overshot
wheels are occasionally to be met with; but I am not aware of any
other instance where the water derived from the under-drainage of
the land, and that alone, has been converted to purposes so valuable,
and where so much ingenuity has been displayed in the adaptation of
means which to a superficial observer would appear totally inadequate
to the production of such important results. The merit of the plan
consists in its originality; there is nothing in the practical adoption
of the principle which suggests difficulty, and every one who examines
it on the spot naturally asks himself, why has not this been adopted
elsewhere?"

The management of Lord Hatherton's farm well merits description. Of the 1,700 acres to which it extends, about 700
are kept regularly under culture; the rest comprise the park,
the irrigated meadows, and some higher improved ground adjoining Cannock Chase, and too distant to be economically
managed in regular cultivation. The quality of the land is dry
green crop soil, with a considerable mixture of sand, in some
parts slightly peaty, and none of it of much natural fertility.
It is managed generally in the four-course rotation. 175 acres
are annually in Italian rye grass and clover, one-half of which
is cut two or three times, and carried home for consumption in
the farm buildings, the other half being folded over with sheep.
12 or 14 two-horse loads of dung are laid on the seeds in spring
as far as there is sufficient dung for the purpose. In autumn,
the land is ploughed with the Bedford two-wheel plough and
two horses abreast, and immediately pressed. At this operation
there are usually nine ploughs, followed by three pressers, and,
a seed-box being fixed on each presser, the seed wheat is dropped

immediately behind the presser at the same time. Early in the season wheat is sown at the rate of two bushels an acre, later two and a half, a smaller quantity of seed having been found to yield a larger proportion of light wheat. The land is harrowed, and on that portion for which there was not dung applied in spring, 4 cwt. of superphosphate are spread and harrowed in between the first and second turn of the harrows. With this force nine acres are ploughed, pressed, sowed, and harrowed daily. A gang of boys follow the pressers, and pick up any tufts of grass which remain on the surface, throwing them into the next open furrow, where they are covered on the return of the plough. The land, being kept clean, is seldom hoed in spring. After wheat it is prepared in the usual manner for turnips. In spring, when ploughed and well loosened by the grubber, it is worked with a revolving harrow, of which there are two kinds used here. The first has strong teeth, goes round with a slow motion, and penetrates the ground to a considerable depth. It is a heavy implement to draw, requiring the power of four horses. It tears up the tufts of roots and weeds, exposing them on the surface to the air and weather. In a day or two the second revolving harrow follows. Being of lighter construction, shorter teeth, and more rapid in its revolutions, it seizes the tufts of weeds, thrashes the earth out of them, and leaves them to be gathered together by light harrows, by which they are readily removed from the ground. Swedes are sown in the middle of May, as soon as the mangold sowing is completed, of which 25 acres are grown annually, and highly valued. On thin soil the yellow globe is sown, on deep soil the long red variety. Of the whole green crop, 175 acres in extent, 25 are sown with mangold, 100 with swedes, and 50 with hybrid and white turnips. For early swedes 10 to 12 loads of farm manure per acre are laid in open ridges 27 inches apart, and over that is sown broadcast, by hand, a mixture, consisting of 2 cwt. of superphosphate, 2 cwt. of salt, and 12 bushels of wood ashes, saturated with liquid from the tank. The land is then split

over the dung and this mixture, with a double mould board
plough, and the turnip seed drilled in on the top of it. For
common turnips the usual application is 2 cwt. guano, and 2
cwt. superphosphate (or 4 cwt. superphosphate without guano),
2 cwt. of salt, and 16 bushels of saturated ashes, all sown
broadcast, and covered in by the double plough, the seed being
then sown on the ridge ; 50 acres of swedes and the greater part
of the white turnips are eaten on the ground by sheep, with cake
and meal, the turnips being taken up before the sheep are admitted, and given to them in troughs, cut. The rest are drawn
for consumption in the feeding houses. After turnips, as much
of the land as possible is sown with wheat, all that is cleared
before the beginning of March being deemed in sufficient season.
Beyond that time, the remainder of the turnip land is sown with
barley, which is not so remunerative a crop as wheat; this land,
under the management it receives, yielding as many bushels of
wheat as of barley per acre. In the month of May, when the
wheat or barley is well up, clover and grass seeds are sown by
an ingenious contrivance, which is found to answer admirably.
On a common light wooden roller a box for seeds is fixed, a
spindle in which takes its motion from the axle of the roller.
The seeds drop before the roller, being protected from wind by
a light board in front, which reaches within a few inches of the
ground. A boy leads the horse, the sowing and rolling is completed at one operation, and no part is missed, as the rolled
ground shows distinctly where the seed has already been delivered.

A great deal of draining has been already done on this farm
and over other parts of Lord Hatherton's estate; and wherever
it is required, it is now being executed. Three-inch tiles and
half-soles are preferred; the drains are generally laid eight yards
apart and three feet deep. For under water, six to eight feet
in depth is gone to, if necessary. Subsoil ploughing, after
drainage, has been repeatedly tried, but discontinued as, on
experiment by weight of the crops produced where the land

R

was subsoiled and where it was not, no advantage was discovered.

An interesting experiment was tried on one field of fair quality, but of light staple, to ascertain whether wheat could be taken alternately for a number of years in succession without deteriorating in quality or produce. In the last fourteen years seven wheat crops have been taken from this field, each year showing progressive improvement, the seventh crop being one-third more productive than the first. The best white wheat is sown on this farm, and a frequent change of seed is found advantageous. Lime is a particularly valuable application on all land in *high condition*, and on newly broken-up land.

A fine herd of Hereford cows is kept in the park, the produce of which are reared and fattened. There are always on the farm about 200 head of cattle, 80 of which are in course of being fattened. Two lots are fed in the year, the feeding being carried on in loose boxes summer and winter. It has been the custom here for some years back to house-feed cattle during summer, and a decided preference is given to boxes over stalls for that purpose. Besides green food (turnips, swedes, mangold, or cut grass, according to the seasons), the feeding cattle get oilcake or corn daily, beginning with 3 lb. each, and increasing to 6 or 7 lb., but never exceeding that quantity. 14 pairs of horses are kept for the work of the farm and plantations, and 14 working Devon bullocks are bought in February every year, worked during the spring and summer at the turnip land, assist in getting in the wheat, and are fattened and sold to the butcher in the course of the winter. The sheep stock comprises 2,000 head of Southdowns, the sale sheep of which are fattened at two years old.

It is impossible, in the space at our disposal, to specify every point of management on this farm deserving of notice. The skill with which everything is conducted, the genius which seized every natural advantage and changed it from an enemy to a friend, is not more conspicuous in great principles than in minute detail. The invention of the revolving harrow, the furrow-presser, the machine for sowing and rolling seeds at one operation, are all due to Mr. Bright, who, within the eighteen years during which he has acted as Lord Hatherton's agent, has seen the whole stock kept on his Lordship's home farm more than doubled, and the acreable produce of each crop increased in the same proportion. The means placed at his disposal were ample, as Lord Hatherton's practical knowledge enabled him to appreciate the value of his agent's plans; and these have spread over the estate, on which the same liberal encouragement is given to the tenantry.

Adjoining Lord Hatherton's estate is Cannock Chase, still containing 14,000 acres of uninclosed ground. One-half of this is believed to be quite capable of profitable cultivation, being chiefly dry turnip land on red sandstone. In the midst of a populous county, within a few miles of Wolverhampton, Walsall, and Lichfield, it seems strange that no effort should be made for the improvement of a tract so extensive.

LETTER XXX.

TAMWORTH.

May 1850.

BEFORE leaving Staffordshire, we visited the estates of Sir Robert Peel, with the view of ascertaining in what manner one who occupied so distinguished a position as a statesman, and who took so conspicuous a part in the establishment of the free trade policy, interpreted practically the relations between landlord and tenant. The deplorable event which subsequently deprived this country of its greatest statesman, will rather enhance than diminish the interest with which any record of the private, no less than the public, transactions of the late Sir Robert Peel will be regarded by his countrymen. We, therefore, deem no apology necessary for introducing *now* to the public a description of Sir Robert Peel's estates written previous to his death.

Drayton Manor, the country residence of Sir Robert Peel, is situated about two miles to the south-west of Tamworth. It is a modern mansion of large extent, built on the site of the ancient manor house, surmounted with towers, presenting a varied and picturesque outline, combined with a uniformity of architectural design. Placed on a rising ground, it looks over a richly wooded English landscape, having ornamental gardens in the foreground sloping down to the edge of a lake, from the opposite side of which the green turf of the park stretches away among the woods. With much natural beauty is united all that the acknowledged taste of its owner could add to it by art. From the Manor House the chief part of the

estate extends westward, though there are detached properties
all round the neighbourhood as far as Shenstone, and Kingsbury
in Warwick, besides that at Blackburn in Lancashire. The
quality of the land is generally good. It is well situated for
markets, being within a distance of four to twelve miles of
Birmingham, in the neighbourhood of Tamworth and Lichfield,
and intersected by railways and canals. It comprises a large
extent of fine meadow on the banks of the Thame, a tributary
of the Trent, which carries in its waters the whole sewerage of
Birmingham, and occasionally deposits these with fertilising
effect on the flooded low grounds. The estate altogether is
very extensive and valuable. A great part of it has been
drained; Sir Robert having taken upon himself the whole
expense, and charged his tenants 4 per cent. interest on his
outlay. The work has been done in a systematic manner, chiefly
under the inspection of Mr. Parkes, entirely to the satisfaction
of the tenants.

The farms on the estate are held on yearly tenures, the tenants
having the utmost confidence in the security of their possession,
many farms having been for a long series of years in the same
family. The farm buildings are substantial, and kept by the
landlord in good repair; and any additional accommodation that
may be found necessary is made, the tenants paying a moderate
interest for the outlay. The tenants are bound to a regular
system of cropping, which is insisted upon only when there is
an apprehension that the farm is being improperly exhausted.
The custom of the county gives to outgoing tenants a modified
compensation for unexhausted improvements.

Public attention was naturally directed to the measures proposed to be adopted by Sir Robert Peel for readjusting his
arrangements with his tenants in accordance with the alteration
in prices likely to follow the withdrawal of protective duties.
The probable necessity for such a readjustment he frankly
recognised in the letter addressed to his tenants on the 24th of
December 1849, in which he states his opinion that the recent

changes of the law will be " to maintain a range of low prices
in average seasons, and to prevent very high prices in seasons
of dearth ;" and because he believes that such will be their
effect, he looks upon these changes as irrevocable. Considering,
however, that some time must elapse before the precise effect of
free trade, in regulating the prices of agricultural produce, can
be definitely ascertained, he proposes to defer for a time that
general review of the relations between himself and his tenants
which he contemplates.   When that is undertaken, it is to be
with reference to the special case of each farm, and to the
effects which the abolition of duties on many articles of import
used by the farmer, and the improved means of conveyance, may
have had in diminishing the cost of production as well as the
price of produce.   Fairly estimating these, he will make an
abatement of rent when satisfied there is a just claim, but does
not intend to make a general and indiscriminate abatement.
Meantime, after pointing out the home competition with which
the farmer, deficient in skill and capital, has to struggle against
his neighbour who is possessed of both, he sets apart 20 per cent.
of the current rent, to be expended in such immediate improvements, on each farm, as will assist the farmer to meet that home
competition successfully.   Besides this he offers to drain, where
still requisite, on the usual terms on the estate, as already
explained, and to consider favourably any proposals for additional buildings which the tenants may think necessary, they
paying a reasonable rate of interest on the outlay.   Where
leases for more than one year exist — and such are very few in
number — he offers to release any tenant who may wish to
withdraw ; and if a lease for years is desired, he is prepared to
grant it on being satisfied of the skill and capital of the tenant
requiring it. Or he is willing to enter into a written agreement
securing the repayment of unexhausted improvements, if such
an agreement is preferred to a lease.   Such is a brief outline of
the arrangements with his tenants which were contemplated by
the late Sir Robert Peel.

Those of the tenants with whom we had an opportunity of conversing, expressed their satisfaction with them, and their entire confidence in the justice of their landlord when the time for the final adjustment should arrive. The 20 per cent. for the past half-year has been nearly all expended in manures, bones, guano, lime, &c.; so that in effect it has been the same with all good farmers as a return of money, inasmuch as they would have voluntarily laid it out in the purchase of such manure themselves; while in the case of bad farmers, if the expenditure on manure is compulsory, it is for their own advantage. The rents of Sir Robert Peel's farms have never been raised by competition, and by the tenants themselves they are considered quite as moderate as those of any of the neighbouring proprietors. Additions have been made in the shape of interest for money expended in draining, &c., but these the tenants can better afford to pay than to be without. Active improvements are going on in the different farms, and increased exertions being made to develope still further the capabilities of the soil. The taste and neatness of the farmhouses and gardens indicate not only confidence in the permanence of their tenure by the tenants, but a degree of worldly comfort to which the tenants of a grasping or needy landlord are too often strangers.

The mode of farming adopted on Sir Robert Peel's estate is much the same as that of the surrounding district, the light land being generally managed in the four-course, and the heavy land in the six. Part of the light land is often allowed to remain two years in grass. The best farmers manure heavily, using artificial manure for their turnip crops, and reserving the farm-yard dung for spreading over their young seeds in autumn. Three to four cwt. of guano, and 20 bushels of bones, per acre, are considered a fair dressing for the turnip, excellent crops of which are grown. The guano is sown broadcast, and the bones and seed drilled in together on the flat. Part of the crop is drawn for consumption in the yards, and part eaten on the ground by sheep, the best farmers giving cake at the same time. A few acres of mangold wurzel are usually grown, to be used when the Swedish turnips are done.

The corn crops are sown by drill, and hand hoed in spring, chiefly by task work. Beans and pease are likewise sown by drill, and partly hand and partly horse hoed; 3s., 3s. 6d., and 4s an acre are paid for hand hoeing the corn crops. Winter beans are being sown here with success, though they are of recent introduction. So good a plant of wheat is scarcely ever got after mixed rye grass seeds as after clover root. We may mention a peculiarity in drilling wheat, which we were told had been practised with great success on a neighbouring estate. Instead of drilling in the usual way at nine inches between the rows, two rows are put in four inches apart, and then with an interval of 14 inches, thus getting the same number of rows in the same space, but with alternate intervals so wide as to admit of very effectual horse hoeing, the smaller interval being cleaned with the hand hoe. A field managed in this way last year, and otherwise in high condition, yielded a very heavy crop of red wheat.

Wheat is reaped, not mown; it is usually done by task work, and on account of the rate paid for it, from the scarcity of labour in harvest, it often costs more than 14s. an acre for reaping and binding. Barley and oats are usually mown and left in swathe till ready to be carried to the rickyard, when they are forked into the waggons loose like hay. Potatoes and other vegetables for sale are usually prohibited to be grown by the terms of agreement, though we should suppose such a condition would not now be enforced, considering the facilities of transport to the large consuming towns possessed by this district, and the readiness with which artificial or town manure can be obtained in return. It has not hitherto been the custom on Sir Robert Peel's estate to interfere with the tenant in the details of his management, although strict enough rules are prescribed by the printed agreement drawn up some years ago for such of the tenants as desired to

have one. The great majority of the tenantry prefer to be
without such agreement (and we think they are right, as we
apprehend it to be a very different thing from that written agreement proposed to be given by the right honourable baronet in
his last letter), holding their farms without any writing whatever, and leaving all claims to be adjusted by the custom of
the country. On the light land of good quality white wheat is
usually grown; on the heavier land red Lammas, Spaldings,
and other varieties of red wheat are preferred.

In some places much inconvenience has been sustained from
the drains on the estate being choked up by an ochrey deposit.
One main drain pipe, twelve inches in bore, was nearly closed
by it, and had to be cleaned out by drawing a long iron wire,
with a bunch of straw tied to it, right up the pipe, for which
purpose it was necessary to sink openings every hundred yards
or so into the drain. After the ochre is loosened by the rubbing
of the straw, the drain is flushed, and the deposit washes away.

The management of stock varies considerably; some farmers
going chiefly into dairying, others part dairy and part feeding,
and all keeping more or less of a sheep stock. The long horn
cow, though still holding her place in individual cases, is giving
way, as a general rule, to the improved short horn; the earlier
maturity of the latter, and its greater aptitude to fatten, recommending it to the pocket of the rent-paying farmer. Where cheese
is made, it is usual to commence making in the end of February
or beginning of March, and to continue till December. About
25lb. is the weight of cheese preferred, at which size they are
found less liable to crack than when larger, and to become
sooner ripe and fit for sale and carriage to the market. They
are usually sold in three lots — the first in August, the second in
October, and the third about the beginning of the following year.
Butter is made from the whey, which is afterwards given to the
pigs. Where the feeding of cattle is practised, a portion of the
cow stock is sometimes set aside for rearing the calves — two
being put to each cow, and turned out with the cow to the fields.

After several months the calves are weaned, and a little dairy produce is got from the cow.

The favourite breed of sheep is a cross between the Shropshire black-faced hornless ram and Leicester ewe, the cross being continued with a pure Shropshire ram. The produce is a very hardy sheep, larger and with more wool than the Hampshire Down, but with a considerable resemblance to that breed — combining the superiority in quality of meat of the Shropshire, with the earlier maturity of the Leicester. It is a very active, shapely sheep, with good carcass and excellent constitution. The wool of the cross breed is superior in quality to that of the pure Leicester. Some farmers sell the hoggets fat at twelve or fifteen months old, others feed them a second year.

In working the land, the farmers almost uniformly yoke their horses in line, both on heavy and light land. Three horses in line are the *minimum* number for a single plough — four, five, and even six being often seen where the land is stiff, or a deeper furrow than ordinary is being taken. On very light land we saw a team of five horses in line drawing a double furrow plough, set with wheels to work without being guided by a man, and turning over two acres a day. On light land three horses are commonly used in a single furrow plough, the driver walking alongside the horses, and never touching the plough, except at the headlands in going out and entering to a new furrow. Three horses in a set of not very heavy harrows, on loose ground, with a driver in front, and a man behind to free the harrows from weeds, is a common turnout. We have had frequent opportunities of observing the teams yoked in this fashion, most frequently five in line; and from the irregularity of the draught, the leader sometimes pulling the whole row, or pulling a lazy horse forward, or himself and his followers leaving the whole work to be done by the horse next the plough, we have no hesitation in saying that, on light land, an immense waste of power is caused by this system. Nor do the farmers feed their horses well or keep them in good condition, it being usual to see one strong horse next the plough, three very so-so horses next in advance, and the leader, a strong horse, in front. Even the strong horses have a dull hanging look, as if badly fed, and out of condition.

The state of the labourers on Sir Robert Peel's property is very satisfactory.   The rate of wages is from 9s. to 10s. a week with beer, which is reckoned to be worth 1s. more.   Cottage rents vary from 30s. to 4l., according to the accommodation, the extent of garden ground, and the locality.   Garden allotments are let at 1s. per rood of 64 yards, in the country, and 1s. 6d. near Tamworth.   The cottages are superior to the general description of labourers' cottages throughout the country, all being kept in good repair at the landlord's expense.   Many comfortable looking cottages which Sir Robert Peel deemed insufficient in point of accommodation, are about to be taken down, and rebuilt on a more commodious scale; and such as become old and uncomfortable, are either thoroughly repaired or rebuilt, the new rents being slightly increased to pay a moderate interest on the outlay.   The labourers are well employed, and are said never at any former period, to have been in more comfortable circumstances than at present.   Irish labourers are employed during the summer at hoeing and other operations at a lower scale than the English, but not to the displacement of the English.   This practice has existed for several years back over this and the adjoining districts.

LETTER XXXI.

CHESHIRE.

CHESTER, May 1850.

CHESHIRE has been long famous for its cheese; and that branch
of agricultural husbandry requiring much experience and mani-
pulative skill, has given a peculiarity of character to the farming,
and to the habits and management of the farmers.   Dealing
with a strong tenacious soil, in a comparatively damp climate,
nature seems to have early pointed out to the farmers that their
produce would be most safely, and least expensively, got
under a system of grass farming.   Much of the old grass land
was then formed into high round ridges, or "butts," as they
are here termed, by which a natural drainage was effected; and
even at this day that form of ridge is adopted by many, in addition to tile drainage, where land is being laid down to permanent pasture.

Dairy farming requiring much personal attention, the occupations are generally small, — the farmer and his wife and
family, with one or two servants, doing all the work themselves.
We think we are not wrong in saying that the farmer's wife in
Cheshire is the most important person in the establishment; the
cheese, which is either made by her, or under her directions,
forming the produce of two-thirds or three-fourths of the farm;

the remaining fraction of which comprises the business of the farmer. The labours of a dairy farm being of daily recurrence, and commenced by the different members of the family at an early age, great skill is acquired by them in the art of cheese making. To the management of the dairy, however, their acquirements are limited. Improvements in agriculture have made very slow progress in Cheshire; and the number of persons brought up in their parents' houses to dairy farming being constantly on the increase, there has hitherto been much competition for farms; not because it was a lucrative business, but because it was the only business with which they were acquainted. While, therefore, there has been little improvement in the land, with an increasing number of farmers supporting themselves upon it, the farmers, as a class, have not made money. Their wives and families work harder than those in most other counties, and content themselves with a very moderate share of the luxuries of civilisation; and yet, with all their hard work and frugal living, the farmers of Cheshire, when they get their rents paid, have, on an average of years, a smaller balance left to themselves than is thought absolutely requisite elsewhere.

Where land is principally in grass, and pastured with a regular dairy stock, there is not much difficulty in ascertaining the annual produce with considerable precision. Two-thirds, or more generally three-fourths, of the dairy farms are in pasture exclusively; the remaining third or fourth being under the plough. Farmers are usually bound to lay the whole of their manure on the grass lands, purchasing what they may require for the arable. The annual produce is divided, and the rent valued on this principle, — that the cheese from the grass land should pay the rent; the rest of the dairy produce, and the corn from the arable land, going to the tenant to pay labour, interest of capital, and profit. Under this arrangement, it is very plain that the landlord gets the lion's share; and we do not hesitate to say that, taking into account the quality of the land, and the industry and frugality of the tenants, we have not yet seen any county in England where wet cold clay yields so much to the landlord, and so small a proportion of the produce to the tenant, as in Cheshire. It is a somewhat anomalous circumstance, that the less fertile parts of the county—the poor cold clays for a circuit of some miles round Nantwich— are said to produce the best quality of cheese.

The eastern side of the county, where it joins the Derbyshire and Yorkshire hills, is of a rather mountainous character; and on the Shropshire side the face of the country is broken and irregular. For five or six miles from Macclesfield, in a north-western direction, it is the same; the surface being undulating and picturesque, though much too thickly clothed with wood for a cultivated district, which this neighbourhood is to a considerable extent. Beyond this, to the westward, the remainder of the county has the appearance of a great plain, nearly four-fifths of the whole lying little more, on the average, than 100 feet above sea level. Drawing a line from Macclesfield by Northwich, to the Mersey below Runcorn, the land to the east and north may be described as the tillage division of the county; that on the south and west being chiefly devoted to dairying. The last, which is the principal feature in the husbandry of the county, we shall now describe; the former, being somewhat exceptional and dependent on the proximity of large consuming towns, may be more briefly treated.

Seldom more than a fourth, and frequently not more than a sixth, of the land in a dairy farm is under tillage. Part of the grass land is never broken up, the *night* pasture adjoining the cow-houses being always preserved. Several of the other fields are taken up in succession, one field being laid down to pasture, and another broken up for a course of tillage. The farmers are very apt to adhere too long, however, to the same piece of land; partly from an aversion to break up old grass, and partly from the difficulty of bringing the tough glebe on these wet clays into a friable condition. For this reason, the landlord or his agent has frequently to interfere, in order to prevent the tenant working the same field too long. The mode of management is then to break up for oats, then fallow, followed by wheat, then oats ; the same course being continued for a round or two, when the field is laid again to pasture, by being sown with seeds from the hayloft, and a small portion of clover. Such is the unimproving course on the stiff clays.

On the more friable lands, potatoes and turnips are substituted for the bare fallow; though, even on these, very little provision is made for the winter feeding of the dairy stock. On a farm of 150 acres, which we examined, 20 acres were in white crop, and 6 in green ; 3 of which were in potatoes for sale, and 3 in turnips for consumption on the farm, the stock on which was 40 cows, besides young animals. The land was foul, and unskilfully cultivated ; and the farmer had no wish to break up more, as he found the dairy the most lucrative branch of his business. The fields, however, were small, and much injured by hedgerows and timber, about which the landlords have hitherto been very tenacious. Though the soil of that part of the county we now refer to — between Macclesfield and Knutsford — is generally of a friable texture, well suited to cultivation, the shade of the trees prevents corn filling well ; there being plenty of straw, but neither wheat nor oats yield on the barn floor anything like what they promised in the field. The small fields and lofty hedgerows render it difficult to save the crops in a wet season ; and as the air is then sluggish, with little wind, much loss from this cause is frequently sustained. The produce per acre varies from 15 to 30 bushels of wheat, and from 30 to 36 bushels of oats.

The management of the grass lands chiefly occupies the dairy-farmer's attention. In arranging the farm, it is usual, in Cheshire, to provide a "night" and a "day" pasture; the former, as already mentioned, contiguous to the cow-houses, the latter more distant. These form the chief dependence of the stock, and are seldom or never broken up for tillage. The rest of the grass land is kept for hay, and for the feeding of the horses and young stock. The hay land should be dressed with dung every second year, but that is not always done. On the banks of the rivers and streams which intersect the county, there is a great deal of mowing ground, which in many cases requires no other dressing than what is left on it by the winter floods. Such meadows, if they were not exposed to injury by summer floods, would be invaluable to the dairy farmer; but as they are not embanked, and in almost every case their course is obstructed by dam dykes for mills, the hay is occasionally swept away entirely. These summer floods occur once in three years on the average; and, even though the hay is not carried off, it is frequently so much damaged, that the cattle which are fed on it become diseased.

The benefits of tile-drainage on the dairy pastures of this county were early appreciated; and a great deal has been done on many estates. Little attention, however, was at first paid to the manner in which the work was executed, the custom being for the tenant to put in, as he chose, all the tiles the landlord gave him. He who could "bury" the greatest number of tiles, accounted himself, and was generally accounted by his landlord, the best tenant. His interest was obviously to put them in at the least cost to himself; and moreover it was not then understood that on level lands a considerable degree of skill was requisite in laying off the drains; while that higher knowledge of principles, so conclusively shown to be necessary by Mr. Parkes, in his essay on the Influence of Water on the Temperature of Soils, was never thought of at all. Better ideas on the subject are now becoming prevalent; and many landlords take care not only that a certain number of tiles are "buried," but that they are laid in the manner likely to be most effectual and permanent. The Marquis of Westminster may be named as an instance of both practices. For some years back he had given to his tenants at the rate of 1,000,000 tiles annually; not only giving the tiles gratis, but also paying about a third of the labour of putting them into the ground. Finding that, from want of care and skill in executing the work, much of the anticipated benefit was lost, the Marquis this last year has taken the whole under his own direction, having had 200 men at work during the winter, under a superintendent, who lays off the drains and sees that the work is properly executed.

Next to tile drainage, of which a great deal yet remains to be done in Cheshire, is the application of bone manure. The wonderful effects produced on the dairy pastures by this substance are supposed to arise from the exhaustion which the soil has undergone through the annual withdrawal from it of earthy phosphates in the cheese which for a long period of years has been sold off the dairy farms. In the milk of each cow, in its urine, and in the bones of each calf reared and sold off, Professor Johnston estimates that a farm parts with as much earthy phosphates as is contained in 56lb. of bone dust. It is therefore quite reasonable to expect much benefit from a manure which restores these phosphates to the soil. There may be something also in the soil of Cheshire peculiarly favourable to the action of bone manure, as in other dairy counties where the same withdrawal of phosphates has been long going on, an equally good effect, it is said, has not followed its application. So certain is it here, that it is a recognised rule on some estates that, by boning and draining four acres of land, an additional produce will be got equivalent to the maintenance of one cow. Tenants readily pay 7 per cent. to their landlords for an expenditure in bone manure, and its effects are said to have raised many a struggling hard-working farmer from poverty to comparative independence. On the stiff clay lands its action is greatly promoted by tile drainage, and on such lands its benefits, when applied to grass, are found greater than on those of more friable texture, though on these, also, it is used with great advantage. The quantity applied varies from one to two tons an acre, ground to a coarse powder in the usual manner. The best time for laying on this manure is in autumn, early enough

s

to admit of a fresh growth of grass to cover it before winter.
On " seeds " which are intended for permanent pasture, it should
be laid on after harvest.   Its effect is to cover the ground
thickly with clover, trefoil, and succulent grasses, in lieu of the
thinly planted and very innutritious pink pointed grass which
previously occupied the soil.   Some farmers told us that it had
doubled the produce, and improved the quality, of their cheese;
and in no instance did we find that its application had injured
the quality of the cheese as we had heard in Leicestershire.
The bones chiefly used are purchased, after being boiled, from
the size manufacturers of Manchester.   They are cheaper and
more easily ground into powder than unboiled bones, and their
effect, though deprived of the gelatine, is believed to be equally
good for dairy pastures.   The benefit is supposed to last for ten,
fifteen, and twenty years; indeed, grass land which has received
a good dressing of bones will never completely revert to its
original sour state.

In the management of cows there is nothing peculiarly good.
They are poorly wintered on straw till after Christmas, when they
get a few turnips, two or three acres of an indifferent crop among
forty cows, and hay till March or April, when they drop their
calves.   From that time till the grass is ready the best farmers
give them a little bruised oats or oilcake, which is discontinued
as soon as they are turned out to pasture.   On that they remain
during summer night and day, being brought home morning and
evening to be milked.   The number of cows kept on the dairy
farms of Cheshire may be accurately estimated from the following particulars, kindly supplied to us.   On thirty-six farms,
containing 6,600 acres, 2,200 of which were in tillage, and 4,400
in pasture and hay, a stock of 1,176 cows, besides the necessary
quantity of young cattle, is kept, in these proportions : —

1st Class   600 acres, at 3   acres per cow, keeps 200 cows.
2d   „   800   „   3½   „   „   226   „
3d   „   3,000   „   4   „   „   750   „

The rent of these farms averages from thirty-two to thirty-three shillings an acre, the first class being of course above that, and the third class below it.   Besides the rent, the tenants pay tithe, land tax, and poor rate.

It is unnecessary to give a detailed description of Cheshire cheese-making.   A few particulars must suffice.   The process is carried on during the day, the preceding evening's milk being mixed with the morning's milk, so that it may be all " set " and made into cheese by one instead of two operations.   It is of much consequence that the milkhouse be sweet and cool, as if the evening's milk is in the least degree sour, the next day's cheese will be sour.   In cold weather it is necessary to warm a portion of the evening's milk before mixing it, but in summer the heat of the morning's milk is generally sufficient to bring the whole to the proper temperature for setting.   Thermometers are scarcely ever used, but the temperature at which the milk is coagulated is believed to range between 75° and 85° Fahrenheit. Before adding the evening's milk a small part of the cream is skimmed off for butter, the froth and bubbles being carefully taken off, as the air they contain is supposed to be injurious. From the morning's milk in the cheese tub, the bubbles are also carefully skimmed off and broken.   In little more than an hour the curd will be ready for " breaking," which is effected by passing the " curd breaker " very slowly through it.   The whey is then carefully taken off, and the curd placed in a basket in which a coarse cheese cloth has been first laid.   In this it is pressed for the further extraction of whey.   This process proceeds until the whey is sufficiently removed to admit of the curd being salted.   The quantity of salt is not very definite, and is regulated by the taste of the dairymaid, though according to Mr. White — whose detailed account of the process of cheese-making, in Vol. VI. of the *Royal Agricultural Society's Journal*, may be consulted with advantage — the average in a first-rate dairy was found to be 1lb. of salt for 40lb. of dried cheese, or about forty gallons of milk.   After the salt has been completely intermixed with the finely broken curd, the curd is placed in the cheese vat, which is put under a lever press, and iron skewers are stuck through the holes in the vat, in which they remain a few minutes, and are then withdrawn to allow the whey to run off.   Passing over the subsequent process till the cheese is finally taken out of the press, it is then to be " dried."   A strong canvass bandage, about two inches broad, is wound tightly round the cheese to keep it in shape and prevent cracking.   In this state it is placed in the drying-house or cheese room, where it is daily turned and wiped with a cloth.   The bandage is kept on the cheese in many dairies till it is sold, being changed and a fresh one put on when it is removed from the dairy to the cheese loft.   The cheese vary from 50lb. to 120lb. in size, the largest size, if of the same quality, bringing the highest price. Butter is made from the whey-cream, which is skimmed off as the whey is slowly scalded.   With this is frequently mixed the portion of cream which has been taken off the evening's milk ; and where the management is good, the butter so produced is of good quality, scarcely distinguishable from the best. Besides the cheese, each cow yields from 15lb. to 20lb. of butter during the season.   The yield of cheese is very various, being dependent on the quality of the stock and the pasture, the supply of winter food, and the skill of the dairyman.   It reaches from $2\frac{1}{2}$ cwt. per cow on the poorest, to 4 cwt. on the better description of farms.

Quite a different system is followed by Mr. Littledale on his well managed farm, near Seacombe, on the opposite shore of the Mersey from Liverpool. His dairy stock, principally of the large Yorkshire breed from the banks of the Tees, are constantly fed in the house, summer and winter. The buildings which they occupy are large and airy, with feeding passages in front of the cattle, plenty of light, and the means of ample ventilation. During the summer they are fed on cut clover, Italian rye-grass, and vetches, receiving a feed four times a day at regular intervals, and not too much at a time. In winter they have brewery grains, turnips, and mangold, half and half of each, with linseed, corn, &c., when the green food runs short. Their fodder in winter is a mixture of hay and oat straw, cut and steamed in a couple of steaming boxes which hold a ton each. These are connected by a pipe with the boiler of the steam-engine, the steam, as it issues from the pipe, being received under a false bottom of the steaming box, of perforated cast-iron, through which it diffuses itself among the cut mixture, which is thereby cooked in less than an hour. Five tons a week of this fodder are consumed during the winter, the horses being likewise supplied with it. The extent of ground last year which furnished food for the stock was as follows: — Six acres of vetches, eighteen acres of Italian rye-grass, part of it cut four times and all of it thrice, having been regularly dressed with liquid manure (collected in tanks from the cow-houses and stables), after each cutting, and twenty-five acres Italian rye-grass cut once, after a hay crop had been previously taken. Thirty-five acres of swedes, part of which are consumed by sheep, and twenty acres of yellow globe mangold, supply the winter food. The stock consists of eighty-three milch cows, and fifteen farm horses. The cows are milked twice a day — 4 A.M. and 3½ P.M. The milk is chiefly disposed of in the neighbourhood — that being the most profitable plan. What remains is kept two days till it is soured, when it is churned. A ten-horse steam engine thrashes, churns, bruises grain, linseed, &c., grinds flour, and cooks fodder. The farm as regards manure is nearly self-supporting, the constant house-feeding affording a great supply. We regret that our space does not admit of fuller details in describing many interesting points in the management of this farm.

The demand afforded by the neighbourhood of large towns, as already mentioned, has in some parts of this county led to peculiarities of management. This is especially the case along the Bridgewater Canal, from near Warrington by Altring-ham, &c., and in some of the sandy tracts westward of Birken-

head.   On these lands it is common to take a double crop of
potatoes in the same year, the early crop being gathered by
the middle of June, and the late crop then planted.   Sometimes cabbages to be sold in the following spring are planted
in November, after the second crop of potatoes, the land
being well manured for the cabbages, and then, after they are
taken off, trenched, and again planted with potatoes.   The
canal affords a ready means of sending the crop to Manchester
and getting back manure.   Carrots are also grown extensively
in the neighbourhood of Altringham.

Some of the small farmers in Wallasey, Morton, and Bidstone
grow early potatoes for the Liverpool market.   The ash-leaf
kidney, previously sprouted three inches or so, is dibbled in
upon well dunged land in January, covered with about an inch
of soil.   The ground is then covered with straw, fifteen to
eighteen inches in depth, which is taken off in fine days and
put back on at night.   By these means early potatoes have
been got by the 12th of April.   They then sell at 1s. 6d.
to 2s. and 2s. 6d. per lb., and the small farmers, who do not,
however, grow more than from a rood to an acre of this early
crop, find them very remunerative.

LETTER XXXII.

LANCASHIRE.

SOUTHERN DIVISION — GENERAL DESCRIPTION — GREAT OCCASION AND LOCAL
FACILITIES FOR IMPROVEMENT — LORD DERBY'S ARRANGEMENT WITH HIS
TENANTS — LORD SEFTON'S ESTATE — IMPROVEMENTS IN TOWNSHIP OF
SPEKE — SLOW PROGRESS ON COLD CLAY SOILS — EXCELLENT MANAGEMENT
OF MR. LONGTON'S FARM — FREQUENT MANURING AND REMUNERATIVE
CROPS — OTHER SIMILAR EXAMPLES — SOIL AND CLIMATE FAVOURABLE
TO GRASS AND GREEN CROPS — MR. NEILSON'S FARM — EARLY CROPS NEAR
ORMSKIRK AND ON THE BANKS OF THE MERSEY — HIGH RENT OF GRASS LAND
IN THE UPPER DISTRICT OF COUNTY — ROTATION OF CROPS MUST BE
REGULATED BY CIRCUMSTANCES — RENT AND RATES.

WIGAN, LANCASHIRE, Oct. 26.

THE population and extent of Lancashire, as well as its peculiar
natural features, render it necessary to adopt a division into
North and South districts, in order to give such a consecutive
description of its agriculture as may be easily understood. For
convenience of reference we shall consider the Southern division as comprising all the country to the south of Preston
and the river Ribble ; while the Northern division will be
held to embrace all the rest of the county to the north of
that river.

The southern division of the county, both from extent and
population, is by much the more important. Manchester,
Liverpool, Wigan, Bolton, Blackburn, Rochdale, Bury, and
many other populous towns are comprised in it. In mere
superficial extent it includes nearly two-thirds of the whole
county. Containing as it does the great coal-fields of Lancashire, it has from that circumstance become the seat of the
cotton manufacture, which in its wonderful progress and importance has of late years exercised so great an influence on the
character and legislation of the British empire.

The eastern side of the county is bounded by a range of

hills dividing it from Yorkshire, composed of millstone grit, the soil on which is generally thin and poor. The southern and western sides extending along the Mersey from the east of Warrington to Liverpool, and thence by Ormskirk to Preston, rest on the new red sandstone; while the coal measures occupy the whole central space. The outline aspect of the country is not picturesque. On the western side next the sea it presents great flats of sand, over which sweep the cold vapours and tempestuous winds of the Irish Channel, unbroken by any mountainous ridge to intercept their severity. Further inland the ground rises very gradually, the numerous straggling fences and stunted hedgerow trees giving strongly marked signs of the rigour of the blast, while large tracts of wet undrained bogs add to the dreariness of the landscape. The coal fields are now reached, and the evidences of a busy population everywhere present themselves. Tall chimnies vomiting forth smoke, long rows of narrow brick houses, heaps of brick and lime rubbish, mounds of refuse on the sites of abandoned collieries, or great banks of coal where these are at work, present themselves on every side. As the traveller winds through some of the valleys on the line of the East Lancashire Railway, especially in the district between Bury and Manchester, his eye lights with pleasure on many spots where wood and water and green pastures are picturesquely blended; but let him leave this line and follow the cross country roads, many of which are still paved streets, and a different aspect presents itself. On either side he finds wide waste spaces, the receptacles of all sorts of rubbish; in the neighbourhood of farmhouses, the site of the dungheap, the better parts of which in this moist climate are washed into the adjoining ditches, which in their turn supply water to the drinking pools for the dairy stock. Through all this part of the country everything looks unfinished; there seems a constant transition, a progress in which agriculture alone appears to have hitherto but little participated.

Within the last thirty years the population of South Lancashire, three-fourths of whom are consumers, has been doubled, yet the improvement of agriculture as a general rule has been neglected, and good farming, of which there are many firstrate examples, is still the exception. Possessing an "insatiable" market at their doors for everything a farm produces, the very flowers in the farmer's garden being convertible into money, and having the advantage besides of inexhaustible supplies of manure easily accessible from the numerous manufacturing towns and villages everywhere scattered over the country, there must have been unusual obstacles to counterbalance such incentives to improvement. The climate and the soil are both against the farmer, unless he has the means and intelligence to turn both to the best account. Compared with Middlesex the fall of rain throughout the year in Lancashire is as 40 to 20, while two-thirds of the soil, in its natural undrained state, is a strong clayey loam upon a subsoil of clay, expensive to improve, and, when improved, requiring much skill and capital to develope its capabilities. Unfortunately the great proportion of the country is held by small farmers who, however industrious, do not possess the intelligence or capital requisite to meet the natural difficulties referred to, while much of it being held on life leases, and the great proportion of the rest on tenures from year to year, there is wanting, also, that permanent interest in the land which forms the chief motive to an improving farmer. On land of this description improvement is expensive and the returns comparatively slow, and the farmers of most means and greatest intelligence accordingly decline to embark them here so long as they can find land of an easier description open to them. Nor have the landlords generally given any material aid to their tenants. Reaping an easier harvest from the mineral wealth under its surface, and from the sums paid to them as compensation by the numerous companies whose railways traverse the district, they could afford to leave the farmer to struggle with his ungenial soil, seldom asking from him any increase of rent, and therefore not deeming themselves called upon to afford him any assistance in its improvement. So long as his scanty crops brought remunerative prices this state of things continued undisturbed, but low prices bring demands for reduced rents, or such outlays in permanent improvements as will enable the farmer to meet the change of times. These demands have in a few instances been met directly by a reduction of rent, in others by permission to remove useless hedge rows (hitherto held sacred), and to break up more grass land, and in the great majority by a hearty co-operation on the part of the landlord in executing drainage and increasing the house accommodation for stock.

On the estates of the Earl of Derby, where a system of enlightened and judicious improvement has been in progress for many years, there has been no general reduction or abatement of rent. Each case is treated on its own merits. When a tenant asks an abatement, a new valuation at "times' prices," by impartial parties, is offered to him, on condition that the rent shall be either reduced or raised according to that valuation; but the alternative is seldom or never accepted, and the rent remains as before. On this estate drainage has been carried on very extensively, a corps of 70 to 100 men being regularly employed under a superintendent who lays off the work and sees its perfect execution. The farmer's intervention is not employed, except that he carts the tiles and pays five per cent. on the outlay. This is understood to be the rule on the estate, though in many cases small farmers receive the benefit of the improvement without any charge. Very substantial and commodious farm buildings are also being erected in several parts of these extensive estates; beautiful Welsh slates, hewn at the quarries of the requisite dimensions, being extensively used in the internal fittings, for mangers, division stalls, &c.

On Lord Sefton's estate, near Liverpool, the same judicious improvements have been carried out — useless fences being removed, and replaced by straight lines of trimly kept hedges ; the fields enlarged, convenient roads of access provided, the farm buildings made suitable for the occupation, and the farmer thus placed in the most favourable position for making the best of his capital and skill.   In the township of Speke, eight miles south of Liverpool, and on the estuary of the Mersey, the land- lord, Mr. Richard Watt, has laid down a railway from the river to the town for the convenience of his tenants, by which they are enabled to transport the Liverpool manure from the barges to the neighbourhood of their farms with great economy and facility.   Other improvements of a permanent character have likewise been executed by the landlord, and, the soil being favourable, there is probably no portion of the country in which there is such a general system of good and remunerative farming, or a more industrious, skilful, and successful set of farmers.

The practice followed by the farmers on the undrained lands, which comprise the larger proportion of South Lancashire, differs very little from the description given of it seventy years ago by Young.   The land intended for summer fallow seldom gets the first furrow till April or May, and after that two or three separate furrows, followed by harrowing, are given at such times as best suit the farmer, seldom with much reference to what best suits the condition of the soil.   The fallows are thus badly executed, the work frequently done in moist weather, and the seed sown at last under unfavourable circumstances. The value of green crops has in many cases tempted the farmer to forego even such a preparation of the soil, but where the land is undrained clay it is very doubtful whether this management is any improvement.   Wheat follows the summer fallows or green crop, then oats, which are sown with clover and grass seeds, and mown for hay the two following years.   " Some crop as long as three seeds are returned after a fallow or fallow crop, and then let it lie for pasture without sowing grass seeds. These farmers are never troubled with fat stock or overflowed

with milk and butter."* The produce of crops under such management is necessarily scanty, and the returns from dairy stock fed
on pastures of such wretched quality cannot be remunerative.
Low prices are telling heavily on a farmer so circumstanced, and
each returning rent day cuts more deeply into the number of his
dairy cattle, the only convertible capital he possesses, and renders more hopeless any prospect of relief from his accumulating
embarrassments.   There would be little benefit in lingering
longer on such instances as these, unfortunately too common on
the colder soils of South Lancashire, and we shall therefore turn
to the description of a farm of the better class, on which capital
and intelligence have been united with industry and perseverance and a ready adaptation to surrounding circumstances.

The farm of Mr. William Longton, at Rainhill, near Prescott,
between six and seven miles from Liverpool, contains 160 acres,
and, with the exception of about 50 acres, is held on a yearly
tenure.   The whole farm has within the last ten years been
drained at his own expense.   The soil is partly a strong loam,
with clay subsoil, and part a sandy loam on a porous subsoil;
the surface gently undulated, and about 100 feet above sea level.
The main drains are laid with tiles and slate soles; the others
are made at intervals of 21 feet apart, and from 32 inches to
3 feet in depth.   These are filled one foot with cinders, which
are got at the glass works at St. Helen's, and cost 2s. a load —
one load sufficing for 80 lineal yards of a drain — and are found
very efficient and permanent, and not half so costly as tiles.
Mr. Longton's system of farming is (1st) green crop, after grass,
(2d) wheat, (3d) barley, (4th) seeds, (5th) grass cut for hay,
(6th) grass again cut for hay or pasture, according to circumstances.   He commences this rotation by ploughing or skimming
his grass land in autumn, with a very light furrow, in which
state it remains during the winter.   As early in spring as the
weather suits, this furrow is cut to pieces by a sharp wheeled
roller being passed across it; it is then well harrowed and

* Rothwell's Agricultural Report of Lancashire, 1850.

torn to pieces; then ploughed with a deep furrow, which, after the surface is thus broken, is easily reduced; and then drawn into ridges 30 inches apart, into which are placed 20 tons of the best town dung per acre; on this the potato is planted; it receives the usual careful cultivation during the summer, and as soon as the crop is removed in autumn the land is ploughed and drilled with wheat. This is sometimes, but not always, followed by barley, though Mr. Longton is decidedly of opinion that barley after wheat is the best management with which he is acquainted. The barley is sown with a mixture of grass seeds and clover, which in the autumn receive a dressing of 15 tons of nightsoil per acre mixed with earth. The seeds are mown the first year twice for hay, which is all sold. In autumn the ground is again dressed with 15 tons of mixed manure, or with guano, and cut once the following season for hay. The aftermath is pastured. If the root appears good it is again dressed in autumn with the same quantity of manure, and again cut for hay; if otherwise, it is pastured. The returns from this management have this year been as follow, viz. : —

1. Potatoes (a short crop), 220 measures of 90 lb. each, per acre, selling at present at 2s. 6d. a measure.

2. Wheat, 40 (Liverpool) bushels, 70 lb. each, of white wheat per acre, and upwards of 2 tons of straw, worth at present 2l. per ton.

3. Barley, 60 bushels per acre.

4. Seeds, first cut 2 tons of hay per acre, second cut 1½ ton, selling at present at 5l. per ton.

5. Grass ley, yielding 1½ ton of hay, and excellent aftermath for pasture.

6. Ditto.

To obtain these returns Mr. Longton purchases annually 800 tons of the best town manure, besides what is made on the farm by the horses and dairy stock, and what is collected of roadside scrapings, old banks, &c. His practice is to sell every-thing his farm produces when it yields him a remunerative price, and to buy in return what is requisite to keep it in high condition. His farm horses are fed on steamed Egyptian beans and hay, each horse when at constant work consuming about a bushel of beans (costing 3s.) per week. The price of the best manure, which used to be 8s. or 9s., is now only 5s. a ton; and this difference is a considerable item where so large a quantity is purchased.

There might be many examples given of farms in South Lancashire equally or even more productive than this. Where the soil is favourable and has been carefully drained, the yield of green crops and grass may be stimulated to any extent by the inexhaustible supplies of manure which Liverpool and the manufacturing towns afford. Mr. Rothwell in his "Report" gives two instances of farms within six miles of Manchester — the first 156 acres in extent, for which 2,000 tons of manure were purchased; the second, 165 acres, for which 1,360 tons of manure were purchased in one year; and in both cases with amply remunerative results. The crops of Swedish turnips produced in this county cannot be excelled in any part of the kingdom — 40 tons an acre, in good seasons and under the best management, being quite common. Such a crop may at this moment be seen on the highly improved farm of Dr. Sillar, of Rainford, though this year the season has not been very favourable and the crop is in general much below an average. The humidity of the climate is favourable to the culture of green crops, the farmer has an ample command of manure, he has markets on every side of him for their sale, and he who has made the most use of these natural advantages has met with the most success.

The farm of Mr. Neilson, of Halewood, exhibits several points worthy of notice. A light tramway with waggons is made use of for taking the turnip crop off the ground in moist weather. The tramway is readily shifted, and the crop is thrown into the waggons, which are then each pushed along by a man, so that the entire crop may be removed from the ground, which thus receives no injury from the feet of horses. The tramway can be constructed for 1s. 4d. per yard, and might be very advantageously introduced on all heavy farms where it is found difficult to take off the turnip crop in moist weather. A gang of men are at present employed on a considerable field of Mr. Neilson's in taking off the turnip crop, which they draw from the ground, fill into the waggons, and convey outside of the gate at the rate of 6s. an acre, shifting the tramway at their own cost. At this work they earn 2s. 3d. a day. A large stock of dairy cows is kept on this farm. They are house-fed summer and winter, receiving in winter a mixture of steamed straw, ground turnips, and 1 lb. per head of boiled Egyptian bean meal poured over the mixture. Besides this and a sufficient supply of turnips and fodder they receive 2 lb. of oilcake daily. A large stock of pigs, 200 in number, is kept on this farm.

On the early, friable, loams in the neighbourhood of Ormskirk and along the Mersey, two crops of potatoes are sometimes got the same year. For the earliest crop the seed is prepared about the beginning of the year, by being sprouted under cover, and planted out into beds as soon as the weather admits. The land is very heavily manured, and great care is taken to preserve the young shoots unbroken. The second crop, the seed having undergone the same preparation, is planted as soon as the first is removed. But the more frequent custom is to transplant swedes after the first crop of early potatoes, and very excellent crops are occasionally obtained in this way.

The higher district of the country along its eastern boundary is chiefly in grass, stocked with dairy cows, for the produce of which the farmers find a ready market in the towns and villages thickly scattered over its surface. For its quality and exposure this land yields a high rent, 2l. to 3l. per acre being often paid. Some of the oat crops of the few fields which are in crop are still unsecured. (October 26th).

The rotation of crops adopted by the best farmers in South

Lancashire will surprise those who have been accustomed to consider any departure from the alternate system of corn and green crops erroneous.   Among many the golden rule of farming is that no two white crops shall follow in immediate succession; but the successful practice of a contrary system in this district may teach us how vain it is to prescribe the same rules for totally different circumstances, the same husbandry for the climate of the eastern side of the island with its 20 inches of rain per annum, as for the western side with its 40 inches of rain.   The true test of any system is its continued success, and the practice of the best farmers in this district, and those whose farms are in the highest state of cultivation, producing crops of all kinds which would astonish some of the wisest sticklers for rotations, combine in attesting the advantage in every point of view of taking a crop either of barley or oats, immediately after the wheat crop.   The four-course farmer takes his crop in this succession—clover, wheat, turnips, barley. The Lancashire farmer prefers it thus — grass, green crop, wheat, oats, or barley; his two green crops following one another, and his two white crops the same.

The rent of land, within six miles of Liverpool, ranges from 40s. to 4l. per statute acre according to quality, condition, and situation.   Within the same distance of Manchester similar land lets at the same rent.   Beyond that distance the rent varies from 20s. to 30s. per acre for the unimproved farms; but where drainage, buildings, and other improvements have been effected by the landlord, it ranges from 30s. to 40s.   This is for land suited to the culture of green crops and wheat, and from which the whole produce may be sent to market.   Besides the rent the tenant has to pay the rates, which, including tithe, land-tax, highway rate, poor rate, and church rate, will vary from 10s. to 12s. 6d. an acre.   The rent of the cold clay soils is much lower.   Within the last 20 years on many of the larger estates there has been little or no change in the rent.   Farms are seldom let by tender, and except when there is a change of

tenancy the farmer is usually left undisturbed. For small
farms, however, there is great competition, and a prudent agent
finds it neceessary to guard himself against being misled by
reckless offerers. The highest offer is seldom or never accepted.
There are instances, however, of greatly increased rents, one of
which may be mentioned. In 1823 the rent of a certain farm
was 150*l.* It is still occupied by the same family, but has
twice changed owners, and at each change been re-valued.
The rent is now 400*l.*; in the course of 27 years the farm has
been greatly improved, and we are assured by a very competent
judge that it is now better worth its present rent than it
was in 1823 at 150*l.* Farms vary in size from 20 to 200 acres,
the great majority are under 100 acres, and very few exceed
200 acres. There is no custom in the county which secures
to the tenant any compensation for unexhausted improvement.
The only right he has is this — that he sows the wheat crop
in October, quits the farm on the 2d of February, returns
to reap the wheat crop at his own expense, and is allowed
half the crop for his trouble. No other compensation can be
claimed except the price of the clover seeds which he sowed
with his last crop. The dung belongs to the farm.

LETTER XXXIII.

LANCASHIRE, — *continued.*

LANCASTER, Oct. 1850.

BEFORE entering on a description of North Lancashire it will
be necessary to notice one point regarding the southern division
of the county for which there was not space in our last letter
—that is, the reclamation and cultivation of peat mosses, or
bogs, the management of which forms a feature in the husbandry
of both North and South Lancashire.

In both divisions of the county these mosses are very extensive. Chat Moss, which lies seven miles west of Manchester,
and is traversed by the railway to Liverpool, is about five miles
long and three miles broad, and is situated at an altitude of 100
feet above the level of the sea. Its surface is composed of a
long, coarse, sedgy grass and heath, beneath which there is a
depth in some places of 34 feet of moss. Under ordinary
circumstances we might expect that this dreary waste, demanding considerable capital and great skill for its profitable reclamation, would long remain untouched by the hand of improvement.
But, with a railway through its centre connecting two of the
most populous and important towns in the kingdom, and
surrounded on all sides by a dense manufacturing population,

offering a ready market for every article of produce, and an equally ready and almost inexhaustible supply of manure, the wonder is, not that anything has been done, but that anything still remains to do. The possibility of effecting remunerative improvement is not a matter of doubt. The thing has been done with eminent success. Yet two-thirds of this great tract still remain waste and unproductive.

There are other extensive tracts of moss in the southern division of the county,— White Moss, near Middleton; Rainford and Kirby Moss, between Knowsley and Ormskirk; Halsall Moss, near Southport; Rufford Moss, &c.; and, as these are in every instance either traversed by railways, or in the close vicinity of canals, they possess facilities for improvement unknown to many other districts of the country in which moss land has been profitably cultivated. In the nothern division of the county there are also extensive tracts of moss. The most valuable of these are in the Fylde, to the north of the Wyre, and are estimated by Mr. Garnett to extend to 20,000 acres altogether.

Besides the advantage of locality which the moss lands of this county possess, they have generally, either beneath their surface or in their immediate vicinity, beds of rich calcareous marl, by the application of which, in conjunction with drainage, they can be converted from worthless unwholesome wastes into rich and productive lands. On Chat Moss the principal improvers have been the late Mr. Baines, of the *Leeds Mercury*, Mr. Reed, Messrs. Evans & Co., Lord Ellesmere, and Colonel Ross. The process of improvement may be briefly described. Large open ditches are cut, into which the covered drains, laid 10 yards apart, are run; but as great subsidence takes place at the first drainage of moss lands, these covered drains are not completed at first; sufficient time is allowed for the land to consolidate, the drains are then cut to the requisite depth, and in many cases laid with the top sod, dried by exposure to the air, being pressed wedgelike into the drain, where it leaves a hollow space for the water some six inches in depth.   The surface plants are then burnt off, and the turf torn to pieces by ploughing and cross-ploughing.   When this has been sufficiently broken, the next great step in moss improvement is begun by laying on the marl, which is most easily effected by the use of a moveable railway.   100 to 150 tons of marl are laid on an acre.   So soft is the moss at this stage of operations, that it is frequently necessary for both men and horses to have flat pieces of wood attached to their feet to prevent them from sinking.   The land is now fit for potatoes, for which it receives a further heavy application of nightsoil and ashes.   Great crops are got, which, till within the last year or two, were nearly exempt from disease.   That is not now the case, and seems to be owing to the presence of marl, as on unmarled moss the potato crop continues to grow successfully. A mixture of lime and salt has been tried by Mr. Evans, which, besides being much less bulky than marl, is found very effective in destroying the surface moss, and preparing it for potatoes.   Turnips, oats, and potatoes are found by experience the most paying and certain crops, though, on some improved mosses, the regular four-course rotation is adopted.

North of the Wyre, Mr. Wilson Ffrance, of Rawcliffe-hall, is one of the most extensive and judicious moss improvers. Of 736 acres allotted to him, 19 years ago, he has drained, made roads, and marled the whole.   Marling, with the assistance of a moveable railway, costs him about 3*l.* an acre ; drainage (with peat turfs), 1*l.* 6*s.*   To this must be added the expense of main drains, roads, and farm-buildings, which, on Mr. Ffrance's allotment, amounted altogether to between 9*l.* and 10*l.* an acre. The produce of potatoes on this improved moss is very great, and as soon as it is ready for being cropped, each portion is competed for eagerly, and lets at high rents.   Some of it fetches as much as 2*l.* per acre, and the whole extent cannot be estimated at less than 1*l.* per acre, which, on an outlay of 7,000*l.* over 736 acres, is a return of rather more than 10 per cent.

On some of the mosses in the manufacturing districts, the whole produce is sold, and manure bought; no stock except the working horses being kept on the land. Large crops of swedes, common turnips, and mangold are grown, and oats frequently yield 60 bushels or more per acre, besides a great bulk of straw, which in such localities can be very profitably converted into cash, and replaced much more cheaply by manure.

Such is a very short sketch of the present condition of the peat mosses of Lancashire. Many interesting details may be learned by a perusal of Mr. Garnett's prize essay on the farming of the county. Our purpose is rather to direct attention to the fact of the existence of such comparatively unopened mines of wealth, for such they are when regard is had to their locality. The bogs of the west of Ireland have been and are being reclaimed to advantage; but here, in the midst of the greatest manufacturing wealth, and possessing the best markets in the world, thousands of acres are still lying waste and unimproved. If left to the native farmers, the reclamation will be slow, for, as a class, they are individually possessed of little capital and no great enterprise, and where allotments are made to them they show no readiness to improve them.

Lancashire, north of the Ribble, differs in many important respects from the southern division already described. The populous manufacturing towns and villages disappear; and, with the exception of the extensive and peculiar district called the Fylde, the country is narrowed into a strip a few miles in breadth, extending from the sea to the mountainous tract which on the east divides it from Yorkshire. To the north, and disjoined from the rest of the county by Morecombe-bay, lies the rich district of Furness. The coal measures are now left behind; red sandstone, millstone-grit, mountain limestone, and clay slate, form the geological features of the district. The outline is on the whole more picturesque, the fields and trees look fresher and greener, though, except in valleys and sheltered

glens or bays protected by bluff headlands, the stunted appearance of the hedgerow trees, and their inclination from the west,
show the effects of the prevailing westerly sea winds. Near
the coast the land is managed chiefly in alternate husbandry;
as it begins to rise towards the hills it is principally in grass,
held by small dairy farmers; and the hills are stocked with
blackfaced sheep.

The extensive district called the Fylde first demands our
attention. It embraces all that low alluvial district lying
westward of the Preston and Lancaster railway, and extending
to the coast. It is divided by the river Wyre, which falls into
the sea above Fleetwood, and is traversed by the Preston and
Fleetwood railway. Along the coast the climate is very mild,
and the towns of Lytham, Blackpool, and Fleetwood are much
resorted to, for sea-bathing, by the inhabitants of the manufacturing districts in the interior. Containing many varieties
of soil, from a blowing sand to a strong alluvial clay, it gives
scope to different systems of management, and, on the whole,
it is naturally a rich agricultural district, requiring only the
hand of enterprise and the judicious investment of capital for
its profitable cultivation. The Earl of Derby, Mr. Clifton of
Lytham, and Mr. Wilson Ffrance, are the principal landowners.
Twenty years ago many parts of this district were nearly inaccessible, and even yet a stranger would find it difficult to get
through some parts of it but for the railway.

In 1831, on Lord Derby's estate, extending to 12,000 acres
of land, in the neighbourhood of Preston, there was not a field
turnip grown; a few for table use might be seen in the farmers'
gardens. Up to that time the only tenure known on that extensive estate was by leases of three lives, with fines for renewal.
These fines varied, according to the extent and value of the
holding, from 100*l.* up to as much as 2,000*l.*, and the right of
property was maintained by a small annual payment, often not
more than 40*s.*, for a considerable farm. In this way the proprietor lost all immediate interest in the property, and the

tenant, even if he had the requisite skill, had probably parted with all the available capital he possessed in paying the fine. Improvement there was none, and though Lord Derby has granted no renewals since 1831, that very circumstance prevents improvement, as the tenants feel that when their leases run out they will be placed under a different tenure. On other parts of the Fylde the same system prevailed, though not up to so late a period; but a considerable portion of this extensive district is still held on long leases, and many years must elapse before the evil effects of the practice can entirely disappear. Instances are known of tenants, through indolence and inactivity, being reduced to poverty when their payments for rent were merely nominal, and of the same men, on the same farms, when compelled to exert themselves to pay a regular rent, casting off their sloth, retrieving their circumstances, and becoming at last comparatively independent.

On the extensive estate of Mr. Clifton, of Lytham, very large and vigorous improvements have been made. One large open drain, emptying itself into the harbour of Lytham through floodgates which exclude the tide at high-water, extends five or six miles inland, and cost the proprietor 3,000l. It has laid dry a great extent of flat country, from which there was formerly no outlet for the water. Mr. Clifton has enlarged several of his farms, and introduced upon different parts of his estate farmers of capital, occupying 400 acres and upwards, on leases of nineteen or twenty-one years. Their farms have been drained at the joint expense of landlord and tenant, and handsome residences and farm buildings have been erected by the landlord. The rotation of crops prescribed on other parts of this estate is, (1) oats, (2) green crop, (3) wheat or barley, and three years in grass, — while a portion is likewise reserved for permanent grass and meadow, as most of the farmers keep a dairy stock. As there is occasionally a good demand and corresponding prices for hay, straw, and turnips, the tenants are allowed to dispose of these, on condition that they bring upon their farms the following quantities of manure : —

4 tons of manure, for 1 ton of hay sold :

3   ditto   ditto   of wheat straw sold :

2*l.* per acre, in addition to manure already applied, on fields whence turnips are sold.

On the sandy lands along the sea-coast on this estate great improvements have been made by marling, a bank of which is very frequently found in convenient proximity. These sandy soils are now the most successful in potato culture, and, being very inexpensive to work, are much sought after by tenants. They let, according to quality, at 21*s.* to 30*s.* per acre, and the better class of soils on this estate at 30*s.* to 36*s.* per acre, inclusive of tithe.

The general class of farms in the Fylde are from 40 to 160 acres in extent. The fields are small, and the fences zigzag about in all directions. In almost every field there is a marl-pit, the application of which, in former times, was the only improvement known to the farmers. It is still practised, but not in the same wholesale manner as formerly. One of the conditions of the holding was generally that the tenant was to lay the whole dung of the farm upon his meadow ; consequently, the little portion they did cultivate could not get any justice. To this condition they were rigorously bound, many instances being known of men being heavily fined for transgressing it. The dairy forms their principal source of income, and in its management they display much cleanliness and care. Their arable is limited to a very small portion, from which it is usual to take one or two oat crops, then a bare fallow, followed by wheat; after which the land is either sown with such hay-seeds as the farmer has, or left to grass itself over; this course being justified on the principle that the old roots had not by their husbandry been destroyed. Draining, to a very limited extent, is carried on ; dried peat being used as a wedge to form the watercourse. Where well executed, these

peat drains are effective, and last many years. The contrast
betwixt drained and undrained land in the Fylde is very perceptible.   Crossing a fence which divides two fields of precisely
the same soil you may step from one not worth 10s. an acre
into another better worth 40s., the whole difference being due
to drainage and improved management.   In this flat country
there is sometimes much difficulty in getting an outfall for the
drainage; but, besides that material difficulty, there is an artificial obstruction at a place called Skippool, where the proprietor of a small meal-mill has the power of damming up the
water to turn his mill, and in this way actually keeps up the
drainage and sets back the water over several hundred acres of
valuable land.   It would appear that the neighbouring proprietors and their tenants, who are injured by it, have no power
to compel the removal of this obstruction, and negotiations for
the purpose of buying up the privilege have hitherto failed.

   After a clean bare fallow, the crop of wheat yields from twenty-
four to twenty-eight bushels an acre; thirty-six to forty of beans
are reckoned a full crop.   From the level character of the
country it is exposed to severe cold winds.   On Mr. Begbie's
farm, boarded hurdles are erected to shelter his ewe flock, which
are kept for the supply of lambs to the coast villages, whence
there is an excellent and increasing demand during the sea-
bathing season.   Two-thirds of the Fylde district are still undrained, and comparatively unimproved.

   To the east of the Lancaster and Preston railway the country
is chiefly under grass, and let to dairy farmers.   The farms are
generally held on seven-years' leases, some on fourteen; they
extend from 40 to 180 or 200 acres, but may average about
80 acres.   They are precisely of that character which most
encourages competition, being chiefly in grass, and believed to
require no great skill and not much capital in their management.   "Railway men," that is to say, men who have made a
little money by railway contracts, are anxious to retire to them,
and when a farm comes into the market there are often as many

as twenty competitors for it, if a small farm, and three or four if a large one. This competition is recklessly taken advantage of to the utmost by most of the landlords of the district, little regard being paid to the qualifications of the tenant, provided he offers the highest rent. Many farms are at this moment being let at an increase of 10 per cent. and upwards on the old rent, without the landlord undertaking any of those permanent improvements by which alone an increased rent, with lower prices, can possibly be realised. It would be wrong to designate such conduct by any other name than the most shortsighted folly on the part of men whose position and information should guard them from acquiescing in the over-sanguine anticipations (if they really do look before them at all) of unskilful and inexperienced candidates.

Where such competition exists it can hardly be expected that much abatement of rent should be made by the landlords. Ten per cent. in one or two instances has been given back, and in others a certain proportion has been allowed for drainage and other improvements. On the estate of Mr. Garnett, of Quernmore Park, a very judicious system is adopted. The landlord drains for his tenant, and buys bones for top-dressing the grass land, for both of which outlays he charges 5 per cent. The bones are applied to the drained grass lands at the rate of one ton per acre, costing 6l. ; and their effect has in every respect been as great as we found it in Cheshire, in all cases improving the quality, and almost doubling the quantity of herbage. By this means a dairy farmer is at once placed in a position in which he may with success meet somewhat diminished prices. At Bleasdale Tower, between 600 and 700 feet above the sea-level, the effect of bones in improving grass land is very striking. A wild and barren country is here being reclaimed for the use of man. Farm buildings of the most substantial and convenient character, and with all requisite shelter for this elevated district, are being erected, and every encouragement to industry afforded, and regard to the personal comfort of his tenants and their families bestowed, by the liberality of the proprietor, Mr. Garnett.

It may be interesting to compare the present rates of rent, labour, and the value of produce with what they were in the same locality in 1770.   An exact comparison cannot be made, as Arthur Young does not mention the exact situation of the places he describes.   Between Lancaster and Garstang a dairy farm seems to have borne something like the following proportions : —

In 1770.	and	In 1850.
Rent, 21s. an acre.		Rent, 41s. an acre.
Rates 3d. per pound.		Rates, 3s. 9d. per pound.
Tithes compounded for.		Tithes commuted, and included in rent.
4-7ths of farm in grass.		4-5ths of farm in grass.
3-7ths arable.		1-5th arable.
Annual produce of a cow, 4l.		Annual produce of a cow, 9l.
Six horses in a plough, and do an acre a day.		Two, and sometimes three, horses in a plough.
First man's wages, 9l. a-year, and his board.		First man's wages, 15l. to 16l. a-year, and board.
Second man 5l. a-year, and board.		Second man, 10l. a-year, and board.
Dairymaid, 3l. and board.		Dairymaid, 7l. 10s., and board.
Bread (oat), 11 lb. for 1s.		Bread, 4d. per 4 lb. loaf, coarse wheaten bread ; 5d. per 4 lb., best.
Cheese 3d. per lb.		Cheese, 5d. per lb.
Butter 8d. per lb.		Butter 11d. to 1s. per lb.
Beef, 2½d. per lb.		Beef, 5d. to 6d.
Mutton 2½d. per lb.		Mutton, 6d.
Labourer's house-rent, 20s.		Labourer's house-rent, 50s. to 100s.

These prices suggest many reflections which it is unnecessary at present to dwell upon, further than to point out that while rent, rates, and wages of labour have more than doubled, the value of the staple produce has increased in a like proportion, and that oat-bread, which then formed the sole bread of the people, is now much superseded, even in the country districts, by wheaten bread, at a price very little higher per pound in 1850 than that of oat-bread in 1770.

The wages of labour throughout Lancashire will be reckoned high as compared with the southern counties. In South Lancashire 12s. to 15s. a week is the usual rate for Englishmen, and 9s. a week for Irishmen. In that district native labour is so scarce that the farmers declare they could not get on at all without the aid of the Irish. Cottage-rents are from 3l. to 5l. a year, according to accommodation, and those on the large estates are always provided with a moderate piece of garden ground. In some cases when labourers' cottages fall into decay they are not rebuilt, and the labourers are consequently driven into the neighbouring towns. But from the denseness of the population this is not attended with the inconvenience we have sometimes witnessed, as in this county a labourer has seldom to walk so much as two miles to his work. In the Fylde labourers' wages are lower, 9s. and 10s. a week being common. To the north and east of the Fylde wages are higher, 12s. and 14s. a week being the present rate. Women are seldom employed in the fields at hoeing or other light work, there being better payment for them in-doors at the factories. It is necessary, therefore, to employ men in this county in many operations for which women or boys are found competent in other districts. And this makes the manual labour on the turnip-crop nearly double the cost in Lancashire, as compared with such counties. Fuel is abundant and cheap.

A report of the agriculture of Lancashire would be incomplete if it bore no testimony to the influence which the two district societies have had in promoting agricultural improvement. In South Lancashire the Manchester and Liverpool Agricultural Society, of which Mr. Henry White, of Warrington, is the secretary, holds an annual meeting in different parts of the district in turn, at which, besides the usual prizes for stock, competition is also invited in the management of farms, including, under separate heads, draining, subsoiling, irrigating, laying

down land to grass, marling, and green crops, besides rewards
to farm-labourers for good conduct and proficiency.   Inspectors
are appointed by the society to examine the different farms and
crops entered for competition, and, besides the direct benefit of
that competition, the reports convey much useful instruction by
describing the best processes which come under the observation
of the inspectors.   In the northern district the efforts of the
North Lancashire Society have been very successful in introducing improved breeds of stock, and in encouraging competition among both landlords and tenants.

LETTER XXXIV.

YORKSHIRE.

WEST RIDING—WOOLLEN AND WORSTED MANUFACTURES — WEAVER-FARMERS —GENERAL TASTE FOR AGRICULTURE AMONG MANUFACTURERS OF THE WEST RIDING — MILK FARMERS — THIS DISTRICT LITTLE AFFECTED BY THE PRICE OF CORN — MR. STANSFIELD'S FARM AT ESHOLT — MANAGEMENT AND PRODUCTIVENESS OF ITALIAN RYE GRASS — RENT — WHARFDALE — VALUE OF A COW'S PRODUCE—RENT AND RATES — HAREWOOD—CONDITION OF LABOURERS — LEEDS — WAKEFIELD — FARMING AT CHEVET GRANGE — DETAILS OF MANAGEMENT OF CROPS AND STOCK — LABOURERS' WAGES — VARIETY OF CROPS NEAR DONCASTER, PONTEFRACT, AND GOOLE — TENURE AND TENANT RIGHT —TURNIP CROPS POOR AND LITTLE VALUED.

WAKEFIELD, Nov. 1850.

CROSSING from Lancaster to the west Riding by the new line of the North-Western railway, the traveller is carried up the picturesque valley of the Lune, whence, after passing through a bleak high country, he begins to descend into the grassy dales of the upland district of Yorkshire. Here in the sheltered valleys of the mountain limestone the fields are completely inclosed, and though for many miles scarcely a ploughed field is to be seen, there is everywhere evidenced a skilful and painstaking management of grass. The mixed breed of short-horns, long-horns, Irish, and polled Galloways, which satisfy the dairy farmers of Lancashire, now give place to the improved short-horns, which with occasional exceptions are the distinctive breed of this Riding. Around Settle the country is all in grass, and continues so to Skipton, eastward of which a few cultivated fields appear, but so small in proportion as to be quite a subordinate feature in the landscape.

As we approach the coal district the factories become more numerous. Entering it at Keighley, we pass through Bingley and Shipley to Bradford and Leeds, still environed by small fields of grass land, generally well drained and in good condition.

We are now in the coal district of Yorkshire, as much distinguished for its woollen and worsted manufacture as Lanca·shire for its cotton. With a population of about a million and a quarter chiefly employed in trade, agriculture, in the most densely inhabited parts of the West Riding, is of secondary importance; and yet it differs from the cotton districts in this —that all classes engaged in the woollen manufacture seem to have a taste for the occupation of land. Besides those employed in the large mills, there is a class called " clothiers," who hold a considerable portion of the land within several miles of the manufacturing towns; they have looms in their houses, and unite the business of weavers and farmers. When trade is good the farm is neglected; when trade is dull the weaver becomes a more attentive farmer. His holding is generally under 20 acres, and his chief stock consists of dairy cows, with a horse to convey his manufactured goods and his milk to market. This union of trades has been long in existence in this part of the country, but it seldom leads to much success on the part of the weaver-farmer himself, and the land he occupies is believed to be the worst managed in the district. Being chiefly in grass, and not permitted to be sublet or subdivided, it has not led to the same evils which some years ago existed in the north of Ireland, under a much similar state of things, among the hand-loom weavers of that country.

Ascending higher in the scale, successful tradesmen buy small properties and cultivate them, while the capitalist manufacturer is in many parts of the West Riding purchasing the estates and taking the position of the old gentry of the country. Some who do not purchase the land occupy large farms as tenants, and into the management of these they carry the same business habits and the same command of capital which gained them success in trade. At Burley, in Wharfdale, Mr. Horsfall, a wealthy manufacturer and landowner, spares neither time nor personal exertion in increasing the produce of his land. By irrigation he has improved his pasture land 50 per cent., and finds, after trial, that by farming himself and attending to details, he can make 7*l.* an acre from the produce of land, which if let would not yield him more than from 3*l.* to 4*l.* He thus feels himself well compensated for his own trouble and the capital he invests as a farmer.

In the neighbourhood of all the manufacturing towns the system of husbandry is chiefly grass farming for the supply of the towns with milk and butter. Besides the "clothiers," there are small milk farmers who carry on a lucrative business of this kind; they give their whole attention to it, and, when their stock is judiciously selected and well fed, they in many cases make an average produce of 20*l.* from each cow, after deducting the loss occasioned by selling a cow whose milk has failed, to replace her with one in full milk. Besides the manure collected on their lands, they can purchase in the towns and villages any quantity, at a cheap rate, for topdressing their meadows.

From what has been said, it will be easily seen that the manufacturing districts of the West Riding have an agriculture of their own, as little influenced by the price of corn, or dependent on it for success, as that of South Lancashire. Small farms are the rule, but as we pass eastwards towards Wharfdale they increase in size, and, at Esholt, the home farm of Mr. Stansfield, M.P., under the management of Mr. White, presents several features of interest. It is managed strictly on the four-course system, but being highly manured and cultivated it is very productive. A certain portion is kept under meadow, which is irrigated, and with great advantage, as the water is enriched by the drainage of a populous neighbourhood. To supply food for the farm horses and other house-fed stock, a small field adjoining the farmyard, three acres in extent, has been kept for the last nine years under Italian ryegrass. The produce of this field is always cut and carried into the stables for consumption. Last year it was cut six times, and after each cutting was dressed with liquid manure from the tank in the farmyard. One half of the field is broken up every second year for

renewal, as the grass after the second year ceases to be productive. The ground is ploughed in September, well torn to
pieces, and all root-weeds picked off and burnt, the ashes being
then scattered over the surface and ploughed in with a light
furrow; after which the seed (which is got from Dickinson of
London) is sown thick, at the rate of from four to five bushels
an acre. Next season it is ready for cutting, and the small
field of three acres so managed yields the entire summer food of
six work horses and five bulls, besides supplying a fresh bite for
the cows twice a day when they are brought in to be milked.

The produce is thus applied solely to the stock of the farm,
but in the vicinity of the populous towns probably nothing
would pay a farmer better than to cultivate Italian rye grass on
a larger scale for sale.   Its growth might be stimulated almost
to any extent by manure, and when given in a fresh and
succulent state it is eaten eagerly by all kinds of stock.   Under
good management the successive cuttings in one year would
yield from thirty to forty tons an acre of green food, which, at
the common price for that article from dairy and horse keepers
in a manufacturing town, could not be less in value than 40*l.*
to 50*l.*

In the management of the arable land the following details
may be interesting : — Seeds are sown every fourth year in these
proportions per acre: 10lb. red clover, 10lb. white clover,
4lb. trefoil, and 4lb. rib grass, being 28lb. in all, and a full
plant is invariably got. From the middle of April till September two acres of seeds yield food for sixteen to twenty sheep,
and one beast.   They are then ploughed, and after being
harrowed the land is sown in October with wheat.   Wheat
yields from four to seven quarters an acre according to season,
the bulkiest crop not always proving the best on the barn floor.
Excellent crops of swedes are grown with superphosphate of
lime, which is highly approved of as a manure. The barley
crop averages eight quarters an acre, but it is not on the whole
so certain a crop as wheat. Labour here is very expensive, 14*s.*

a week being the present rate for farm labourers. The harvest could not be accomplished without the aid of the Irish, and even with their assistance the expense of cutting a heavy crop of wheat is as much as 16s. an acre, and other corn crops may average 12s. an acre. Cow cabbage, or kohl rabi, is cultivated both here and by Mr. Horsfall, of Burley, with advantage. It is found to yield as much per acre as the Swede, and when given to cows it does not affect the taste of their milk. Land *lets* here at from 30s. to 50s. an acre, according to the quality and situation, and as the farms are small there is great competition for them; but that is not taken advantage of, as in north Lancashire. The best man is invariably made choice of, and he gets the farm at a fair valuation. There are few leases, the holding being chiefly from year to year; but a good tenant is never disturbed, and even a bad one is long borne with.

Proceeding down Wharfdale by Burley, Otley, and Arthington, to Harewood, a rich country is passed through, a large proportion of which is in grass, stocked, as at Farnley, with fine herds of short horns and Leicester sheep. The arable land does not indicate more than ordinary management, and cannot yield so large a net produce to its occupier as the neighbouring grass lands. On the sides of the valley near Otley and the plains beyond it the farms are chiefly arable. They average 200 acres in extent, and are managed on the four-course system. No artificial manures are used for green crops, the farmer depending solely on his farm for sources of reproduction. They all keep dairies, the produce of which is made into butter, and sold in the manufacturing towns. A cow's produce is reckoned to be worth 12l. yearly. Rents vary from 10s. on the poorest arable farms to 3l., inclusive of tithe, for the best grazing land on the banks of the river. Poor and highway rates amount to 1s. per pound in the country districts, and rise to 2s. 6d. and 5s. near manufacturing villages, where sometimes as much as 10s. per pound has been paid by the farmer for poor rates alone. The farmers as a class are not possessed of much capital, and their cultivation does not bespeak great energy or skill. The price of wheat to the farmers here at present is from 42s. to 46s. a quarter.

As we approach Harewood the cultivation improves, and a fine tract of rich, well managed land is traversed. The farmers are busily engaged with wheat sowing, for which in many places the land, after being ploughed and harrowed, is prepared by being "ribbed" with a light single-horse plough, which produces much the same effect as the presser, the seeds falling into the ruts made by the plough, where the roots take firm hold, and come up in rows as when drilled. Farm labourers are paid 12s. to 14s. a week, their cottage rents vary from 4l. to 5l. The landlords have made no abatement of rent, and any farm that comes into the market is eagerly sought for.

In the neighbourhood of Leeds the land is well farmed, whether occupied as arable, pasture, or market garden. On the strong loams near Wakefield, heavy crops are grown by good management in the market gardens here situated, and which supply several of the thickly peopled districts of the Riding. In other parts of the coal district the local value of produce, and the facility of getting manure, have enabled the farmer to grow a succession of crops, which, without these advantages, he could not continue long with benefit. The following rotation is not uncommon in such localities : — 1. turnips ; 2. wheat; 3 and 4. seeds, eaten on the ground; 5. wheat; 6. wheat or barley, according to the nature of the soil. In some cases wheat is taken every alternate year, and Mr. Charnock, in his report of the West Riding, mentions an instance, on the banks of the Calder, of wheat having been taken from the same field for 30 years in succession, with only four exceptions — one of these having been a bean crop, another barley, the third fallow, and the fourth potatoes. The crops during the whole period are said by him to have averaged 39 to 42 bushels per acre.

The farm occupied by Mr. Johnson of Chevet Grange, between Wakefield and Barnsley, may be taken as a favourable

sample of the arable farming of the lower district of the
West Riding. It contains 280 acres of sound good land, capable of growing good crops of wheat and barley, and dry
enough for eating the turnips off with sheep. One-third of
the farm is, by the common rule of the district, kept in permanent grass and meadow : the other two-thirds — 180 acres
in extent — are managed in the four course rotation. Previous to the seeds being ploughed up for wheat, the ground
is sown with a few cwts. of salt, for the purpose of killing
snails. It is ploughed in October, and, after being harrowed,
eight to ten pecks of seed are drilled at seven inches apart.
The seed is a mixture of Spalding's red and Australian white,
in the proportion of three of the former to one of the latter.
The two kinds ripen together, and the mixed sample sells considerably better than if it had been all red, while the produce
is believed not to be diminished. It is not found necessary to
hand or horse hoe the wheat crop, as the land is never allowed
to get foul. The average produce is five quarters an acre. The
preparation of the land for the turnip crop is the next process,
and for this the wheat stubble is ploughed, with three horses
abreast, a deep strong furrow, and the only very deep furrow
given in the course.

In spring the land is wrought to a sufficient degree of fineness
by repeated ploughings and harrowings, or "dressings," as the
conjoint operation is termed here; after which six loads an acre
of manure from the farmyard are spread over it, and lightly
ploughed in. Lime is applied once in eight years at a cost of
36s. to 40s. an acre. The seed is then drilled on the flat, the
drill at the same time depositing a mixture of eight bushels of
bones and two cwt. of superphosphate per acre. The rows are
nineteen inches apart, and when the turnips are ready they are
hand hoed twice, and horse hoed. About one-fifteenth of the
crop is swedes, the rest white, and other soft turnips. Almost
the whole is eaten on the ground by sheep, which are confined
by nets, and shifted from space to space as the crop is consumed.

The sheep eat the turnips from the ground, the scooped-out bottoms being afterwards "dragged" up to be eaten. Boxes with cut straw chaff and a little salt are placed for the sheep to eat, as this is found to keep them healthy, and they are not on this farm put on the turnips until these have become ripe, which is indicated by the leaves beginning to decay. When put on at an earlier stage young sheep are apt to die, and are, at all events, very subject to scour. The turnip crop is reckoned to keep eight or ten young sheep per acre for twenty weeks. In spring the land is ploughed and sown with barley, of which six quarters are considered an average crop. Red clover is sown on one-third of the land, mixed seeds on another, and the other is left unsown, to be followed by a pulse crop (pease or beans drilled at nineteen inches apart) in the following spring. By this arrangement red clover comes only once in twelve years. The seeds carry from three to five sheep an acre when pastured. A part is cut for hay, and in autumn the course is completed. Twenty to thirty beasts are wintered in the yards on straw, and 4lb. each of oilcake daily. Eight work horses are kept on this farm, the thrashing being done by horse power. In winter they are fed on a mixture of oat and wheat straw, and a small portion of clover hay, cut together into chaff. This is placed in the horses' manger, and then slightly damped with water, after which about a quart of bean meal is strewed over it, which, being well mixed by hand, adheres to the wet chaff, and makes the whole a palatable and nutritious feed for the farm horse. At night the horses are all turned loose into a yard, where they are supplied with straw in racks. In summer they are put on the old grass land.

A good many pigs are fed on Egyptian beans, which have been previously steeped twelve hours in cold water, and then, after having lain twenty-four hours longer, to soften and germinate, are found an excellent and economical grain for pig feeding. The pigs receive as much as they can eat, and get nothing else.

Four ploughmen and three extra men are required for the labour of this farm, besides additional labour during harvest. They are paid 14*s.*, 13*s.*, and 12*s.* a-week, according to ability. The farm is very compact, and the different operations are carried on with much neatness. The fields are divided by closely trimmed and straight lines of hedges. The farm buildings comprise a huge barn, with stables for the horses, and sheds and open yards for the cattle. In the stackyard the crop is secured in high long stacks, in the Lincolnshire fashion, — the largest of which is estimated to contain not less than 1000 bushels of wheat. Besides this farm, Mr. Johnson occupies other extensive farms in the neighbourhood, managed on the same plan.

It is impossible, within the limits of a single letter, to include all the interesting details connected with the farming of this part of the Riding. Near Doncaster the land is considerably lighter, and there sheep husbandry is more exclusively followed. In the neighbourhood of Pontefract licorice is cultivated, the roots of which are from two to three feet in length, and require a great depth of soil for their successful culture. The roots are the article of commerce. Flax, teazle, woad and chicory are also largely cultivated on some of the alluvial soils which most favour their growth. On the warp soils, near Goole and Selby, potatoes are grown both for the London and local markets, for either of which the navigable rivers present ready access to the farmers. On the better class of these soils two crops of potatoes are in some instances got in a year, while in others they are grown alternately with oats or wheat; and where the soil is particularly favourable, and the command of manure sufficient, they are grown in succession year after year.

The prevailing breed of cattle throughout the West Riding is the improved short-horn, of which there are several celebrated herds in the district. The draught horses are light and active, and excellent hunters are reared in the neighbourhood of Doncaster, Wetherby, and Ripon. The pure Leicester sheep is being superseded by a cross with the Shropshire ram and Lei

cester ewe, the produce of which yields a much better mixture
of fat and lean meat, and commands 1*d*. per lb. more in the
market.   Some feed the hoggets fat at a year old by giving cake
with the turnips, and this is done where the land is naturally
poor and requires this high feeding to enrich it for the following
corn crop.   Where the land is itself of superior quality, this
mode of feeding, it is said, would make it too rich for the corn
crop; and on such lands accordingly the hoggets are not fed fat
the first year, but after being grazed during the summer, they
are fattened on the turnips the second winter.   The breed of
pigs is very superior.

On the southern and eastern sides of the West Riding, where
the large arable farms are situated, leases are not very common;
but the farmers are sufficiently protected by the system of
tenant-right, or compensation for unexhausted improvement,
which prevails.   We shall more particularly refer to this when we
have seen its effects in other parts of the county   In many instances temporary abatements of 10 per cent. have been made on
account of the low scale of prices, but no permanent arrangement
has yet been entered upon for the future.   The better class of
tenants consider it unfair to press their landlords for a permanent
reduction until it shall be seen what the future range of prices
may be; and a good feeling exists between landlord and tenant in this part of the Riding.   The landlords are looked upon
as "steady" men, who have never shown any proneness to take
advantage of their tenants in times of competition.   Corn-rents
have not been introduced, though many farmers express a wish
for them.

On the whole, the farming of the agricultural division of the
West Riding cannot claim a very prominent place.   Eminent individual exceptions there are, but the great extent of imperfectly drained land, the foul stubbles, the very light, and, in
many cases, carelessly managed turnip-crops, show that much
yet remains to be done by both landlord and tenant.   Of so

little value is the turnip-crop accounted, both intrinsically and
from the mode in which it is here consumed, that the selling
price at present is only from 40s. to 45s. an acre ; and, taking
the district from Barnsley to Church Fenton, we should say
that for one field of swedes which averages 10 tons an acre,
there are five which do not reach half that amount.

LETTER XXXV.

YORKSHIRE.

POTATO COUNTRY ROUND GOOLE — MESSRS. WELLS' FARMS — PROCESS OF WARP-
ING — STIFF CLAY THUS COVERED WITH "GENTLE" LAND — COURSE OF CROPS
— FALLOWS ALMOST SUPERSEDED — BUILDINGS, AND MANAGEMENT OF STOCK
— ECONOMY OF STEAM POWER — HEAVY APPLICATION OF MANURE — GUANO
HAS LESSENED THE COST OF GROWING POTATOES — COMFORT OF THE PEOPLE
ON LORD BEVERLEY'S ESTATE — RENT, RATES, AND FARMERS' CAPITAL. —
HOLDERNESS. — SOIL AND COURSE OF CROPS — AGRICULTURAL MANAGEMENT
— SUNK ISLAND — RENT, RATES, WAGES — WANT OF DRAINAGE AND BUILD-
INGS IN HOLDERNESS — PATRINGTON — MR. MARSHALL'S IMPROVEMENTS —
FLAX MILL.

DRIFFIELD, EAST RIDING, Dec. 1850.

BEFORE quitting the alluvial districts which lie along the tidal
rivers flowing into the Humber, it may be interesting to give a
brief sketch of the system of management adopted on one of the
best warp-land potato farms. Previous to the potato failure of
1846, the country round Goole (the greater portion of which is
embanked land, situated four feet under high-water level of
spring tides) sent annually to the London market about 30,000
tons of potatoes. Since 1846 the crops have greatly diminished,
not in extent, but in yield; yet the price with a short crop is
sufficiently tempting; and the farmer is further encouraged to
persevere by the opening of the Goole and Pontefract Railway,
which gives him access to the markets of the manufacturing
districts of the West Riding and Lancashire, while he also possesses his former outlet to the east by sea to London. Fully
one-half the potatoes of the district now find their way to the
west.

The farms of the Messrs. Wells, of Booth Ferry and Airmyn
Pastures, are situated in the township of Airmyn, near the con-
fluence of the rivers Aire and Ouse, and about two miles westward of Goole. Together they extend to 800 acres, and are

managed alike; but not being contiguous, we shall confine our
description to the Pastures Farm.　Fifty years ago much of
this farm was under water, an excellent breeding place for
wild ducks, in pursuit of which several men in the village
earned their principal livelihood; the rest of it yielded cran-
berries and ling, and abundant crops of rushes.　There are now
400 acres, all under crop, with the exception of from forty to
seventy acres, which are still in process of reclamation by warping.　The details of that process were described at some length
in a former letter; and it is only necessary here to say that the
soil of this farm consists of an alluvial deposit, several feet in
thickness, gained by admitting the muddy tidal waters of the
river; and that the art is now so well understood that the expert
warp-farmer, by careful attention to the currents, can temper
his soil as he pleases.　When the tide is first admitted, the
heavier particles, which are pure sand, are first deposited; the
second deposit is a mixture of sand and fine mud, which, from
its friable texture, forms the most valuable soil; while, lastly,
the pure mud subsides, containing the finest particles of all,
and forms a rich, but very tenacious and expensive soil to
manage.　The great effort of the warp-farmer, therefore, is
to get the second or mixed deposit as equally over the whole
surface as he can, and to prevent the last from effecting a
lodgment.　This he does by keeping the water in constant
motion, as the last deposit takes place only when the water is
suffered to be still.　On that portion of this farm which is at
present of a stiff, tenacious texture, it is Mr. Wells's intention
to apply the warp until he gets the whole surface converted
into what is here appropriately termed " gentle" land.　Warping costs 10l. an acre, the outlay being made by the landlord,
who receives 5 per cent. from the tenant; and as three years
are spent in the process (one year warping, one year drying and
consolidating, and one year growing the first crop), the landlord
foregoes one and a half year's rent, and the tenant pays as
much.　The first crop is generally seeds, hoed in by hand, as

the mud at this time is too soft to admit of horse labour. These
are pastured one or two years; and the land is then drained
with tiles, not less than three feet in depth, if practicable, after
which it is in a proper state for the course of husbandry pursued
on the rest of the farm. As soon as all the farm has been made
"gentle," or suitable for green crops, one half of it will bear
crops for sale, and the other half crops for consumption on the
farm. There will then be annually 100 acres of potatoes and
100 acres of wheat for sale, and 200 acres of seeds, turnips,
rape, tares, swedes, and beans or peas to be eaten on the farm
by live stock. The system adopted at present is as follows : —

1. Potatoes, manured with 20 tons of manure.

2. Wheat, sown out with mixed seeds.

3. Seeds, part mown for green food, and part made into hay.

4. Seeds, top-dressed with 3 cwt. of guano per acre, part
mown and part depastured.

5. Potatoes, manured in the ridge with 7 cwt. of Peruvian
guano.

6. Wheat.

The variations from this rule are these : — On the strong
land, where potatoes cannot be profitably grown, the seeds are
ploughed up after being mown the second year; and then, after
being ploughed five times, the land is drilled with wheat. When
the land gets foul, instead of a bare fallow, turnips or tares are
sown, which are followed by potatoes, two cleansing crops thus
following in succession. Bare fallows are very seldom resorted
to, even on the strongest land, the practice being to sow winter
tares on the wheat stubble, and to clean the land well after
they are mown in May and June, or to winter fallow the land,
and, after repeatedly ploughing and cleaning it in spring and
the early part of summer, to sow rape upon it, which is either
eaten on the ground with sheep and ploughed for wheat, or, after
being eaten down, left for a crop next summer, and the land then
prepared for wheat. The corn crops are sown with the drill,
and the potato and turnip crops grown on ridges twenty-eight

to thirty inches apart. Where the land is sufficiently clean, a crop of turnips, the same season, follows the winter tares, and both are succeeded by potatoes.

Commodious and well-arranged farm buildings have been erected on this farm, in which the whole stock are housed under cover, while there are two large yards, hollowed in the centre for storing the manure, and into which all the stock are turned for a few hours daily to exercise. Besides the work horses, ten well-bred young horses, four of which are sold annually at the great horse fair at Howden, are accommodated with sheds and yards. Between fifty and sixty cows and bullocks are fed on the farm, and a large stock of pigs, which receive barleymeal, and refuse potatoes, steamed together. The barn is fitted with a horse thrashing mill, and grinding and cutting machines for bruising corn and linseed and chopping straw and hay. At Booth Ferry, Mr. Wells has introduced a seven-horse power steam engine for these purposes at an additional cost of 300*l.*, which, from its superior efficiency and economy, he finds, on accurate computation, to yield him a profit, besides interest, of 20 per cent. on this original difference of cost, besides the advantage of being able to thrash any quantity to meet a favourable turn of the market, without at the same time stopping necessary field operations. The cost of thrashing and dressing corn by steam power at Booth Ferry averages $7\frac{1}{2}d$. per quarter; by horse power on the other farm, 1*s.* $7\frac{1}{2}d$. per quarter.

About 1000 tons of manure are made on this farm; besides which 200*l.* are expended in guano and other manure, and 200*l.* in linseed cake for feeding. The introduction of guano has greatly lessened the cost of the potato crop, the large application of 7 cwt. an acre costing only 3*l.* 10*s.* against at least double that sum formerly paid for town manure, besides the manifest saving of labour in carrying on and spreading over the land 7 cwt. of guano in comparison with 20 to 25 tons of manure. The crops of potatoes formerly grown on these rich warp-lands were very remunerative. Since 1846 they have

been extremely precarious, and it has been necessary to substitute an inferior white potato for the York red (so long a
favourite in the London market), on account of its greater
liability to disease; while only six tons an acre of whites are
now got where ten tons of reds used formerly to be reckoned an
average crop.

The township of Airmyn, containing 3,600 acres, principally
of warp-land, is the property of the Earl of Beverley, whose
liberality to his tenants, and considerate treatment of the
labouring class, merit remark.   The conditions on which the
land is re-warped have been already mentioned, but Lord
Beverley likewise makes a large expenditure in the erection
of improved farm buildings for his tenants, in many cases
without remuneration, and supplies tiles for drainage, and posts,
rails, and quicks for fences.   The farmers are allowed to kill
the game, and the timber on the estate is confined to plantations, none being kept in the hedgerows at their expense.   The
labourers who live in the village have each a house and garden,
and a "cow-gate," which comprises one acre and a-half of the
best pasture land in the district, and adjoining the village, and
one acre and a-half of mowing-ground for winter food, at the
very moderate payment for house and "cow-gate" of 7l. a-year.
The owner, with the clergyman and farmers of the township,
have united with the labourers in establishing a "cow club,"
the funds of which are employed in replacing any of the
labourers' cows which may die by accident or disease.   The
farmers live at moderate rents, they farm well, and give employment to all the labourers; the rates are kept low, and, by
the wise and judicious management sanctioned by the proprietor, a bond of union and good feeling, such as is too seldom
met with, extends through all classes in this township.   The
rent of land ranges from 30s. to 35s. an acre, the rates and
taxes 4s. 6d. an acre, and the capital required for its proper
cultivation is estimated at 11l. an acre.   A canal and railway,
and a navigable river, traverse the township, which afford ready

access both to the manufacturing districts in the interior and
to the London market.

Crossing the Ouse at Booth Ferry we enter the East
Riding, the least of the three divisions of the county. Its
population is about 200,000, which is not quite a sixth of
the population of the West Riding. It is usually divided
into three districts, — 1. Holderness, stretching from the sea-
coast to the eastern foot of the Wolds; 2. the Wolds, occupying the elevated chalk district in the centre; and, 3. Howdenshire and the Vale of York, extending from the west
side of the Wolds to the rivers Ouse and Derwent, on the
southern and western boundaries of the Riding. It is altogether
an agricultural district, there being no manufactures of great
importance carried on in it.

Beginning our narrative with Holderness, this is a low-lying
country, seldom rising higher than 50 feet above sea-level, and
many thousand acres of it formed by a deposit from the waters
of the Humber, and lying some feet below its level at spring
tides, from which it is protected by embankment. The greater
portion of the district, however, is gently undulating land, very
favourably situated for drainage, part light enough to grow
turnips as a regular crop, but the most of it in its undrained
state too strong for turnips, and still managed with bare fallows.
On the light land the usual four or five course system is practised. On a well-managed strong land farm of 400 acres we
found the course of cropping as follows : — (1) fallow, (2) wheat,
(3) seeds, (4) wheat, (5) oats, (6) beans. To avoid the total
loss occasioned by a dead fallow, winter tares are sown in succession on a portion of the land, which, after these are mown
off in June, is then repeatedly ploughed, dragged, and cleaned,
and prepared for being sown with wheat in October. Another
portion is winter fallowed, well wrought and cleaned in spring
and summer, and in the middle of August sown with mustard,
which is eaten off by the ewes in the end of October, and then
ploughed in for wheat. This is found an admirable preparation

for wheat, besides affording useful feed at that season for the ewes. Previous to ploughing up the clover root, 3 cwt. an acre of salt is sown to kill the slug, and, after it is ploughed and well harrowed, the land is completely pulverised and prepared for the drill by the use of a "shim," which cuts the tough edges of the furrow and greatly facilitates the regular action of the drill. The wheat after seeds is followed by oats, which are always a good crop, and these are succeeded by beans, which are more precarious, but form an excellent preparative for the following fallow or green crops. A portion of the fallow division is cropped with swedes and white turnips, and these are followed by April wheat. The fallow division is either manured with farmyard dung, or limed at the rate of four tons an acre, at a cost of 32s. to 36s. Lime is found an essential requisite in the clay soils of Holderness, and is repeated every twelve or sixteen years. Tile-draining, where tried, has been attended with the greatest benefit; yet not above one-third of this farm is drained. The whole stock on the farm at present is 8 cows, 20 to 30 wintering beasts kept in the yard to tread down the straw, 108 ewes, 13 workhorses, and about 12 young horses.

On Sunk Island, now containing within its banks 4700 acres, and connected with the main land by a bridge, the management is usually — fallow, wheat, oats, beans. The wheat crop is said to yield five to six quarters; oats, eight to twelve; and beans, which are uncertain, two to five quarters an acre. The rent may be 32s., and taxes 2s. to 4s. an acre more. The stock is limited to a few cows for the use of the family and servants, and a lot of wintering cattle to tread down the straw. Labourers' wages are from 10s. 6d. to 12s. a week.

On the warp soil between Hull and Patrington, rape for seed is sometimes grown on the fallow division, in preparation for which the land is well manured; and when the crop is removed in the beginning of July, the land is ploughed and prepared for wheat. The wheat is followed by drilled beans, well horse and hand hoed, after which a second wheat crop succeeds.

This district, as at present cultivated, is altogether dependent on the price of corn, the quantity of stock kept on each farm forming comparatively quite an inconsiderable object. It is natural that the farmers should complain of present prices, yet farms, when they are offered to be let, are still eagerly applied for. Some landlords have made temporary abatements, and some assist their tenants with drainage; but this necessary improvement, the foundation of all others, is still very far behind, as two-thirds of Holderness are believed to be undrained, or very imperfectly drained. Drainage and increased accommodation for live stock appear to be the chief defects in the farming of the district, — drainage, which will render an additional expenditure of manure a profitable outlay; and better housing for stock, in which the increasing breadths of green crops may be consumed with economy both of the substance of the animal by shelter and warmth, of labour by convenient arrangements to facilitate the operations of the feeder, and by preventing waste of the roots and other expensive substances employed in feeding. An extension of this system will of course lead to larger home supplies of manure, and consequently to heavier crops of all kinds; and there can be little doubt that the farmers of Holderness will be compelled, by a lower range of prices of corn, to direct their attention more than they have hitherto done to increasing their returns from live stock, whether in beef, mutton, and wool, or in cheese and butter.

At Patrington the influence of capital and the energy of the manufacturer have converted the quiet of a retired rural village into a scene of bustling industry. Some three years ago about 1000 acres of land here were purchased by Mr. W. Marshall, of Leeds. He instantly began the work of improvement, and nearly the whole estate has already been tile-drained, under the superintendence of Mr. Parkes. About eighteen months ago the foundation of a new and extensive

suite of farm-buildings was laid. The whole is now completed, and occupied by stock, while the barn is flanked by
a goodly row of large wheat-stacks, the produce of the farm.
Straight lines of well-made roads lead to the different fields,
and give easy access for getting home the crops and taking
out the manure. A steam-engine of eight-horse power occupies
the centre of the barn, within whose capacious roof are fitted,
(by Crosskill, of Beverley), in different compartments, every
imaginable machine for converting the corn and vegetable
produce of the farm into food for the sustenance of man and
beast. The thrashing-machine thrashes and dresses the corn,
and then delivers it in the granary, where it is either stored
or passed to the grinding-loft, whence it descends to the lower
storey, after being ground into flour and dressed, and is there
received in sacks and packed aside ready for the baker. From
the end of the thrashing-machine the straw is carried by an
endless web to another loft, where it is passed through the
chaff-cutter, and reappears below as chaff. Other machines,
conveniently arranged, break beans and oats for the horses,
oilcake for the cattle, and linseed for mixing with the cut
chaff. The root-house is situated at one end of the under storey,
opening by large doors to the farm road, by which the roots are
brought in. Elevators, moved by the steam-engine, lift the turnips rapidly up to a turnip-cutter, placed at such a height that
the cut turnips fall into a truck, whence they can be conveyed
on a railroad throughout the whole of the feeding-houses. A
different compartment contains the cooking apparatus, where,
by steam from the boiler, cooked food of various kinds is
prepared for the pigs and other farm stock. Underground is
a great arched tank, into which all the rain-water that falls
on the buildings is conveyed by spouts and pipes. From this
the engine feeds itself with water, and likewise pumps up
water to a tank on the highest part of the barn, whence it
supplies by pipes all the different divisions of the farm build-

x

ings; and, in case of fire, could be readily turned to good
account.   The engine pumps the liquid manure of the farm to
another tank (in rather too close contiguity to this), whence,
by applying a gutta-percha hose, the liquid may be dispersed
over the manure heap.

The cattle-houses are situated in parallel lines, at right
angles to the barn.   Each animal has its comfortable box,
twelve feet by ten, with a supply of fresh water in one corner,
and a manger for its food in the other.   Between every double
row of cattle a railway is placed, on which the trucks, with
their food, are easily pushed along.   A covered manure-pit
receives the dung, when it is carried from the cattle, and
protects it from the influence of rain and weather.   The
mode of cultivation to be hereafter adopted on the farm we
did not learn, as, in the absence of the manager, there was
no one to communicate such information; but, as the same
spirit and energy will no doubt be manifested in the field,
it will soon be necessary to pack the animals more closely
together in the cattle-houses, as the green crops of a farm
of this extent, if principally consumed at home, will suffice for
three times the number of animals for which accommodation
is now provided.   By converting the boxes into stalls the room
at present occupied by one will suffice for three; and, as
all other arrangements may remain unchanged, the charge
for interest will then fall lighter on each.

At the entrance to the farm Mr. Arthur Marshall, of Leeds,
has erected extensive works for the rotting and scutching of
flax.   In these he at present manufactures the crop of 300
acres, but the works are sufficient for 500.   The farmers of
Holderness, however, do not seem to go very readily into flax
culture, and Mr. Marshall is therefore obliged to hire the land,
sow the seed, provide people to weed and pull the crop, and
the farmer then carts it to the works, where it is stacked
till required.   Mr. Marshall pays 8*l.* an acre for the use of the

land, the farmer undergoing no risk of failure of crop, and no outlay for seed or labour. The average yield of dressed flax per acre is five hundredweight, at present worth 70*s.* a hundredweight, besides two quarters of seed, worth 50*s.* a quarter. The employment given in these works, and in the extensive improvements at the farm, has raised the rate of farm wages for men, women, and children, in the parish of Patrington, from 12 to 15 per cent. above that of the surrounding district.

LETTER XXXVI.

YORK, Dec. 1850.

WHILE on the subject of flax culture at Patrington, it may be well to advert to one or two points which are sometimes overlooked by the advocates for its extension in England. An article, of which the wages of labour in the first stage of its manufacture form fully 60 per cent. of the whole value, must be mainly dependent for profit on the scale of wages. The difference of that scale between most parts of Ireland and the East Riding of Yorkshire is not less than 150 per cent., and with such a difference the manufacturer in this district cannot afford to pay the farmer for the raw material so remunerative a price as in Ireland. Other articles of agricultural produce here are of considerably more value per acre than in Ireland, and that likewise limits the extension of flax culture, as the price offered by the manufacturer is not, when compared with the value of other crops, sufficiently tempting. Flax is acknowledged by all who have tried it to be an exhausting crop; yet that of itself is not a good objection to its culture, though it is an excellent reason for a farmer declining to grow it unless he can obtain such a price as will not only compensate him for the usual expenses of cultivation, but will also enable him to restore to the soil that fertility which the flax crop has removed. So long as the difference of wages gives the Irish manufacturer so great an advantage over the manufacturer in the East Riding, the extension of flax culture here is not likely to be very rapid. But there are other counties in England whose soil is equally well adapted for the growth of flax, and where the establishment of flax works would be a great boon to the unemployed labourers. In many parts of Essex and Suffolk, for instance, where labourers' wages do not exceed 1s. 2d. a-day, and where the rates are at certain seasons heavily burdened for the support of the unemployed labourer, the soil is well adapted to the culture of flax. And it is very obvious that a lower range of prices for corn will make this crop and others of a similar character more worthy of the farmer's attention. The farmer was much better paid with wheat at 60s. a quarter than with flax at 8l. an acre; but if the future price of wheat shall range between 40s. and 50s. he may find it his interest to grow flax.

The sugar beet is another article of culture, which is highly deserving of notice. The growth of beet-root for sugar has become of late years the most profitable crop of the farmers in the north of France and Belgium. It has been tried both in England and in Ireland with such a degree of success as may undoubtedly lead to the establishment in this country also of an important agricultural manufacture. On the heavier class of soils in the Eastern counties rents are extremely low; farmers are discouraged by the low price of corn, and are making nothing by their business; labourers are irregularly employed and at inadequate wages, and for that reason poor rates are unusually heavy; and yet on such soils 14 tons of beet to the acre are often grown without the application of a particle of manure. In such circumstances it would be well for the parties interested to inquire into the trustworthiness of the statements which have been put forth on this subject, as they could establish a strong claim on the Legislature for permission to manufacture freely the produce of their own fields now that they possess no exclusive privilege over the produce of the foreigner.

Leaving Holderness, we entered the Yorkshire Wolds from Driffield, crossing them by North Dalton to Pocklington. The general aspect of the country is much more picturesque than the Downs of the Southern counties or the Wolds of Lincolnshire. It presents a very uniform and gradually inclined plane, joining the low ground on the south-east, and rising to its greatest elevation on the north, about 800 feet above sea level, whence it gradually falls southward to an altitude of about 500 feet. The country is all enclosed, generally by thorn hedges; and plantations, everywhere grouped over its surface, add beauty to the outline, while they shelter the fields from the cutting blasts of winter and spring. Green pasture fields are occasionally intermixed with corn, or more frequently surround the spacious and comfortable homestead. Large and numerous corn ricks give an air of warmth and plenty, whilst the turnip fields, crowded with sheep, make up a cheerful and animated picture. The large corn fields, 30 to 70 acres in extent, attest by the evenness of the stubble the correct manner in which the drill man does his part, and the neatly trimmed hedges and well built ricks show that the labourer is expert, and that the farmer likes to have his work well done. Two horses abreast in the field are yoked to a light wheel plough, with which an acre and a half can be as easily turned over, as one acre where the plough, without wheels, is used. The horses walk at a swift, active step, and in the pole-waggon, in which two are yoked abreast as in a carriage, with the driver on his saddle as postilion, the horses trot briskly along.

The farms are from 300 to 1,300 acres in extent, and the farmers are probably the wealthiest men of their class in the county. For this there are several reasons. In the first place, the extent of their farms required that they should be cultivated by men of capital, and these from their social position were men of education and liberal mind, alive to the improvements

which recent times have produced, and with a command of
capital which enabled them to take advantage of those improvements.   Next, though their farms are situated at a considerable altitude, there is something in the calcareous nature
of chalk soils in all parts of the country which renders them
very "true" in their yield of wheat and barley, and very sound
for sheep and cattle.   Then, besides corn, the Wold farmer
depends mainly on mutton and wool, and these two articles
command at present a higher relative price than any other
produce the farmer has to sell.   Combined with these advantages, his land is light and easy to work; it is laid out in
large square fields still more to facilitate operations; it requires
little or no underdraining or ditching; and its bulkiest crops
are consumed on the ground where they grow, without any
expense in hauling home the turnips or carrying back the
manure.   The size of the farms has also limited the competition
for them, though it is somewhat peculiar to Wold farms that
among men of capital there is a greater competition for a farm
of 800 or 1,000 acres than for one of 300.   This arises from the
fact that there are plenty of men of adequate capital to compete
for such farms; and as the same superintendence is requisite for
a small as a large farm, the large one, under equal circumstances, will be the more profitable.   This competition, however,
does not lead to exorbitant rents, for there is an obvious difference in the results of a competition between men, all of
whom have capital and look for a fair return from it, and
that where a man of capital is pitted against a man of straw
who has everything to gain and nothing to lose by getting possession of a farm.

The system of farming pursued on the Yorkshire Wolds is of
two kinds, — one practised on the lower range of Wolds, and
the other on the higher level.   On the lower Wolds the soil is
chiefly a light calcareous loam, from 5 to 10 inches in depth.   A
farm of 500 acres, which we visited, rented at 20s. to 25s. an
acre, is managed in the following manner: — (1) seeds, (2)

wheat, (3) white turnips or rape, (4) wheat, (5) swedes or hybrid turnips, (6) oats, sown out with seeds. We begin with "seeds," which are eaten with sheep, from Midsummer till September, the sheep getting ½ lb. of cake daily. Eight double horseloads of well made yard dung are then spread over the ground, which is ploughed about four inches deep in October, and towards the end of the month drilled with red creeping wheat. It is of much consequence that the land at this stage should be thoroughly consolidated; and that is effected by using the presser before sowing the wheat, or by rolling the land completely, after it is sown, with either Cambridge's wheel roller or Crosskill's clodcrusher. If thought requisite, the ground is again rolled in spring, but the crop receives no farther cultivation till harvested. The stubble is then ploughed 6 inches in depth, in preparation for the next crop. In spring the land is ploughed, harrowed, and cleaned, and towards the end of June drilled with white turnips and rape; along with which the turnips receive 12 bushels of dissolved bones, and the rape 8 bushels per acre. The rape is eaten on the ground, between Michaelmas and Martinmas, by sheep, which also receive ½ lb. of cake daily. As the crop is consumed, the ground is ploughed and sown with red wheat. After that the white turnips form the food of the sheep till February, and on this division white wheat is sown in February. The wheat stubble, in preparation for swedes, receives a dressing of dung over the greater part of it, and 12 bushels of bones in addition, when the seed is sown in May. When the white turnips are finished, the sheep, last of all, are put on the hybrids and swedes, and at this stage they receive ½ lb. of cake from Candlemas to the end of April. The land as it is ready is ploughed, and sown most frequently with oats, though barley is occasionally taken on a portion. One half of this division is also sown with the following mixture of seeds: 10 lb. red clover, 10 lb. white, 7 lb. trefoil, and 3 lb. rib grass; the other half receives 14 lb. white clover, 14 lb. trefoil, and 7 lb. rib grass per acre. Red clover is thus repeated only once in 12 years. Ryegrass is never sown in either mixture, as it is believed to produce or encourage twitch. Besides the manure already mentioned, 80 to 100 acres of the wheat, on the poorer land, are dressed in spring with 6 quarters of soot, or 2 cwt. of guano per acre. The yield under this management is 28 to 32 bushels of wheat, 48 to 64 bushels of oats, and 32 to 40 bushels of barley per acre. The produce of 200 Leicester ewes kept on the farm is annually sold at 16 to 18 months old, averaging 220 in number, and this year fetched from 40s. to 45s. each. Four cows are kept for the use of the farm, and 15 beasts are bought in autumn, and fed fat during the winter in the farmyard. They consume on an average half an acre of turnips and 15 cwt. of oilcake each. Some Wold farmers sell their sheep at 12 to 13 months old, carrying them on with cake during the whole period. This season 28s. each was the price of such stock, besides the yield of wool — 7 lb. each, at present worth 1s. per lb.

On the higher Wolds the farms are larger on the average; the rents range from 12s. to 20s. an acre, and the farming on the whole is neither so good nor so profitable to the farmer. The quality of the produce is inferior, oats being from 3lb. to 4lb. a bushel lighter, the sample of barley low priced, and the extent of wheat in proportion to other crops more limited, while in quality it is also inferior. The best farmers, from the difficulty of growing red clover, now alternate the four-course with the six, which begins with (1) oats after seeds, (2) turnips or rape, (3) wheat, (4) swedes and hybrids, (5) barley, (6) seeds. Manure is applied to the different green crops as has been already detailed on the lower Wolds, but not in such liberal quantity nor so often repeated. Wheat is seldom taken oftener that once in the course. The yield may be 24 bushels of red wheat, 36 bushels of barley, and 48 bushels of oats per acre. The soil is often from 18 inches to 2 feet in depth, and frequently with an admixture of clay. It is thus more expensive to work than the light friable loams of the lower Wolds.

Sixty or seventy years ago, when the Wolds began to be cultivated, the system adopted by the farmer was a bare fallow, followed by two corn crops; and when that had been repeated as long as any thing would grow, the land was permitted to clothe itself with such natural herbage as it produced, and left for a few years to regain by "rest" the power of reproducing corn. When turnips and seeds were introduced, this system was in some degree departed from; and when, in addition to these, bone manure became known, a regular course of husbandry developed itself. The four-course then was considered the rule of good farming; but further experience has shown the impropriety of binding a farmer to any continued course of cropping, however excellent that course may for a certain period have proved. In the Wolds of Yorkshire, as in the best farmed districts of Norfolk, where indeed the four-course had its origin, all the best farmers have been compelled to change. Red clover refused to grow every fourth year; it was found precarious even every eighth year, and can now be calculated on as a certain crop only once in 12 years. Turnips were therefore substituted for a portion of the seeds, and these in their turn are found too frequently repeated on these light lands. To obviate this new difficulty peas have been introduced, and the turnip crop is thus repeated at greater intervals. At the same time it has been found that the higher system of farming, now within the reach of the man of capital, has enabled him to grow wheat much more frequently than he formerly could do with success; while on many of the Wold farms he finds that he can not only grow as many bushels per acre of wheat as of barley, but he can repeat the wheat crop in alternate years with less exhaustion to the land than if he followed his green crops with wheat and barley alternately. In truth, each year's experience is adding to our knowledge, and extending the means at our disposal for improving the culture of the soil; and the landlord who continues to bind his tenant down to a prescribed routine from which he must not under a penalty deviate, inflicts upon him a very serious injury, without any corresponding advantage.

The third district into which the East Riding is divided comprises Howdenshire and the vale of York, extending from the western side of the Wolds to the rivers Derwent and Ouse. This country embraces a great variety of soil, from the rich warp lands along the Ouse to stiff cold clays and thin moorish sand. In some places there is much difficulty experienced in obtaining the consent of adjoining proprietors for the deepening of outfalls ; and at no great distance from Howden the drainage of a considerable tract has to be carried several miles, which could, with a better outfall, be got rid of by a much shorter course, if any means existed by which owners could be made to agree.

The cold infertile clays of this district are exceedingly unprofitable. They are let at 12s. to 15s. and 20s. an acre, but, being generally undrained, no good can be done in them at any rent. The old system of two crops and a fallow continues to be the common course; but the scanty produce and the low price compel the farmer to restrict his labour, and thus each year adds to the utter hopelessness of his condition. Many farmers who have still something to lose are giving up their farms, several have already emigrated, and the number who remain on these undrained clays will soon be limited to those who have no means to take a farm elsewhere. To meet the difficulty some landlords are making abatements of rent, while a few are beginning in earnest to drain the land. The owners of this description of land have no chance of safety except in the vigorous adoption of this course. Where they have capital of their own, they cannot make a better investment of it than in laying these lands dry, and so giving their tenants some chance of a living for themselves after they have paid their rents; and where they have not capital, they should not for a moment delay to avail themselves of the Government drainage loan, as without drainage in one way or other their estates will soon be valueless. When the land is drained, we should advise that it be gradually laid to grass; under which, by judicious top dressing of bones and other substances, it may probably yield a better rent and less precarious return to the tenant from dairy produce and stock than it is hereafter likely to do under corn crops.

The sand land is of various quality, — some of it very fertile and highly remunerative, and much of it barren and waste. It is generally situated at a low level, well sheltered, and capable of great improvement. Near the rivers and lines of railway, where manure can be readily procured, potatoes are grown on this soil extensively, and more free from disease than on the richer warp lands. Chicory is also cultivated to a considerable extent. A very great proportion of the land is still undrained, although where drainage has been judiciously executed, it has been particularly beneficial. Enjoying many advantages of position, and being light and inexpensive of management, the sand lands of the East Riding exhibit in their culture a less degree of spirit on the part of both owners and occupiers than any other district of the county. There are many eminent exceptions, but by contrast they still more clearly prove the rule. Small inconvenient inclosures, with much hedgerow timber, are serious obstacles to good farming.

There is no want of competition for farms in the East Riding, except on cold clays and springy undrained moor. Many, however, on these two kinds of soil are being vacated. A desirable farm in any other part of the Riding is at once applied for, if from any chance it becomes vacant. Farms are never let by tender, nor are the landlords at all prone to raise their rents by unduly taking advantage of competition. The best man is chosen, and the rent named to him. Indeed, so great is the confidence placed by the tenants in some landlords, that they trust entirely to the latter to fix the rent. It must not, however, be concluded that farming, as at present conducted, and with present prices, is a remunerative business, merely because

good farms still find tenants and old tenants continue to hold
their land. The low prices of stock, seed, and other articles
which an entering tenant must buy, enable him to invest his
capital with advantage ; and the very same reasons influence a
tenant who at present holds land to continue with it, as he must
otherwise part with his stock and crop at a disadvantage. Corn
rents are unknown in the district, and leases equally so. The
farms are all held from year to year, and there is no wish on
the part of the tenants for any other arrangement. In practice
this scarcely ever leads to inconvenience; and in many cases
farming of the most spirited character may be met with where
the tenant has no other security than his confidence in his
landlord. A rather notable instance occurred quite recently,
which shows that this security cannot always be depended on.
By the death of the proprietor of one of the most extensive and
valuable farms in the East Riding the estate changed owners.
The tenant, a wealthy man, who had occupied the farm for
upwards of half a century, and to whose good management it
owed much of its increased value, did not find himself comfortable under the new arrangement, and resigned his farm.
It was immediately re-let to another farmer of capital and experience, at an advance of 350l. per annum.

The condition of the labourers throughout the East Riding is
very satisfactory. They are well employed, fairly paid, and
comfortably lodged.

The following table shows the rate of wages and prices in this
district at three different periods: the first being taken from
Mr. Leatham's Report in 1794, the second from Mr. Strickland's
in 1811, and the third from our personal inquiry in 1850 : —

RATE OF WAGES AND PRICES IN THE EAST RIDING FOR THE
YEARS 1794, 1811, AND 1850.

	1794.	1811.	1850.
Farm servant with board -	£10 to £13	£20 to £26	£14 to £16
Day labourer without board, average of year - - -	2s.	3s. 3d.	2s.
Thrashing wheat, per quarter	2s. 2d. to 3s. 4d.	5s. 6d. to 7s.	hand 3s. 6d. horse 1s. 7½d. steam 7½d.
Reaping wheat, per acre -	6s. to 8s.	10s. to 11s. 6d.	7s. 6d.
Wheat, per bushel - -	5s. 9d.	9s. 6d.	5s. to 5s. 6d.
Barley, ditto - - -	4s. 1½d.	3s. 9d.	3s. 3d.
Oats, ditto - - - -	2s. 6d.	2s. 9d.	2s.
Beans, ditto - - -	4s. 6d.	4s. 9d.	3s. 9d to 4s.
Wool, per lb. - - -	—	10d.	1s.
Beef, ditto - - - -	3¼d.	7½d.	5d.
Mutton, ditto - - -	4d.	8d.	5d. to 6d.
Butter, ditto - - -	1s.	1s.	1s.
Milk, per quart - - -	3d.	2d.	2d.

It does not appear that the high prices of 1811 compensated
the farmer for the increased cost of production; or, with wheat
at 9s. 6d. a bushel, that he was less distressed than at present.
Mr. Strickland says in 1811, —

"If means are not adopted to relieve the farmers, either by lowering their taxes and the prices of labour, or by increasing the value
of their produce, by destroying monopolies and giving them an open
market, they must shortly be found unable to answer the demands
made upon them.   The advance in the profits of agriculture has by
no means kept pace with the advance in its expenses; and farmers
had a better prospect of realising a competence seventeen years ago
than they have at present."

LETTER XXXVII.

YORKSHIRE — *continued.*

COSTS OF CULTIVATION, AND PRODUCE, ON EIGHT FARMS IN YORKSHIRE AND LINCOLN — THE NORTH RIDING DESCRIBED — THE GREAT DEGREE OF ATTENTION AND SKILL NECESSARY TO SUCCESSFUL BREEDING OF STOCK — THE MESSRS. BOOTH — FARMING AT BAINESSE — THE PRACTICE AND ADVANTAGE OF AUTUMN FALLOWING — FEEDING OF STOCK — PRODUCE OF CROPS — PROPORTION OF SWEDES TO OTHER TURNIPS TOO SMALL — THE VALE OF CLEVELAND INJURED BY BEING BROKEN UP FROM GRASS — GOVERNMENT DRAINAGE LOAN — TENANT RIGHT IN THE WEST RIDING — DOES NOT PRODUCE SUPERIOR FARMING — INDISTINCT CHARACTER OF THE AWARDS OF VALUATORS.

NORTHALLERTON, Nov. 1850.

THE different items which make up the costs of cultivation are found to vary considerably, influenced partly by the nature of soil and climate, partly by the system of husbandry pursued, and chiefly by the greater or less command of capital and energy possessed by the cultivator. It is difficult to ascertain with precision the amount of these costs, and any trustworthy facts on this point are, therefore, the more valuable. We have been favoured with such information as has enabled us to compile the following table, showing the sums which make up the cost of cultivation, with the yield of corn per acre, in the after-mentioned districts.—(*See next page.*)

The column of " total cost " does not include interest of capital, insurance, or tenant's profit; and under the head " yield" is included only the corn, as we had no sufficient *data* by which to estimate accurately the annual produce of stock. The table will, however, we trust, supply to practical farmers and landowners many interesting *data* for comparison and future reference, by which they may test their own expenses

TABLE SHOWING THE COSTS OF CULTIVATION, AND THE YIELD OF CORN PER ACRE.

DISTRICTS.	Rent. s. d.	Tithe. s. d.	Public and Local Taxes. s. d.	Manual Labour. s. d.	Horse Keep. s. d.	Tradesmen's Bills. s. d.	Manure purchased. s. d.	Cake, &c. purchased. s. d.	Total per Acre. s. d.	Wheat.	Oats.	Barley.	Beans.	Proportion in bare Fallow.
Howdenshire Farm, East Riding — strong clay, 149 acres arable, 25 acres grass -	13 9	3 2	2 4	24 2	10 5	5 8	5 6	1 9	66 9	20	28	24	—	1-6th.
Holderness — fertile clay on red clay sub-soil, 230 acres all arable -	27 0	Free.	3 9	27 9	9 9	2 2	1 8	6 8	78 2	32	48	—	20	1-5th.
Holderness — 96 acres arable, 17 acres grass	29 0	1 0	4 8	26 4	12 2	5 0	4 5	5 5	89 0	35	55	—	19	None.
Yorkshire Wolds — 218 acres arable, 46 acres grass -	29 6	Free.	5 3	22 5	8 5	4 5	4 9	3 4	78 1	26	48	32	20	None.
Yorkshire Wolds — 400 acres, all arable -	25 0	Free.	5 3	29 0	11 2	4 7	8 6	11 0	84 6	32	48	36	28	None.
Lincolnshire Wolds — 1,017 acres arable, 500 acres grass -	35 0	2 3	1 11	37 4	11 7	7 0	4 11	5 6	105 6	36	64	44	—	None.
Lincolnshire Wolds — 524 acres arable, 130 acres grass -	21 4	Free.	2 3	25 2	10 6	6 9	3 9	7 6	77 3	26	56	36	—	None.
Warp Land Potato Farm —400 acres arable -	30 0	Free.	4 6	32 6	13 10	4 0	10 0	10 0	104 10	33	64	36	—	6 tons potatoes; previous to 1847 10 tons per acre.

and produce, and ascertain in what respects they excel or fall short of the examples here given.   It is proper to explain that the three first columns are averaged on the whole extent of each farm, arable and grass; while the rest are charged on the extent of arable land only.   This is not strictly correct, but it appeared the most just criterion we could adopt, as the comparative extent of grass land on each farm was so exceedingly variable. The figures may be relied on as a statement of facts as they exist — not estimates of what might be, but what they at this moment are.

The North Riding is bounded on the south by the East and West Ridings, on the north by the river Tees, which separates it from Durham, and is traversed by the York and Newcastle Railway from York to the borders of the county near Darlington. The central and northern portions of the riding, situated along the valleys of the rivers, contain the greatest extent of valuable land.   To the eastward lie the Yorkshire moors, comprehending an extent of 400,000 acres, and rising from 1,000 to 1,500 feet above the sea level.   On the west the country rises to the highlands on the borders of Westmoreland, with rich valleys of pasture land skirting the streams which drain that limestone country and flow through the picturesque dales of the North Riding.   Much of the best land in the lower portion of the district is kept in permanent pasture; and in the neighbourhood of Northallerton, Catterick, and along the southern bank of the Tees, are found the most celebrated herds of short horn cattle which now exist in England.   The opposite bank of the Tees, which gave the name to the breed of Durham cattle, has lost its principal men, — the Collings, — and the best herds may now be said to be confined to the south side of Teesdale.   Men are still to be found there who have been bred from their childhood to study the peculiarities of form and symmetry which, combined with early maturity and great weight, have given the improved Short horn its celebrity.   Seldom leaving home, often the first to see their stock in the morning, and the last to visit them

at night, making the health of each individual of the herd
a study, and enabled by constant attention, and particular
management, to encourage the development of such points as
they think requisite, while everything else on the farm is made
subordinate to the stock — these men have acquired a fame
which is the result of such earnest application, and cannot long
be maintained without it.   They succeed; and to all parts of the
United Kingdom is diffused, from the hands of not half a dozen
men in the North Riding, the blood which has improved, and
continues to improve, the native breeds of every district into
which it is introduced.   Two, three, or four hundred pounds
for a bull is no uncommon price; and a cow of rare form and
breeding has been bought by a farmer for 300 guineas.   Ireland
and Lincolnshire borrow for a season the best animals which
leave the district; and the Messrs. Booth have at this moment
bulls which bring them in 100*l.* a year each, and are let to the
same parties at that rate for a succession of years.   But these
prizes are only to the most successful; for many, tempted by
these and similar rates, try the system without counting the
cost of patient study, constant application, and liberal outlay
by which such success has been achieved.   It is not merely the
first outlay, however lavish, that will place the beginner in the
rank of a first-class breeder of Short horns.   He must be prepared to sacrifice every other consideration on his farm to their
welfare; and after he has collected his herd, and fed and
watched them with the utmost care, he must stand the risk and
uncertainty attending their management.   If too fat, they cease
to breed, or they produce dead calves; if too lean, they lose
caste, and their produce sells at second rate prices.   One of the
most experienced men in the district — himself an eminent
breeder and first-rate judge — informed us that one season 34
of his high-priced and high-bred cows missed having calves;
and so great are the risks attending this business, that it is every
year narrowing itself into fewer hands.   Men of station and
wealth embark in it frequently as a hobby, and some, like the

late Lord Spencer and the present Lord Ducie, are successful.
But such cases are exceptional.

The farm of Messrs. Outhwaite, of Bainesse, near Catterick,
may be taken as a very favourable specimen of agricultural
management in the North Riding. It is bounded on one side
by the river Swale, and its soil on the lower division is a good
friable turnip loam on a thirsty gravel, and on the higher side a
deeper and stronger soil, now found, after being thoroughly
drained, the most valuable land on the farm. It comprises 461
acres, eighty of which are permanent grass. The four-course,
extended into a six by introducing beans instead of a portion of
the seeds, is the system of farming adopted, with this peculiarity
— that oats or barley are taken on the clover leys, wheat and
barley after turnips. Wheat is also sown after beans and
potatoes, Spalding's red being sown from October till the 20th
of November, beginning with seven pecks of seed and increasing
to nine. Early in spring, as the swedes are consumed, the land
is ploughed and drilled with Hunter's white wheat; the sowing
of which continues till the middle of March, after which barley
is taken on such land as has not at that time been cleared of
the turnip crop. The oats and barley are also drilled in at
nine inches apart, and the whole of the corn crops on this
farm are repeatedly horsehoed in spring. Beans are sown
upon manured land with the drill, and are likewise well
horsehoed.

The great aim in the culture of the farm is the early preparation of the land intended for the turnip crop; to this all
other work is postponed after the corn crops have been secured
in autumn. The stubbles are then stirred in one direction
by Biddle's "scarifier," the sharp-pointed tines being used
in this operation, and the ground torn up to the depth of five
or six inches. After the field has been gone over once, the
"scarifier" is fitted with the broad share tines, and made to
cross the former stirring at right angles, thus tearing the ground
to pieces, and disengaging the stubble and roots of weeds and

twitch, which are drawn together on the surface by the harrows, then gathered by the horse rake, and laid in a heap to be carried home for littering the cattle yards. The land, now thoroughly pulverised, is ploughed with a clean deep furrow, and in that state is left exposed to the influence of the weather till spring, when it receives one furrow more, and is found in fine condition for vegetating the seed of the turnip crop. The theory on which this early culture is recommended is, that twitch, immediately after harvest, is comparatively weak, and has not extended its roots far beneath the surface; but as soon as the corn crop is removed, and the twitch so permitted to grow without obstruction, it spreads rapidly along the surface, and penetrates deeply beneath it; and every week that it is left undisturbed renders its extirpation more difficult and expensive. Tear it up early, and the seedlings are at once shaken out entire from the tender soil; leave it to strike deeper root, and every broken fibre that remains strikes afresh, and, gaining strength throughout the winter and early spring, gives the farmer at that busy season the expense of a second fallowing. The advantage of this early preparation is attended with this further benefit, that only one furrow is requisite in May, and, the ground not being deprived of its moisture at that season, the turnip seed is almost sure to vegetate at once. So successful have the Messrs. Outhwaite found this management, that they are now enabled to sow their entire extent of turnip land with swedes; and the preparation of the soil, besides being so much more early and effectual, is not nearly so expensive as under the common system.

The swedes are sown on the ridge, twenty-eight to thirty inches apart, eight loads of well-rotted farmyard dung and $1\frac{1}{2}$ cwt. of guano per acre being previously applied and covered in on the weakest land, on which the crop is afterwards eaten with sheep. On the better land 14 tons of dung and $1\frac{1}{2}$ cwt. of guano per acre are applied, the crop in this case being all drawn for consumption by cattle. The sheep are penned upon the turnips,

and receive ¼lb. of oilcake each daily. In the beginning of
May, after being shorn, the most forward hoggets—being those
bred on the farm—are sent to market, and weigh 24 lb. per
quarter. The rest follow as they get ready, after being a month
or six weeks on clover. About one sheep per acre is wintered
on the farm, and seventy beasts, one lot after another, are
fattened in the yards and stalls. The oxen are fed in sheds and
yards loose, six or eight together; the heifers are tied up in
stalls. They all receive cut swedes daily, and for the last ten
weeks 7lb. to 8lb. each of oilcake and bean meal mixed together.
1,400 to 1,500 bushels of beans and barley, the produce of the
farm, are annually consumed on it, the whole of the beans being
so applied, and all the barley which does not sell for malting.
Twelve work horses are kept, and they consume all the oats
produced. The wheat crop scarcely averages thirty-two bushels,
the oats sixty-four, and the barley forty-eight bushels an acre.
The tenants have no lease nor tenant right, and their rent and
tithe together amount to 1,200*l.*

Such is a specimen of the best arable farming in the most
fertile district of the North Riding, the Messrs. Outhwaite
having taken a prize offered for the best managed farm within a
circle of twenty-five miles round Richmond. There are many
other farms in that district which may equal them in annual
yield; but none, that we saw, come near to them in their extensive culture of the swede—confessedly by far the most
valuable and nutritious of our turnip crops. The general proportion of swedes to white and other soft turnips is quite
inconsiderable, and the scanty crops even of soft turnips, which
are everywhere to be seen in this district, do not bespeak very
liberal management. The quantity of manure applied is too
limited, but the land is considered high rented, and the landlords do not in many cases afford their tenants such sympathy
and aid as they think themselves justly entitled to with present
prices. In the neighbourhood of Catterick and Richmond,
several farms have in consequence been given up, the landlords

refusing to submit to any abatement; and one nobleman is said
to have as many as eight or nine farms at present on his hands.
On other estates in the same neighbourhood there are no farms
vacated, the land having been moderately let, and the landlord,
(as in the case of the Duke of Northumberland at Stanwick,
where the rent of good land runs from 18s. to 36s. an acre,
tithe free), also draining the farms for his tenants at a charge
of 5s. an acre, and building them suitable accommodation for
their stock without making any charge for that outlay.

The vale of Cleveland, comprehending the low-lying district
which extends from the York moors to the river Tees, forms
the next prominent feature in the agriculture of the North
Riding.   It is generally a cold strong clay, resting on the blue
lias, by far the greater portion of it undrained and badly
farmed.   Formerly this district was celebrated for its cheese
and horses; but the latter are now scarcely to be met with as a
distinctive breed, the farmers having been tempted to part with
their brood mares at high prices, and the best stock having thus
in process of time been taken out of the country.   Much of
the old grass land has been broken up to support the otherwise failing rents of needy landlords, and as it was then
"called on" as long as it would carry anything, without being
either drained or manured, it is rapidly passing into the same
sterility as the other parts of the farm for whose rescue it was
broken up.   The scanty stock thus yearly becomes diminished;
there being little or no green crops cultivated, the hay and
straw consumed by half-fed cattle is converted into wretched
manure: the crops annually fall off in produce as well as price:
and the only kind of produce — milk and butter — that keeps
its price, is got in lesser quantity from an abridged extent of
grass land.   The system of management generally followed
is what is here denominated "two crop and fallow," a bare
summer fallow being given every third year, followed by wheat,
and that, after receiving a dressing of such manure as the farm-
yard produces, is succeeded by oats.   To vary the crops, beans

are sometimes taken on part of the land instead of oats; and
occasionally a portion of the oat crop is sown out with clover.
On some farms one half remains in permanent grass, and on
these the farmers are enabled to keep more stock and to work
their land at less cost comparatively. But generally the
quantity of stock is inconsiderable; and the quality as well as
the quántity of manure being inferior, from the deficiency of
winter food, the arable land is year after year becoming less
productive, and the tenants, as a matter of course, less able to
meet their engagements. Many farms are being given up, and
even with liberal abatements of rent it is impossible for the
tenants, under such circumstances, to go on. Like all cold wet
clays, there can be no chance here for either tenant or landlord
without effectual drainage. With that and some timely support
there is much to encourage an enterprising farmer. Milk sells
at 2*d.* a quart in the populous towns along the river and on the
coast, and the demand for dairy produce at remunerating prices
is constantly increasing. The farms are small, and the tenants
generally not an enterprising class, nor possessed of adequate
capital for the extent of land they occupy.

Many landlords in Cleveland are availing themselves of the
drainage loan, and within the next two years a very great
extent of drainage will be accomplished; indeed, so much alive
are the proprietors of land in Yorkshire to the necessity of this
operation as the foundation of all other agricultural improvements, that they have already applied for a greater aggregate
sum than has been allotted to this county, and each individual
is therefore restricted to a certain proportion of the sum for
which he has applied. The present time is most favourable,
in every view, for proceeding with the work. The general
employment which it diffuses over an agricultural district
comes at a peculiarly appropriate time, when the farmer's
necessities, in many cases, lead him to dismiss part of his
labourers. There is likewise a great reduction in the cost of
pipes for drainage, arising from the introduction of better

machines and greater skill in the workmanship. Two-inch pipes, which very recently in this Riding cost 25s. per 1,000, can now be supplied for 15s. Drains are made from 3 to 4 feet in depth and from 18 to 36 feet apart, according to the nature of the soil, and at an average expense altogether of from 3l. to 4l. per acre. A very few years ago the same operation would have cost double these sums. The Government loan is repayable in 22 annual instalments of 6½ per cent., which repays both principal and interest. A few landlords charge their tenants 5 per cent. of this annual sum, and themselves pay 1½. Most frequently the tenant is bound to pay the whole, and, in addition, to cart the tiles free of charge. And we are sorry to say that more than one instance exists in Yorkshire, where the landlord charges his tenant 7½ per cent., thus putting into his pocket 1 per cent. besides securing a permanently higher value for his land by an outlay to which he does not contribute a single farthing. This grasping conduct, so utterly at variance with the intention of the Legislature, is quite unworthy the character and position of a respectable landlord.

In our letter from the West Riding we referred to a custom, existing in the southern part of it, of compensation to the outgoing tenant for certain acts of husbandry and unexhausted improvements, or, as they are more briefly termed, "tenant right." In no other part of Yorkshire have we met with this custom; and we have not the slightest hesitation in saying, that any dispassionate observer who will compare the state of farming in that part where it exists with the general average farming of the East and North Ridings, where it never has existed, will at once affirm that it has not produced a better style of farming. On the contrary, the farming of the southern division of the West Riding is not to be compared in any single point with that of the wolds of the East Riding, or the better farmed lands along the Ouse and Humber, or in Holderness, or the North Riding. And we were assured by an extensive farmer of much experience in the West Riding,

who has himself had to pay this tenant right, and is therefore
familiar with its operation, that it leads to frauds of every kind,
— which in truth cease to be counted frauds, inasmuch as the
party who suffered at his entry feels himself justified in retaliating on his successor.   Instances have been known of toll-men
being bribed to sign for false quantities of manure as having
passed through their bar ; and it is quite common to secure the
services of a valuator not according to his character for skill
and justice, but mainly in reference to his skill in getting up
and carrying through a "good" valuation.   One absurdity of
the system is, that five "dressings," or preparatory ploughings
and harrowings, are as a matter of rule charged against the last
turnip crop, though very possibly two or three such dressings
at the utmost are all that a skilful farmer would himself
bestow.   So sensible are the valuators of the haphazard nature
of their awards, that they, in rendering their account, specify
each item for which a charge is made; but, to prevent unnecessary questions, they put down no sum opposite to that
item, contenting themselves with a single and lump sum for the
whole at the last.   In what other branch of business would
such a blundering system be tolerated ?   The best farmers are
now desirous of having certain points restricted, and believe
that it would be a benefit to their class if the landlords would
purchase up and put an end to many of its vexatious exac-
tions.   An entering tenant who has to pay down in cash a
considerable portion of his whole capital for a doubtful benefit,
the return from which he cannot reap till he himself quits
the farm, is greatly crippled in his means at the very outset;
and it is notorious that some farmers are become so expert in
the trade that they make a business of taking a farm for a few
years, and then quitting it with a high valuation.   We repeat
that to whatever other consequences this custom may lead,
whether to landlord or tenant, it has not in the southern division of the West Riding conduced to superior farming.

LETTER XXXVIII.

DURHAM.

RAPID INCREASE OF POPULATION — APPEARANCE OF THE COUNTRY — GREAT
CAPITAL INVESTED IN COAL MINES — SLOW PROGRESS OF AGRICULTURAL, AS
COMPARED WITH MINING ENTERPRISE — COURSES OF CROPS — DENTON FARM
— TWO-CROP AND FALLOW SYSTEM DETAILED — REDUCTION OF RENT NOT SO
EFFECTUAL HERE AS AN INCREASE OF CROPS — EXPENSE OF DRAINAGE —
COAST-SIDE FARM — EXCELLENT AND PRODUCTITE MANAGEMENT — PRO-
DUCE OF CROPS — STOCK — ADVANTAGES OF THE SYSTEM.

DURHAM, Dec. 1850.

THE county of Durham is not much more than half the area of
the North Riding of Yorkshire, while its population is a third more
numerous. In 1841 it exceeded 324,000, and, from the great
extension of the collieries since that time, there has no doub
been a rapid increase of population, the increase during the
preceding ten years having been 27 per cent. — with one exception the greatest in England. Its surface is of an irregular and
very hilly character, except along the north bank of the river
Tees, from which, widening towards the sea, a considerable
tract of rather level country extends. Its peculiar conformation
is favourable to picturesque beauty; and though to the railway
traveller its chief features may seem a succession of engine
chimneys, lines of coal waggons, great fires of coal waste, and
numerous shabby tile-roofed villages — the roads through the
different valleys of the county skirt along streams, not always
limpid, but often shaded by venerable woods, encircling the
ancient feudal castles of the nobility, from the highest tower of
which they still display their banner. Raby, Wynyard, Lambton,
Lumley, Ravensworth Castles, and others, occupy sites of great
beauty, generally placed half-way up the hill, backed by wooded

heights, and commanding a prospect of the cultivated valleys beneath them.

The geological features of the county comprise the red sandstone on the north bank of the Tees, next the magnesian limestone, extending from Darlington in an easterly direction to Hartlepool, and thence north along the coast to Tynemouth. Within this comparatively narrow strip lie the valuable coal measures, which commence near Staindrop in Durham, and extend northwards to the mouth of the river Coquet in Northumberland. The whole western boundary of the coal measures is formed by a tract of millstone grit; and beyond this lies the mountain limestone, the green hills of which yield excellent pasture for sheep. The larger proportion of the arable land of the county is of a tenacious character, sometimes a thin infertile clay, and nearer the banks of the rivers a deep strong loam. Along the coast the soil is of a more friable nature, yielding sound crops of excellent potatoes, turnips, and other vegetables, which find a ready market in the neighbouring seaport towns.

Unlike Lancashire and the West Riding, the coal fields of Durham have not led to the establishment of a great manufacturing population, the coal being wrought principally for export to London, and to the east coast and continental ports. In this business large capitals are employed, partly by speculators, but principally by the great landowners themselves, whose incomes, like those of Lords Londonderry and Durham, are chiefly derived from coal. To give an idea of the capital employed in this business, we may mention that Lord Londonderry, for the convenience of his own trade, constructed the harbour of Seaham at a cost little, if at all, short of 300,000*l.*, and to this harbour he has about forty miles of railroad leading from his different collieries. The return from capital invested in working coal has been so much more remunerative than land, that improvements on the latter have been comparatively neglected, and the skill and enterprise so abundantly lavished below ground form a very marked contrast with the absence of those

qualities and the evident defect of capital everywhere too conspicuous on its surface. To show the different progress made
in the two departments within the last eighty years, we may
mention that the system of farming practised in 1770, viz.,
"two crop and fallow," as described by Arthur Young, with
a yield of sixteen bushels of wheat, and thirty each of oats and
barley per acre, is exactly the common practice of the present
day, with a yield rather diminished than increased. Let us
turn to the coal trade, and see its progress. For the coal waggon
roads then "the track of the wheels was marked with pieces of
timber let into the road, for the wheels of the waggons to run
on, by which one horse is enabled to draw fifty or sixty bushels
of coals." These roads led to the shipping ports. As in process of years the trade increased, "Keels" were employed, with
light draught of water, enabling them to penetrate many miles
up the tidal rivers to wharfs at no great distance from the
collieries, to which the wooden tracks (gradually superseded by
iron rails) conducted. The progress of business, and the discovery
of steam power have in their turn nearly done away with the
keels, and the consequent expense of transhipment; and thousands of tons of coal are now more cheaply and expeditiously
taken from the mouth of the coal pits and laid in the holds of
sea-going vessels, than hundreds, or probably even tens of tons,
were slowly and laboriously dragged by horses along the wooden
tracks described by Arthur Young.

The four-course system of husbandry is common on the
friable soils on the northern side of Teesdale, though on many
farms of this description, owing to the indolence of the farmer,
a large proportion of the green crop division is managed with
bare fallow. The stiff land of the county, which, as already
mentioned, forms its greatest proportion, is managed on the
"two crop and fallow" system. The coast lands, and the farms
within the influence of such towns as Sunderland, Shields, and
Newcastle, are probably the best managed in the county. We
shall describe the practice on a farm in each of these various

localities, in order to present to the reader, as concisely as possible, a view of the present state of farming in Durham. The first and last examples are confessedly much in advance of the average in their respective neighbourhoods.

The farm of Denton, some six miles west of Darlington, was purchased by the Duke of Cleveland, two years ago, from the late Mr. Culley. It has been occupied by the present tenants, Messrs. Heslop, for the last fifteen years, and contains 490 acres of land — 380 of which are under cultivation, and about 110 in grass, 80 acres of which are prime feeding land. The four-course, lengthened to a six by introducing a pease crop, on account of the uncertainty of red clover, is the system adopted over the farm, a portion, which is strong land imperfectly drained, being annually in bare fallow. The Northumberland five-course has been abandoned, as the second year's grass was found greatly to encourage the growth of "twitch." Two-thirds of the turnip land are now sown with swedes, which receive, in the ridge, ten tons of dung and two cwt. of guano per acre. These are all drawn for consumption in the yards and boxes. On the weakest land white turnips are sown, which are eaten on the ground with sheep. Wheat follows the pease stubble, the land receiving ten loads of dung per acre; it is also sown after bare fallow, and after white turnips, and such part of the swede land as is cleared early enough in the season. Barley and oats are sown in spring after swedes and clover ley. Wheat yields from 32 to 40 bushels, barley 40, and oats about 60 bushels an acre. A changing stock of cattle and sheep are kept on this farm, part of the cattle being bought in April, and sold fat at Christmas; and a second lot, kept on till June and July, are sold into the West Riding, where a good demand at that time exists for the "Feasts" which are then held in the manufacturing districts. Four hundred half-bred Leicester and Cheviot hoggets are bought at Stagshaw in April, fed on "seeds" during the summer, and finished on white turnips in autumn, whence they are sold in lots as they become ready. Between wool and increase of carcase these sheep leave for eight months' keep 20s. each on an average.   They receive no cake or extra feeding.

The cattle are grazed in summer on the old grass lands, and part fed in open sheds and courts, and part in loose boxes in winter.   In the latter the cattle are found to make decidedly more progress, with less waste of food than in the open yards and sheds.   But the boxes, though very commodious, are extremely costly, 300*l.* having been spent in the construction of boxes, with turnip and straw houses, for twenty cattle.   Water is provided by pipes for each animal, and in future it is intended to occupy the boxes in summer as well as winter with house-fed cattle.   This farm was taken in 1835 by the present tenants at 900*l.* a-year, wheat then selling at 4*s.* 6*d.* a bushel.   In 1839 prices had risen, and the rent was then increased to 1,100*l.*   The farm is tithe free.   On this rental it was bought by the Duke of Cleveland two years ago ; and there are many farms on his Grace's old estate in the neighbourhood, of equally good quality, let for a third less rent to hereditary tenants, who have hitherto raised their easy rent with little exertion ; but who, content with that, and secure in their holdings, have made no endeavour to improve their farms ; and, while they have neither enriched themselves nor their landlord, have done nothing to enlarge the field of employment for an increasing population of labourers, nor contributed any greater produce to the extended requirements of the country.

The " two crop and fallow " system is that which prevails over all the strong undrained land of the county ; and draining, we regret to say, is still greatly neglected.   The system is of two kinds — either simply (1) fallow, (2) wheat, (3) oats, or of that and the following *improvement* combined : (1) fallow, (2) wheat, (3) clover.   This combination gives (1) fallow, (2) wheat, (3) oats, (4) fallow, (5) wheat, (6) clover; or one-third of the farm bare fallow, one-third wheat, one-sixth oats, and one-sixth clover.   The land intended for fallow is seldom ploughed before February, by which time the ameliorating

effects of severe frost on this heavy soil must be lost, or nearly
so.   After receiving the usual repeated ploughings and harrowings during summer, the land is commonly limed, and then ridged
up in 10-feet mounds, well gathered and rounded, to carry off the
water.   On this the wheat seed is sown broad-cast in autumn,
and receives no further attention till the following harvest.
During next autumn and winter, the manure from the farm-
yard, such as it is, is spread over the stubble and ploughed in.
In spring the land so prepared is sown with oats.   The oat stubble lies till February, when it is ploughed; and the same routine
of bare fallowing is pursued during the summer.   The wheat crop
this time receives no manure ; and, in spring, clover seeds are
sown with it, which next year are mown for hay.   The clover
root is broken up in February, again to undergo a bare fallow.
No roots are cultivated, and no purchased manure or food made
use of.   The farms are small in extent, the farmers hard-working
and industrious, but without means, and strongly prejudiced in
favour of their old ways, though these have yielded them nothing but ill-requited toil.   They keep very little stock, which
being ill fed, the manure made on the farm is merely rotted
straw.   The yield of their wheat crop may be from 12 to 20
bushels an acre,— 15 being a full average for the undrained
lands ; and their oats from 20 to 30 bushels.   But it is quite
obvious that even these meagre crops are likely, under a con-
tinuation of the same management, to become gradually lessened ;
for there is a constant abstraction from the soil without any corresponding return of manure.   The rent of such land varies
from 10s. to 16s. an acre; and the tithe and rates may be
3s. 6d. an acre more.   A simple calculation would show that
such crops cannot pay, even at considerably higher prices than
the present, and that here the difficulty is not one of rent,
but of produce.   A shilling a bushel added to the price of
wheat would only increase the farmer's returns 15s., whereas
an increase of produce from 15 bushels the present crop, to
25 bushels an acre, would make a gain to him of four

times that sum. It must therefore be a wise course for a judicious landlord to promote, as much as he can, the improved
culture of his estate; for he can have no hope of any balance
being left for rent until such an increase be attained. The first
step to this is thorough drainage; and, fortunately, clay can be
got in all places where it is most needed, and the cost of coal
for burning it for drainage pipes, in Durham, is a mere trifle.
We have seen instances in the county, where cold clay land, laid
up in high, crooked ridges, has been completely drained by the use
of two-inch pipes, placed from three and a half to four feet deep,
and the distance between each drain eight yards, the drains
being carried in parallel lines, quite irrespective of the ups and
downs of the old crooked ridges. The cost of this operation
over an extent of 2,600 acres, on one estate drained during the
last two years, has averaged 4*l.* 10*s.* an acre.

Our next example is one of the coast farms, where the management and results present a very remarkable contrast with
the "two crop and fallow" system just described. Seaham Hall
farm, near the harbour of Seaham, is occupied by Mr. At-
kinson, and extends, with the adjoining glebe-land, to 480 acres,
about 60 of which are in permanent grass. Part of it is still
undrained, a stiff strong soil, which in that state is managed
with bare fallow; but as soon as the drainage is effected, fallows
are dispensed with, and the regular system of the farm carried
out there as on the more friable soils. On one field of this
heavy land, which was only drained last spring, there is now
growing an excellent crop of swedes, estimated by the tenant
at 28 tons an acre. The rest of the farm, which is naturally
well adapted to green crops, or has been rendered so by
drainage, is managed in the following manner : — (1) clover,
(2) potatoes, (3) wheat, (4) turnips, (5) potatoes, (6) wheat,
sown out with "seeds." The clover root is ploughed up in
autumn, cross-ploughed in spring, and, after being harrowed,
the land is drawn out into ridges, 28 to 30 inches apart, into
which from 16 to 22 loads of dung per acre are placed, and

the potato seed planted. The land is wrought in the usual manner during summer, and the crop is raised and stored in long narrow heaps, covered with straw, and a light coating of earth to exclude frost. The potato crop is followed by wheat, the ground being first ploughed, and the seed either drilled in or sown broadcast in little ridges formed by the single plough, in which it grows up very much in the same manner as when drilled. The action of the single or ribbing plough, with which one man and horse can go over an acre and a half of ground per day, is believed to pulverise the soil better, and secure for the seed a more genial bed than any other preparation. The seed — red and white wheat, mixed in equal proportions — is sown at the rate of six pecks an acre, between the 20th of November and Christmas, the latter period being the favourite time for wheat-sowing along the coast. One and a half to two cwt. of guano is applied, and harrowed in with the seed. The wheat crop is succeeded by turnips, for which the land is prepared in the usual manner. Two-thirds of this division are sown with swedes, one-third with white turnips for early consumption; 12 to 15 loads of dung, $6\frac{1}{2}$ cwt. of guano, and 10 bushels of dissolved bones (mixed with sawdust), are all applied together to each acre of the turnip land. The crop is drawn for consumption in the yards and stalls. The next crop is potatoes, for which the land is very easily and cheaply prepared after the turnip crop. But as a good crop of turnips is supposed to exhaust the land more than clover, the potato crop receives a somewhat heavier dose of dung after turnips than after clover. In other respects the management is the same. Wheat again follows the potato crop, receiving the same dressing of guano as before; but the downy Essex seed is now used, as it is found not to lodge, and is therefore more favourable for the clover seeds which succeed it. The mixture of "seeds" sown is 8lb. red clover, 2lb. white, and two quarts of rye grass per acre. These are sown among the growing wheat crop in spring, and mown the following year

z

for hay. The average produce of wheat is from 32 to 40 bushels an acre; of potatoes, 5 to 6 tons since the disease; and of swedes, 28 to 35 tons an acre. The potatoes and wheat are sold off the farm, the swedes and clover consumed on it. Seventy head of cattle, 50 of which are fed fat, are kept on the farm, and 300 sheep, principally half-bred hoggets. The farm is managed with seven pairs of horses. The rent and tithe amount to 650*l.*, and the rates are 1*s.* per pound.

The leading features of this management are that four-sixths, or two-thirds, of the land are in well manured green crops, and two-sixths, or one-third, in white corn crop. The successive green crops keep the land clean and friable, and render the farm comparatively cheap, both in manual labour and horse work; seven pairs of horses could not under other circumstances accomplish satisfactorily the work of a farm of this size. Then, not only are the green crops heavily manured, but the intervening wheat crops also. In five out of six years the soil receives an annual application of manure. The farmer is thus enabled to grow the most valuable crops — potatoes and wheat — for sale, and swedes and clover for consumption. He can sell two-thirds of the annual produce of his arable land without injuring the farm, because he restores to it a full equivalent in manure. Comparing the returns of this six-course with those of the " two crop and fallow " system, there is a difference more than adequate to meet the increased charges of higher rent, labour, and manure, and, when all these are deducted, a handsome balance remains for interest and tenant's profit; whereas the whole produce under the latter system cannot, at present prices, pay the expenses of cultivation, without leaving a farthing for rent or tenant's profit.

LETTER XXXIX.

DURHAM — *continued.*

DURHAM, Dec. 1850.

THE usual system of cultivation practised by the clayland farmers of Durham is, as already mentioned, a " three-course," viz. : — (1) fallow, (2) wheat, (3) one-half oats and one-half clover. Nearly the whole of the fallow is managed as a dead fallow, there being very little green crop cultivated. Occasionally this rotation is prolonged by pasturing the clover a second year. The live stock is quite inconsiderable. Three cows and six young cattle to 100 acres may be about an average stock for the clay farms. As this stock is badly wintered ($2\frac{1}{2}$ acres of inferior turnips per 100 acres being the average extent of the turnip crop), the home supply of manure can be neither rich nor plentiful. To meet rent and the expenses of cultivation, the farmer's sole dependence is on his wheat crop, a little also being received from that portion of the hay crop which he sells off the farm. As a general rule no manure, except lime, is purchased. That is laid upon the bare fallow in preparation for wheat. The system is very exhausting; a bare fallow, stimulated by lime, is sown with wheat, which is followed by oats or hay. Each return of this rotation further reduces the soluble properties of the soil, as these are not restored by the small quantity of inferior manure applied in nearly the same proportion in which they are abstracted. The

same farm which thirty years ago averaged from 20 to 24
bushels of wheat, and 30 to 36 bushels of oats per acre, is now,
under this process, reduced to 14 bushels of wheat and 18 to 20
bushels of oats. One farmer assured us that his oats did not
last year average more than 10 bushels an acre.

Diminishing produce and lower prices are producing their
natural effect. The rents vary from 11s. to 16s. an acre; tithe
and rates 3s. 6d. an acre more. The evil here is not high rents,
but defective produce. If the farmer paid no rent, he could not
continue this system with present prices, and have a profit. In
a rotation of six years he has,—

	£	s.	d.
Two crops of wheat, 14 bushels an acre each, 28 bushels, at 5s.	7	0	0
One crop of oats, 20 bushels, at 2s. 4d.	2	6	8
One crop of hay, 1½ ton, at 3l.	4	10	0
Two bare fallows	0	0	0
	£13	16	8

His expenses, exclusive of rent, will be,—

Six years' tithes and rates, at 3s. 6d an acre - £1 1 0
Bare fallow, viz. : —

	£	s.	d.
Ploughing and harrowing, five times, at 11s.	£2	15	0
Lime, once in 12 years, proportion for three years -	0	10	0
Seed wheat, two bushels, at 5s. -	0	10	0
Harvest, inclusive of carting and stacking -	0	14	0
Thrashing and marketing -	0	5	0
	£4	14	0

This process, repeated twice - - £9 8 0
Oat crop, seed, labour, and harvest expenses - 2 2 0
Clover seed, and labour - - 1 7 6
                                                    13 18 6

Deficiency, besides rent - - - 0 1 10
Add rent, six years, at 15s. an acre - - 4 10 0

Total deficiency on six acres - - £4 11 10

An increase of price to the extent of 2s. 6d. a bushel on his wheat, and 1s. 2d. on his oats, is necessary to make good this deficiency, so that neither a reduction of rent, nor an increase of 50 per cent. in the present prices, will make this farmer's business profitable. There is no remedy possible here but a better system of farming and an increased produce.

To this point Lord Londonderry particularly directed the attention of his tenantry, by a public letter addressed to them in January last. After pointing out the inevitable consequences of an adherence to the common system of two crops and a fallow, he recommended a change, the main principle of which was to get rid of successive corn crops, and to substitute green crops for bare fallow. In order to accomplish this, he proposed, — 1st, to drain the land in the best manner, charging 5 per cent. on the outlay; 2d, to improve the buildings and foldyards, so that the stock might be kept under cover, and their provender be economically consumed; 3d, to make liquid manure tanks to receive the drainage of the houses and folds; 4th, to give his tenants, gratis, from one to two cwt. of guano, or an equivalent of dissolved bones, to be applied, in addition to the manure made upon the farm, to green crops; 5th, to provide a supply of bones and guano for sale to the tenants at cost price, and to erect a bone mill and apparatus for dissolving bones; the use of which was to be given to such of the tenants as chose to avail themselves of it.

Such measures as these, if zealously carried out, cannot fail to be attended with the best permanent effects on the interests of both tenant and landlord. We saw them in operation on the farm of Barmston, on the river Wear, a strong poor clay, reduced by the system already described to such a state of sterility that it was abandoned to the landlord as utterly hopeless. One field of the last tenant's pasture still remained to attest its condition. On this there had been no stock during the whole summer, as it was considered incapable of feeding anything to advantage. The whole summer's grass, therefore, was still on the ground, and a very miserable, white, scanty, innutritious herbage it seemed to be. Lord Londonderry has taken this farm into his own management; and under the direction of Mr. Gibson, his able agent, every field is being thoroughly tile-drained. The drains are uniformly made in parallel lines, 8 yards apart, and from $3\frac{1}{2}$ to 4 feet in depth, the soil being of nearly uniform quality, and no regard is had to the old crooked lines of ridge and furrow. The land is then ploughed as deeply as two powerful horses can move it ; and, after being wrought and cleaned in spring, it receives 20 loads of ashes and 6 cwt. of guano per acre, and is then sowed with swedes. The crop on the ground after this management was not less than from 20 to 25 tons an acre. These are drawn for consumption in the buildings, and are followed by wheat, which is sown out with clover. The wheat crop of this year yields 32 bushels an acre, on land which three years ago had all the appearance of having been reduced to a *caput mortuum.* By this change a heavy crop of swedes is substituted for the equally expensive, but totally unproductive, bare fallow ; these are profitably consumed by cattle, which leave a large supply of rich manure to increase the productive powers of the soil : the one year's wheat crop of 32 bushels is more valuable than the two wheat crops of the former system ; in short, the ascending scale of fertility is begun, and the ruinous descending scale of exhaustion is abandoned.

A necessary supplement to the substitution of green crops for bare fallow, on this description of soil, is increased house accommodation, as turnips cannot be eaten on the ground on these strong lands. The house accommodation at present is inferior and inadequate. Where so much has to be done, it is very important that some economical mode of construction be adopted, for the expenditure recommended by most of our book authorities would swamp a landlord altogether. Whilst we certainly should desire something of a more permanent character, we subjoin the particulars of an estimate and specification drawn up by Mr. Gibson, which may be useful to landlords, as exhi-

biting a cheap method of affording increased accommodation to
their tenants. With care this may last a considerable number
of years, until a landlord is gradually able to get over his whole
estate with buildings of a more permanent and substantial
description. The system of stall-feeding is adopted as the
most economical in first cost, and believed to be equally profitable as compared with any other in the progress of the stock.
Close wooden sheds are proposed to be erected, 15 feet wide
inside, with a feeding passage in front, and a cleaning passage
behind the cattle. The sheds are to be made of home-sawn
wood, and roofed with the same, coated with coal tar. Inside
they are to be fitted in the usual manner, with stalls, mangers,
doors, &c. The whole may be so erected at a cost of 10s per
head, where the timber is got free of expense on the estate. If
the value of the timber is added, the cost will be 30s. per head.
A shed 70 feet long, by 15 feet wide inside, affording accommodation for 20 cattle in stalls, 7 feet to each pair, will cost as
follows : —

	£	s.	d.
3400 superficial feet 1 inch deal, at 12s. per 100 -	20	8	0
50 larch posts, at 8d.	1	13	4
40 couple sides at 8d.	1	6	8
20 baulks, at 10d.	0	16	8
170 feet wall plate, at 1d.	0	14	2
170 feet runners, at ½d. -	0	7	1
2 barrels coal tar, at 5s., in Durham	0	10	0
Nails	1	10	0
Workmanship	2	14	0
	£30	0	0

On Lord Durham's estate an attempt has-been made to introduce the Northumberland, or five-course system, but the
tenants do not take to it kindly. The farms average 200 acres
in extent. During the last 10 years 14,000l. have been expended in drainage by the landlord, the tenants being charged
5 per cent. on the outlay. The average rent may be from 25s.
to 30s. an acre, and tithe 5s. to 6s. an acre. Lord Durham

last year allowed 20 per cent. of the rent to be expended in drainage, buildings, and manure.

The four-course is the common system on Lord Ravensworth's estate, which extends some miles westward from Newcastle, on the south bank of the Tyne. The land is generally of superior quality, and is let at rents varying from 40s. to 3l., and as much as 4l. an acre, in the vicinity of the town. The landlord executes drainage at a charge of 5 per cent. on his tenants. The farms vary in extent from 50 to 200 acres ; they are held from year to year, but the same families continue in their farms for generations. The demand for milk in this populous neighbourhood is good, a cow's produce being reckoned to be worth 20l. Other articles of farm produce, such as potatoes and other vegetables, are much in demand, and at remunerative prices. Manure in any quantity can be purchased at a moderate rate, the best quality costing 3s. 6d. per two-horse load, and a second quality is delivered on the estate by railway at a cost, including carriage, of 3s. per double-horse load. Notwithstanding these advantages, the tenants are not in a very prosperous condition. Some buy manure extensively, and use their opportunities with spirit. The most improving tenants on the estate, and the men of most enterprise, are said to be innkeepers and butchers from Newcastle, who carry their business habits and intelligence into the management of their farms. Contiguous farms of the same quality and rent vary in their produce many bushels an acre, according to the energy and command of capital possessed by their occupants. Lord Ravensworth has been giving temporary deductions to his tenants; but it is said that he now contemplates making a general permanent abatement of 12 to 15 per cent.

Besides the population of the large towns in this county, — Durham, Sunderland, Shields, Gateshead, and, on the other side of the Tyne, Newcastle,—there are very numerous populous villages scattered throughout the eastern side of the county, all of which are occupied by well-paid colliers, good consumers of

produce, and convenient customers for the farmer.   One extensive coal owner pays about 10,000*l.* in wages monthly, the greater
part of which is spent in bread, meat, dairy produce, and beer,
all in one way or other the produce of the farmer, who has thus
every encouragement to exertion which good markets can give.
But, besides this demand, he has likewise a ready sale for hay,
large supplies of which are required by the numerous horses
employed above and below ground at the various collieries.

The horses employed at the collieries are of two classes — pit
and waggon horses; the first a small compact horse, for working
in the pits under ground; the other a larger and more powerful
horse, for drawing the heavy waggons on the surface.   The
usual feeding given to these horses is three bushels of oats and
twelve stones of hay per week to each pit horse, and three and
a half bushels of oats and fourteen stones of hay to each
waggon horse.   Lord Londonderry's agent, Mr. Gibson, has
effected a very considerable saving by giving the horses their
food in a prepared state.   The hay is cut into half-inch chaff,
oats and beans are crushed, and, in addition to this dry
mixture, a certain proportion of each, with linseed, is given
steamed for the evening meal.   In this way two bushels of corn
and seven stones of hay per week are found sufficient for the
pit horses, and two and a half bushels of corn and eight stones
of hay for the waggon horses.   We can testify to the good condition and spirit of the horses in this establishment; and, as the
same feeding is given to the farm horses on Lord Londonderry's
farms, we think it may be useful to mention in detail the daily
allowance of the

PIT AND FARM HORSES.

Dry feeding consists of cut hay and crushed oats and beans, all
mixed together : —

Hay for each horse per day	-	-	11lb.
Oats and beans	-	-	-	12
			—	23lb.

PIT AND FARM HORSES — *continued.*   23lb.

Steamed feed : —
   Hay   -   -   -   -   -   3lb.
   Beans   -   -   -   -   -   2
   Linseed   -   -   -   -   -   1 — 6

           Total daily allowance   -   29lb.

WAGGON HORSES.

Dry feeding : —
   Hay   -   -   -   -   -   12lb.
   Oats and beans   -   -   -   -   14
                                — 26lb.

Steamed feed : —
   Hay   -   -   -   -   -   3lb.
   Beans   -   -   -   -   -   3
   Linseed   -   -   -   -   -   1 — 7

           Total daily allowance   -   33lb.

The saving effected by this simple change in an establishment like Lord Londonderry's, where 300 horses are constantly employed in the collieries and farms, must be very great; according to our reckoning, considerably more than 1000*l.* a year.

The Durham breed, or improved short-horns, are, of course, the prevalent cattle of the county. The north bank of the Tees is not now, however, so famous for this breed as it was once rendered by the celebrated Messrs. Collins : the Yorkshire side of the river now bears the palm. Small West Highland cattle are grazed in considerable numbers in the county. On the extensive farms which Lord Durham holds in his own management, a large number of 1½-year Highland heifers are bought at about 50*s.* each in autumn; they are crossed next year with a short-horn bull; and the following year, after suckling their calves, they are fattened and sold at about 7*l.* each. Another class — stots — are bought at the same age, and, after being kept two years, are sold fat in November at from 10*l.* to 11*l.* each. The cross-bred calves, after being suckled by their dams, are put on good keep, and are turned out prime fat at three years old, the oxen then averaging fifty to sixty stones imperial, which, from the superior quality of the meat, sells at the highest figure per stone.   The heifers, though of equal quality, are much smaller in size, and do not bring, within some pounds, the price of the oxen.   We had an opportunity, at the great November cattle fair at Darlington, of seeing a large collection of the cattle of the district, and, though there were several superior lots, there were also too many of a description quite inferior to what might have been looked for in the immediate neighbourhood of Teesdale.

The relations between landlords and tenants in this county present some very instructive points.   The Duke of Cleveland's estate in Durham comprehends the greater proportion of the country from the borders of Cumberland along the north bank of the Tees to within a few miles of Darlington.   Within these limits are included many varieties of soil and climate, from the rich lowland arable farms at Denton and Pierce Bridge, to the exposed mountain grazings in Teesdale Forest.   This extensive estate was valued 50 years ago, and during the period which has since elapsed the rent then fixed has undergone no change.   There are no leases, but the tenants are hereditary, the same families, in direct descent, occupying the same farms for centuries.   One of the best farmers on the estate has in his possession a lease of the land he now occupies, granted to one of his ancestors in the reign of Queen Elizabeth.   Though the Duke has never allowed the scale of his rents to be changed, and admits no competition or interference with his hereditary tenantry, he has not neglected the improvement of his estate. He keeps a drainage bailiff to lay off and superintend the whole drainage on his farms, whether executed by himself or the tenant.   He supplies tiles free of charge, the tenant being at the cost of putting them into the ground ; or, where stones are the material used, he pays two-thirds of the whole cost.   He does not object to the removal of useless fences and hedgerow

trees when they are shown to be injurious, and has encouraged
the enlargement of arable fields where it promotes economy of
labour.   The cottages on the estate are generally held directly
from his Grace, who in that case keeps them in good repair.
The rents vary from 2s. to 2l. a-year, with gardens in all cases;
and new cottages of a more commodious description are let at
from 2l. to 4l. a-year.   Allotments not exceeding a quarter of
an acre, and now being more limited, are let at from 32s. to
48s. an acre, all rates being paid by his Grace, and fences and
roads kept in good order at his expense.

The average size of the Duke's arable farms may be about
150 acres; the largest on the estate does not exceed 500 acres.
When the rent was fixed the valuation was low, the rent of
very good arable and pasture land running from 15s. to 26s. an
acre tithe-free, and an inferior quality from 9s. to 15s. an acre.
To the latter is generally annexed a right of pasturage on the
adjoining moors, on which the Duke reserves the game, but
gives his tenants the use of the pasture for stock rent-free.
On this great estate, during the last 50 years, there have not
been a dozen changes of tenantry.

With so many favourable circumstances, one might have
reasonably expected that the Duke of Cleveland, from his fine
old feudal castle of Raby, would have looked down on a
contented and prosperous tenantry, disturbed by no complaint,
but gratified by a reciprocal endeavour on their part to improve
the estate, and render more fertile and remunerative to themselves the annual produce of their farms, in the entire benefit
and enjoyment of which they are so amply and ungrudgingly
secured.   Truth constrains us to say that this is not so.   Their
easy rents have been made during a period of comparatively
high prices with little exertion.   The certainty they felt that no
additional rent would be exacted, and that the son would, as a
matter of course, succeed to his father on the same terms, led to
an indolent feeling of security.   Lower prices have found them
even less prepared than their more highly-rented neighbours;

and the Duke, in declining to make abatements, is not more exempt from complaint than other landlords who have not the same excuse.

When we consider the circumstances under which this fixed rental has been unchangeably continued during the last 50 years, we shall be better able to appreciate the propriety of the arrangement. During that period the average price of wheat rose as high as 119s. 6d., and fell as low as 39s. 4d. a quarter. The population of the whole country has doubled, and that of this particular county had, within the ten years preceding 1841, made a more rapid increase than that of any other county in England. The demand for all articles of consumption produced by the farmer, besides corn, must have kept pace with the increase of the population. In the midst of this activity and industry we find a great estate standing nearly still for half a century, the landlord declining to avail himself of the natural and legitimate benefits of his property, the farmer letting slip the opportunities he possessed, the labourers increasing in numbers, but finding from agriculture little or no increase of employment, and the increasing population forced to seek from abroad those supplies which the land in their own neighbourhood has failed to yield in sufficient abundance. However much we may admire the beneficence and liberality and unselfishness of the Duke of Cleveland, we cannot acquiesce in the wisdom of this arrangement. The principles by which the amount of rent regulates itself according to the varying circumstances of a country cannot, more than any of the natural laws, be laid aside with impunity. And though we should much more deprecate the system of recklessly screwing a tenantry by inviting unfair competition and adopting every means which the present state of the law affords to unprincipled or heedless landlords for unduly enhancing their rents, we yet deem it right to state the circumstances of a case of a contrary character to show that the real interest of all — landlords, tenants, labourers, and the public — are injured by any practice which fails to keep pace with the progressive improvements of the country.

LETTER XL.

CUMBERLAND.

Eastern District.—humidity of climate—grass farming therefore most remunerative — sir james graham's estate — consolidation of good land into large farms — encouragement of good tenants — great improvement of the estate — to which the tenants largely contributed — tenure — system and expense of drainage — wood-land — rent — course cf crops — mr. birrel's farm — pig feeding — practice of "sowing out" without a crop much approved — lord mansfield's arrangements with his tenants — industry of the cumberland farmers and their families.

Carlisle, Jan. 1851.

The lower lands of East Cumberland chiefly rest on the red sandstone formation, and the upper district on mountain lime-stone. In the plains along the banks of the rivers and streams —the Eden, the Esk, the Irthing, the Caldew, and the Line—the soil is generally a fertile alluvial loam ; on the low-lying ridges which divide the several plains, it varies from a strong retentive soil to good friable turnip land ; and near the Scottish border, on the Sark and towards the Solway, there is an annually-diminishing extent of unimproved bogs or peat moss. The lands along the valleys are very liable to be flooded by the sudden rise of the streams, after heavy rains in the moun-tainous district on the eastern border of the county. The quantity of rain which falls during the year in this county is, in the most favoured parts, nearly twice as much as on land at the same elevation on the east coast, while the greater frequency of rainy days imparts a character of humidity to the atmosphere much more beneficial to the growth of grass and green crops than corn. This humidity of climate has given to those who cultivate the soil with a wise desire to enlist nature on their side, instead of vainly trying to supersede an influence which may be modified but cannot be controlled, a preference for stock over corn farming. By keeping this in view we shall be better able to appreciate the value of the advice lately given by Sir James Graham to the tenantry of Cumberland, to " plough less and graze more ;" a corroboration of which is found in the management adopted by all the best farmers we visited in this county. The fact that agricultural labourers here are not in excess, as in some of the southern counties, removes of itself any objection (if such is really well-founded) on the score of diminishing employment ; while all attempts which have been made to introduce on a large scale the corn system of the east coast, however successful for a time, have in the end been found unremunerative. A wet autumn has interfered with the proper season for sowing the wheat crop; or if got in during a favourable time, and after giving every promise of an abundant increase, the continued rains of a wet summer "lodge" the crop, and the farmer, when he thrashes out his bulky stackyard, is greatly disappointed in the quality and quantity of the yield. Mr. Curwen tried the system forty years ago on a large scale, and failed; and though, for the introduction of the best breed of short horns, and the spirit he infused among all classes for agricultural improvement, his name is still gratefully recollected, as a corn farmer his example has not been followed. The future range of the price of corn is likely still further to limit any desire for its more extended cultivation in this county ; and the fact that the men who have really made money here have done so as breeders or feeders of stock, has become so generally understood, that to the development of that branch of their business the attention of the best farmers is now chiefly directed. The farms, for example, of Mr. Ferguson, of Harker Lodge, a considerable breadth of which used to be kept in cultivation, are now being laid to grass with such success that, on an extent of 700 acres, 3,000 hoggets and 200 cattle are fed during the summer, the hoggets being sent off to the market as they become fat, after yielding an increase of 10s. per head, on the average, for their keep from the end of April till November.

Netherby, the estate of Sir James Graham, occupies the north-western extremity of the county on the Scottish border, and extends from the Solway for seven or eight miles up Esk-dale. It includes in one compact and undivided property the whole of the land between the lower part of the river Line and Dumfriesshire, and comprises altogether nearly 30,000 acres of land, between 2,000 and 3,000 acres of which are wood, much of it fine old timber. Sir James succeeded to the estate up-wards of thirty years ago; and since that time he has been unintermittingly engaged in its improvement. Neither time nor large expenditure have been spared, to make it—what it is now confessed to be — the best-conditioned estate in Cumber-land. Time and money alone could not have done this, had not both been expended with judgment; and it may surprise our readers to hear that a statesman who, during the greater portion of that time, has occupied so conspicuous a place in the councils of the nation, is more minutely acquainted with the details of arable farming and the general management of land than many men of inferior capacity who devote their whole lives to the business.

The leading feature in the management and improvement of the Netherby estate has been the timely consolidation of the good land into large farms, and the proper subdivision and enclosure of the inferior lands. An idea may be formed of the extent to which this principle has been acted on, and the con-sequent diminution of an overgrown agricultural occupying population, from the fact that the number of rent-paying tenants holding land in 1820 was 340, and in 1850 only 165. Fine farms of 300 and 400 acres, now occupied in one holding by an enterprising tenant, were then held in seven or eight separate possessions. The demolition of useless clay buildings and super-fluous hedges caused an immense saving of horse power; and, as one great feature in the management seems to have been the careful selection and encouragement of good tenants and the unsparing weeding out of bad ones, Sir James was at every step assisted by his tenants in the further improvement of his property. He had not to work single handed against ignorance or indifference, but enlisted on his side both energy and capital. In planting sheltering woods to enhance the value of his farms, he was at the same time laying by an improving capital; and in erecting new farm buildings, in draining, removing and replanting fences, making open waterways and embankments, and constructing roads, he was assisted by his tenants, who contributed in labour a large part of the cost. The buildings on the estate, which were then chiefly of mud and thatch, have all been replaced by substantial stone and slate. Year after year sees a diminishing extent of moss, the landlord contributing the material for drainage, and the tenant performing all the other cost of the reclamation. During the currency of his lease he enjoys the benefit of his industry, but at the close of it the landlord participates in the increased value caused by the improvement. In constructing new fences the same principle is adopted, and the tenant is bound to maintain them constantly in good order. This is strictly attended to, and the neatly trimmed thorn fences along all the lines of road traversing this extensive estate mark its boundaries on every side. Great though the expenditure of the landlord has been, it could not have effected so much without the aid of an enterprising and industrious tenantry, wisely directed it is true, but still rendered in addition to the rent.

The farms are all let on lease at money rents for a period of fourteen years, free of all manner of tithes. The tenant enters on his farm at the term of Candlemas, and pays his rent at two terms in the year — Whitsuntide and Lammas. He pays all taxes, rates, and burdens, already imposed, or that may be imposed, by law upon farmers during his term of possession. The stipulations as to management are very stringent; but we were assured that a good tenant is never interfered with.

With regard to drainage, the former custom on the estate was for the landlord to furnish the tiles free of charge, and the tenant to put them in.   This system was commenced many years ago, but the drains were then made too shallow, and a great portion is now being taken up and relaid at greater depth.   Sir James now executes all drainage at his own cost, the tenant performing carriage and paying 5 per cent. on the net outlay.   Twoinch pipes, with a flat side to lie on, are the size chiefly used; they are thirteen inches in length when burned, and cost at the kiln 16s. per 1,000 for all sizes as required (a due proportion of large mains being furnished at the same price), the cost of coals, inclusive of cartage, being 10s. per ton.   The drains are made, according to the nature of the soil, from three to four feet in depth and seven to ten yards apart, and the highest charge for interest to the tenant is 3s. 6d. per acre.

In the management of his woods, Sir James does not fail to take advantage of a new outlet afforded by the increasing wants of our manufacturers.   We heard that he is at present in treaty with a thread manufacturer of Manchester for the erection of a steam-power mill at Longtown, at which the small wood of the estate is to be cut into bobbins.   This trade is already largely established at Windermere, and supplies an excellent market for the tops, boughs, and rubbish which used formerly to be burnt to get them out of the way.   For this purpose beech, hazel, alder, birch, and ash coppice are all suitable, and are now sold, where the trade is fully established, at 1s. per cubic foot. At a sale lately in that district a coppice of this description brought 30l. an acre, free of all expense of labour, to the owner of the land, and in about fourteen years more the same coppice will be again ready to cut.   Such prices cannot, of course, be looked for at first; but when the trade is fully organised, the manufacturers compete with each other, and as higher rates call further on their ingenuity, powder factories are established, where all the wood under one inch in diameter, and which is, therefore, unsuitable for the bobbin mill, is turned to profitable account.   Besides the direct benefit to an estate in affording a good and accessible market for the small wood, a mill of this description employs fifty or sixty hands, and in process of time it may lead to the manufacture of the thread as well as the bobbins.

There has been no abatement of rent on the estate, nor are there many complaints. The prices of corn are low, and the returns from cattle have not been remunerative; but Sir James Graham's tenants draw a large proportion of their annual receipts from the feeding of sheep and pigs, both of which have been paying well. Considering the condition of the farms as regards drainage, fences, and buildings, the rent appears to be fairly charged, and the tenants have no doubt that Sir James will not permit them alone to bear all the burden of unforeseen low prices, should they continue. The present leases were entered into in 1843. Rents vary considerably, according to the quality of the soil ; from 20s. to 26s. an acre may be the average for arable land, about a tenth of which is reclaimed moss. The highest rent for a large farm of excellent land, completely drained, fenced, and housed, is 36s. an acre, the tenant also paying rates, the whole of which do not together exceed 1s. 6d. per 1l.

The usual course of husbandry adopted on the estate is the five course, the land remaining two years in grass. A good farmer who desires to change the system is at once permitted to do so ; and on land of superior quality it is understood that the landlord would not object to such a tenant taking wheat every alternate year if he found such a practice advantageous.

As an example of the mode of farming practised on the estate, we may shortly describe that of Mr. Birrel, of Guards. He occupies the extreme north-western boundary of England, the land lying a few yards from the shop of the famous Gretna blacksmith who binds for ever the runaway lovers who present themselves at his forge. In extent the farm comprises 475 acres, sixty being reclaimed moss, for which the landlord supplied the tiles, and the tenant expended the rest of the cost of the reclamation. This was a condition of the lease, the landlord providing tiles, and the tenant binding himself to reclaim a certain number of acres annually. The farm is divided into three natural divisions, (1) of moss, (2) weak land unsuitable for wheat, and (3) good land fit for the growth of any kind of crop. These are each subdivided into five fields, and, the rotation being a five course, there is thus annually a field of each quality of the farm bearing the same kind of crops — there being three fields of oats, three of green crop, three of wheat, barley, or oats, three of "seeds," and three of pasture. This, with steady management, insures as nearly an uniform result as can be attained. The green food and fodder bear a pretty constant proportion to the quantity of stock, and the amount of horse and manual labour are regulated in the same manner.

The rotation begins with oats, which are sown broadcast on one furrow, and yield twenty-seven bushels an acre on the inferior land, to thirty-six and forty-eight on that of better quality. The oats are followed by green crop, one third of which is potatoes, one third white and yellow turnips, and one-third swedes. The potatoes are manured with the best dung of the farm, and yield from $6\frac{1}{2}$ to $7\frac{1}{2}$ tons an acre of the Prince Regent variety, which is at present selling at 3l. per ton. The turnips and swedes receive about 20 tons of dung and 2 cwt. of guano per acre, all that have received dung being drawn for consumption in the feeding houses. The turnips which are to be eaten on the ground by sheep receive no dung, but have 3 cwt. of guano per acre. The swedes are all taken up in December and stored for use in spring, as the farmers here are quite sensible of the injury done both to the soil and the root by leaving it in the ground during the winter and early part of spring. Two-thirds of the land, after green crop, is sown with oats, and one-third with wheat, the average yield of which is thirty bushels an acre. Twenty fat cattle, reared on the farm, are sent to market annually, and 300 Cheviot lambs are bought in September, and, after being wintered on turnips, are sold fat off the grass during the summer. They receive cake or corn from the end of January till sold, and leave an average advance of

20s. each. Sixty pigs are bred and fattened annually on the produce of the farm, and realize about 200l. They are bought by dealers, who take them by railway to Leeds, Nottingham, and other midland towns, whence a brisk demand has arisen, accompanied by paying prices. Five pairs of horses, assisted occasionally by two young ones, do all the horse work of the farm. They are chiefly fed in the stable, except when during the summer they are turned out nightly on a good pasture.

So little do the farmers wish to increase the extent of their arable land, and so sensible are they of the great advantage of laying their land to grass with a full plant of "seeds," that it is not uncommon to see them of their own choice sowing out their fields without a corn crop. This practice is attended with much success. On the fine farm of Mr. Gibbons, of Burnfoot, a large extent of strong land is now being laid to grass in this way. It is fallowed and thoroughly cleared of weeds, and in the month of June or July, as soon as it can be got ready, it is sown with a mixture of grass, rape, and clover seeds, in these proportions — 5 lb. perennial rye grass, 7 lb. white clover, 7 lb. cow grass or perennial red clover, and 3 lb. or 4 lb. of rape seed per acre. This is stocked with sheep the same autumn, the 40-acre piece which we saw having yielded upwards of eight weeks' excellent keep for 380 sheep. It is now a deep rich green, very refreshing to the eye at this season, and will keep and feed a heavy stock of sheep during the coming summer.

Bordering Sir James Graham on the Scottish side is the estate of Lord Mansfield, who has lately concluded an arrangement with his tenants, which is said to have given complete satisfaction. The basis of that arrangement is, that the rent during the last leases of fourteen years is converted into a grain rent, at the average price of the county for these fourteen years, and, that being ascertained, 10 per cent. is deducted, and the balance, as a fixed money rent, becomes the future rent of the farm. On these terms the whole of the tenants have willingly entered into new leases. By this arrangement an improving tenant reserves to himself the whole benefit of his improvements, whereas a new valuation taxes the improving tenant unfairly, and lets the sluggard go free.

The tenants of East Cumberland are an industrious, hard-working, and economical class of men. Their families are brought up to industry, the young men working in the fields, and the daughters assisting in the dairy and the house. The in-doors work of a Cumberland farm house is a serious matter, as all the farm servants, married and single, receive the whole of their food in the farmer's kitchen. They have bread, porridge, and milk to breakfast — broth, meat, and bread to dinner — and milk-porridge and bread to supper. Besides preparing all this, the daughter of a substantial farmer or independent yeoman may be seen on market-day at Carlisle selling her poultry and dairy produce, while her father or brother is disposing of and delivering his corn and potatoes.

LETTER XLI.

CUMBERLAND — *continued.*

WEST CUMBERLAND. — AN IMPORTING COUNTRY — "STATESMEN" — GRADU-
ALLY BEING BOUGHT OUT — SIZE OF FARMS — LORD LONSDALE'S ESTATE —
NO LEASES, YET PERFECT CONFIDENCE — HUMIDITY OF CLIMATE—REQUIRES
MORE FREQUENT DRAINAGE AND LARGER PIPES — TILE WORKS — GROWTH
OF SWEDES — MR. TURNER'S FARM NEAR WHITEHAVEN — HIS PLAN OF
"SOWING OUT" WITHOUT A CROP — YIELD OF CORN CROPS — HAY — LIVE
STOCK — EXCELLENT MANAGEMENT OF MILCH COWS — CONTRASTED WITH
THAT OF GLOUCESTERSHIRE — SHORT HORNS INTRODUCED BY MR. CURWEN
— COVERED DUNG HOUSE AND CORN RICKS AT GILGARRON — PROSPEROUS
STATE OF LABOURERS — RENT OF LAND.

WHITEHAVEN, Jan. 1851.

THE western division of Cumberland presents a much greater
variety of soil and surface than the eastern. Its geological
features comprise granite, clay-slate, trap, limestone, red sandstone, and coal. On its eastern boundary it is shut in by the
lofty mountains of the lake district, from which it slopes in
undulating ridges of greater or less elevation to its western
boundary on the sea. With a sea-coast line of nearly fifty
miles in length, it possesses numerous shipping ports, the principal trade of which is the export of coal. From Maryport to
some miles beyond Whitehaven, coal is raised close to the coast;
and at Whitehaven one of the best seams is worked for a considerable distance under the sea. The populous towns of
Whitehaven, Maryport, Workington, and others, along with the
mining population scattered throughout the district, consume
more agricultural produce than it yields, so that West Cumberland is an importing country.

A great change has been effected within the last half cen-

A A 4

tury by the inclosure of the commons, which before that time comprised nearly half the lowland of the district. Another peculiar feature of this division — not indeed confined to it, as it extends over the rest of Cumberland and the adjoining county of Westmoreland — is the gradually diminishing numbers of the class called "statesmen," or yeomen proprietors of small estates, from 40 to 100 acres in extent. This class of men, formerly very numerous, have been located on their patrimonial estates for many generations. Their original possession is said by some to have been granted to them on condition that they should be ready to follow their lord, or the warden of the marshes, in re-pelling the border forays. The young men of this cla s are in many instances zealous improvers, but the older generation are strongly prejudiced in favour of old systems, and generally very unwilling to advance with the progress around them. They are comfortably housed, and, as a class, most industrious and economical. But they cannot easily accumulate wealth, as the eldest son gets the patrimonial estate, and the rest of the family the savings, increased very generally by small annuities payable out of the estate. These little properties are seldom subdivided, but every year they are being absorbed into the larger ad-joining estates, either by unavoidable sale arising from accu-mulated embarrassments, or by the offer of a tempting price which the " statesman" thinks it imprudent to decline. In point of general intelligence the " statesmen " are not superior to farmers occupying the same extent of land, and are much in-ferior in education and enterprise to the more considerable farmers, though as regards real property they may be the wealthier of the two. There are many known instances in which " a statesman," paying no rent, has become hopelessly embarrassed, and a farmer succeeding to the occupation has both paid a fair rent and made a profit.

The great proportion of the arable lands is held in farms of from 50 to 150 acres in extent. Some farmers occupy from 200 to 300 acres, and a few as much as 400 acres. Many hold on leases of 14 years or more ; but the principal estate in the district that of the Earl of Lonsdale, is chiefly let by verbal contracts from year to year, without any stipulation whatever as to the mode of farming. Notwithstanding the absence of leases, the farmers on this estate are a very enterprising class; and as the most perfect confidence subsists between the landlord and tenant, the latter most liberally invests his capital in the cultivation and permanent improvement of his farm. When a farm, from any cause, falls into his Lordship's hands, a good young tenant is carefully sought out; and as this has long been the practice of the estate, the farmers are mostly selected men, vieing with each other in the management of their farms. They have no tenant right or repayment for unexhausted improvements, but they know that they are dealing with a family which has always felt its own interest identified with the prosperity of the tenantry.

The climate of West Cumberland is of a peculiarly moist character, the rain guage showing an annual fall of from 47 inches of rain at Whitehaven to as much as 160 in some parts of Borrowdale. As this falls upon a soil in many cases of very impervious character, it may be readily conceived that thorough drainage here is not only a necessary but a most difficult operation. There is so little uniformity in the nature of the soil and subsoil, that many varieties are met with in almost every field. Within the last twenty years an immense extent of tile drainage has been made, great part of which has proved comparatively of little effect, from having been done too shallow. Eighteen to thirty inches was then the usual depth, and many thousand roods of drains put in at that depth are now being taken up and relaid at from three to five feet deep, with the best effects. This greater depth was introduced into the county about four years ago by Mr. Parkes, who was employed by Lord Lonsdale to superintend the drainage of his estates. But Mr. Parkes, probably not sufficiently adverting to the difference in the quantity of rain-water to be carried off by the drains in this county as compared with some of the drier districts of England, adopted then a uniform

rule of small pipes and wide intervals apart.   To use the
forcible language of Mr. Pusey,—

> "How can a fixed rule be laid down for the depth or distance of
> drains or the size of the pipes, when one county has 25 inches of rain,
> and another has 50 inches, to be carried off by these drains?   If a
> man living in Oxfordshire said that inch pipes would drain his land
> well, a voice from Cumberland might exclaim that it was absurd to
> use less than 1½ inch pipes, which he found far the best.   Yet the
> smaller pipe might be more competent to its duty in one place than
> the larger one in the other."

It never can be safe to act altogether in defiance of local experience until we have had time to mature our own ; and, accordingly,
further experience has shown that very wide intervals and excessive depth will not do for the soil of Cumberland.   Four feet
deep and seven to ten yards apart are now the standard on Lord
Lonsdale's estate, an additional drain having been in many fields
put in between the wide intervals which were at first unsuccess-
fully adopted.   There are now established in this division of the
county twenty-seven drain-tile works, which are estimated by
Mr. Dickenson to have produced, since their establishment,
116,000,000 of draining tiles and pipes — a quantity sufficient
for the drainage of 58,000 acres of land at seven yards' distance
betwixt each drain.   A great deal, however, yet remains to be
done.   We have not seen in any other district so great a number of wheat fields with water oozing down their open furrows.

The chief excellence in the farming of West Cumberland is
the successful management of the swedish turnip crop.   For
this the soil and climate seems to be peculiarly suitable   Forty
tons an acre are said to be sometimes got, and twenty to thirty
tons are reckoned an average crop.   The manure used is twenty
carts of good farm-yard manure, and two cwt. of guano, per acre.
This is put in ridges about thirty inches apart, on which, after
being closed in by the double plough, the seed is sown from the
30th of April till the 30th of May, the earlier the better.   About
the middle or end of October the crop is taken out of the ground,

and stored in long slightly-thatched narrow heaps for winter use. The tops are usually ploughed into the ground as manure for the wheat crop, which is then immediately sown.

Without further entering into a general statement, we may detail the management of a particular farm, which will better exhibit the peculiarities of system than any abstract description. The farm of Moresby Hall, within three miles of Whitehaven, on the estate of Lord Lonsdale, contains 340 acres of land, in the occupation of Mr. Turner. He has no lease, no prescribed rotation of crops, and is never interfered with by his landlord as to the management he thinks it right to adopt. The lea is broken up and sown with oats, which yield forty-five to forty-eight bushels an acre. The oats are followed by swedes and yellow turnips in about equal portions, the land receiving twenty loads of good dung and two cwt. of guano per acre. The swedes yield from twenty to thirty tons, the yellow about twenty-five tons an acre. The swedes are all stored early in November, and the land then sown with wheat. On the high land, not suitable for wheat, half the turnip crop is drawn, to be consumed by sheep on the adjoining grass land in wet weather, and the other half is fed off on the land during periods of dry weather. This prevents the soil — which, though drained, is a moist clay—from being "poached" by the trampling of the sheep. The great object on this farm being to provide rich food for a large head of stock, there follows a peculiarity in the management at this stage, the benefits of which are daily becoming more generally appreciated. Instead of taking a white corn crop after the turnips, and laying it down with "seeds" in the usual fashion, the land is laid to "seeds" without a crop. It is ploughed when dry, well harrowed, and rendered smooth on the surface, and then sown with the following mixture of grass, clover, and rape, two cwt. of guano having been previously scattered over each acre, and slightly covered by the harrows,— two pecks Italian rye grass, two pecks perennial rye grass, 4 lb. rib grass, 5 lb. white clover, and 3 lb. rape. The seeds are covered by the roller, they grow rapidly, and are ready to be stocked with sheep in July. A ten-acre field of poor land, sown last April in this manner, kept and fed 100 clipped hoggets from the 20th July to the beginning of November. It is now (January) a rich deep green, and will be early ready for a heavy stock during the present season. This lies two years in pasture, and is then ploughed for oats, which, from the high condition of the land, cannot fail to be bulky and productive. The wheat on the better land yields about thirty bushels an acre of the old English white variety, which, from having been long grown in the district, has become acclimated, and is found to stand a moist season best. The wheat is sown with the usual mixture of grasses, part of which is mown and part depastured. Whatever has been mown is uniformly ploughed up the next spring for oats, as it does not afterwards yield good pasture, and the best farmers in this district find it their interest to have nothing in pasture that cannot keep a full and well-fed stock.

The management of hay next deserves attention. Mr. Turner every year mows thirty-five acres of old land of fine quality, the same fields being mown every year. One half of this is top-dressed annually with twenty loads an acre of good manure, which is laid on, either immediately after the hay is got, or in the months of October and November — at all events while the grass is growing. The aftermath yields an abundant pasture for a large herd of short-horn cows. The crop of hay weighs about $2\frac{1}{4}$ tons per acre. Great attention is paid to managing it with the utmost expedition ; and by the fourth day, if the weather be favourable, it is carted to the hay barn, where it is at once stored, as it is got, for winter use. The hay barn is a large loft over the cowhouse, and contains the whole of the hay given to the cows throughout the winter. Over the stable, for the farm horses, is a similar large loft, in which their winter supply of rye grass hay is stored in the same manner. The hay secured in this manner is of the finest quality, and proves the advantage of careful management.

During the summer this farm feeds 80 cattle, 40 of which are large short-horn dairy cows, and 300 sheep. In winter it keeps 40 cows, 20 cattle, and 150 sheep. The sheep are chiefly Cheviot lambs, bought in September, which are fed, as already noticed, during the winter, and, after being shorn, are sent off to the fat market in summer and autumn as they become ready. Between wool and carcase they leave an increase of 20s. to 25s. each.

The dairy cows, 40 in number, are kept for supplying White-haven with milk and butter. This number is regularly main-tained in milk throughout the year; those which have become dry being either fed off, or sent to another farm and more moderate feeding, till they are ready again to take their place among the milking stock. The best heifer calves are reared to keep the stock good. The mode of feeding the cows is as follows: — On the 1st of November the winter management begins. The cows are then kept constantly housed, except being turned out, two or three at a time, for a few minutes daily, to the drinking pond. They get turnips twice a day, two stones' weight at each time. They receive likewise a cooked mixture of oats and tares, grown together for that purpose, and cut by the chaff-cutter, then boiled with chaff, and given twice a day, a bucketful to each cow at a time. The boiled mixture is placed in a stone trough twelve hours, to become cool before being given to the cows. They also receive a small handful of the best old land hay four times a day, the forty cows consuming nearly a ton each during the winter season. The hay is conveniently let down through a trap-door from the hay barn into the cow-house as it is needed. The cows receive a little oat straw the last thing at night. By the 1st of May they go to grass; they are milked daily at five a.m., a portion of them again at one p.m., and the whole of them at five p.m. They are then all turned out again to a pasture field near the house till nine, when they are brought in and kept in the house all night. They are thus protected from the chills of damp cold nights, and require no food till again turned out to their pastures after being milked in the morning.   The morning and mid-day milk goes to Whitehaven for sale, the evening milk is made into butter.   Milk sells at 2*d*. per quart for new, and 1*d*. for skim, and butter from 9*d*. to 11½*d*. per lb.

The annual produce of each cow is very considerable, and the farmer finds it his interest to give his cows throughout the year the best and most nutritious food.   What a contrast does the winter feeding of this stock present to the starving system of the dairy farmers of Gloucestershire, and how different the quality and quantity of the rich manure produced as compared with the little dried heaps of miserable droppings, which they scatter sparingly over the land!   The horse work is done on this farm by four horses in winter and seven in summer.

The farm of Mr. Jefferson, of Preston Hows, is conducted much in the same style, the grasses which he has laid down without a corn crop being of first-rate quality.   The swedish turnip crop on this farm in 1847 was found to weigh 40 tons an acre.

The stock principally kept in the arable parts of West Cumberland is the improved short-horns.   These were first introduced by the late Mr. Curwen, who spared no expense or pains to get good blood.   The produce of his stock still maintains its fame, and some of the best herds trace their descent from his.   In some quarters the polled Galloway is preferred, and on the farm of Mr. Rigg, of Abbey Holm, very fine specimens of that breed may be seen.   The sheep are chiefly bought at the lamb fairs on the Scottish border, and pure Cheviot seem to be preferred to the half-bred Leicester and Cheviot.   A few flocks of well-bred Leicester ewes are to be met with in different parts of the county.

The continued rains which at certain seasons fall in West Cumberland are most injurious in their washing effects on manure heaps exposed to their influence.   Captain Walker, of Gilgarron, has erected covered sheds over his dungheaps, and constructed a capacious tank to receive all the drainage from his feeding houses. This is pumped up and applied to the manure heap, which it keeps moist, and prevents from too rapidly heating or decomposing. It has been attended with the best effect, the superiority of the manure so treated showing itself to an inch when applied to the land. The sheds are cheaply constructed of light wood, covered with M‘Neill's patent felt. Sheds of the same description are erected in two parallel rows, in which the whole of the corn is housed. The platform on which the stack rests is raised sufficiently above the ground to render it proof against vermin, while it also admits circulation of air. The cost of one of these sheds, capable of holding 700 or 800 shocks or stooks of wheat, complete in every part, is 40*l.* They are an excellent contrivance, and, in this moist situation (between 500 and 600 feet above sea level), of peculiar utility. Wheat is grown at this elevation at Gilgarron, of fair quality.

The condition of the agricultural labourer in West Cumberland is very satisfactory. For Englishmen employed as day labourers the present rate is from 11*s.* to 15*s.* a week. Cottage rents are from 2*l.* to 3*l.* 3*s.* per annum. Fuel is every where plentiful and cheap. The most common mode, however, of paying and feeding farm servants, both married and single, is by engaging them for the half-year, and giving them their food in the farm-house. It is the same practice as that common in East Cumberland. The best men have 8*l.*, and ordinary men 6*l.* for the half-year; boys, 2*l.* to 3*l.*; women, 2*l.* 10*s.* to 5*l.* and their victuals, which are abundant and good. Barley bread, which formerly was chiefly used by the labouring population, is gone out of use, as wheat bread is preferred, and its price now brings it within the reach of all. The poor rates throughout the district are generally exceedingly moderate.

As a class, the farmers of West Cumberland are plain, industrious, and intelligent. Their sons and daughters are brought up to habits of industry and economy. Agricultural complaints are not much heard, and farms, when they come into the market, are eagerly sought after. Their system of sheep-feeding on turnips has not yet received all the aids which the turnip-cutter and the cheapness of cake and corn have afforded to the Southern farmer. The stubbles are not always so clean as they ought to be. So that while we gladly accord to them in several points superior merit, we cannot acquiesce in the claim, laid by more than one of them to us, that they are the best farmers in England.

The rent of good arable land near Whitehaven is from 25s. to 30s. an acre, the highest about 35s., all tithe free.

LETTER XLII.

NORTHUMBERLAND.

RENT OF LAND ON TYNESIDE — GENERAL DESCRIPTION OF THE COUNTY — SOUTHERN DIVISION — MUCH CLAY LAND INDIFFERENTLY FARMED — DEPRECIATION IN VALUE OF THIS KIND OF SOIL — ARRANGEMENTS MADE BY THE LANDLORDS IN CONSEQUENCE OF FALL IN PRICE OF CORN — MODE OF LETTING LAND — FARM MANAGEMENT ON DUKE OF PORTLAND'S ESTATE — SYSTEM ON HEAVY LANDS — INCREASE OF VALUE BY DRAINAGE — EXPERIMENTAL TRIALS OF FEEDING AT HOWICK GRANGE — ADVANTAGE OF HOUSE FEEDING IN SUMMER — MR. SCOTT'S FARM AT BEAL — ADVANTAGE OF GIVING CAKE TO CATTLE ON GRASS — BEST MODE OF DRAINAGE — DOUBTFUL BENEFIT OF LEVELLING DOWN OLD CROOKED RIDGES.

MORPETH, Jan. 1851.

ENTERING Northumberland on the west by the railway from Carlisle, we passed from Hexham to Newcastle, down the valley of the Tyne, which is highly picturesque and fertile. A broad sparkling river — now shut in by high banks clothed with venerable woods, now confined to its channel by embankments protecting rich arable holms, here peacefully gliding past the graveyard and the parish church, there more impetuously breaking over the rocky bed which forms the base of some old feudal castle — mingles itself with the tide a few miles above Newcastle, and then exchanges its picturesque character for that of commercial bustle and activity. Along the river the land is of excellent quality, let at from 40s. to 50s. an acre, and principally under cultivation. The holms which, on the banks of the Wharfe or the Tees, form the rich feeding lands of the district, are here applied to the production of corn. Out of the valley the country is on both sides of inferior quality, much of the south side being worth little more than 7s. an acre, the land cold and undrained, and the farming of a very ordinary description.

B B

Between Newcastle and Morpeth the soil is generally a strong clay.   Northwards of Morpeth and to the west it is a poor infertile clay.   Along the seaboard and the line of railway past Warkworth, Alnwick, Belford, and to the border of the county, it consists of strong wheat land, more or less fertile, but generally of superior quality.   Between the Cheviot Hills and the ridge which forms the eastern boundary of Tillside there stretches a tract of excellent turnip land, held in large farms by intelligent cultivators, and where the five course, or Northumberland system of farming finds its best illustration. To the westward of this the country is high and unenclosed, and stocked with Cheviot and half-bred sheep.   The geological features of the county comprise the coal measures, millstone grit, mountain limestone, and greenstone.

The agriculture of the southern division of the county is not at all superior to that we have described on the cold clay lands of Durham.   The farms are many of them larger in extent, but the land is chiefly undrained, and, being nearly all under the plough and very indifferently stocked, it appears to be in poor condition, and at present prices must be bearing hard on the capital of the occupying tenants.   Very many farms have been surrendered to the landlords, who are now becoming thoroughly impressed with the necessity in which they stand, either to meet their tenants liberally by encouraging drainage and affording them temporary relief, or to find their whole estates abandoned to them in wretched condition.   The appearance of the farm horses, and the quality of the scanty c k, sufficiently evidence the straits to which many of the smaller farmers have already been reduced; while the less carefully prepared fallows and diminished accumulations of manure afford no promise that the returns of next year will better their circumstances.   The reliance has been so wholly on grain that any abatement of rent, which is unaccompanied by a wise expenditure in improvement by drainage, so as to encourage the keeping of stock and the more certain action of manure, will only postpone for a little the inevitable crisis. It must surprise many who have hitherto been led to consider the agriculture of Northumberland as a model for the rest of the kingdom, to learn that a great portion of the county, extending from near Newcastle on both sides of the railway as far north as Warkworth, is as little drained and as badly farmed as any district we have yet seen in England, and that the occupiers of the small farms can only eke out a scanty subsistence by careful parsimony, and by employing no labour except that of themselves and their families.

The larger farms in the district of which we are now speaking are better cultivated, but the farmers are even less hopeful. In the best times they assert that thay never could calculate on making a gross return from their farms equivalent to two rents, and that now most of them can scarcely make even one. Many of them occupy more than one farm, and, as the farms of this class extend to from 400 to 700 acres each, it is not uncommon to find one holding of 1,000 to 1,500 acres altogether. The business they declare not to have been a good one ever since the war, and, the management continuing much the same, it is now worse than ever. A great reduction in the rent, we were assured by one extensive farmer, would not adequately meet his present difficulties. A case in illustration was mentioned to us of a strong clay farm of good wheat land, 600 acres in extent, the rent of which in 1825 was upwards of 800*l.*, besides 100*l.* more for tithe, but which, notwithstanding a large expenditure by the landlord, would not at present bring much more from a tenant than half the former rent. Such statements must be received with considerable caution, though on the whole they show that the amount of rent has hitherto been out of all proportion in excess of the gross produce as the farming is at present conducted, and that the two must in future be made more justly to approximate.

Nearly all the landlords in the district are now satisfied that to keep this land in cultivation they must be ready to assist their tenants with drainage and farm buildings. Many have also given abatements of rent, and it is believed that those who have hitherto declined to do so are accumulating arrears much heavier in reality than the loss they would suffer by any voluntary abatement. All the agents who have been authorised to make abatements are ready to give you instances in which arrears have been at once paid off, in consequence of this liberality, and of the confidence which it has given to the tenants to go on with such improvements as form the best guarantee to the landlord for the maintenance of his future rental. We may here describe the arrangements which the leading landlords of the county are at present adopting.

The Duke of Northumberland, whose great estates in this county are, with the exception of one or two farms, let on holdings from year to year, offers a re-valuation to every tenant who finds his farm too dear. This is made by two gentlemen in the district of high character and competence, and is binding on both landlord and tenant. If the latter, however, prefers to quit the farm, he is permitted to do so. The valuators assume a certain rate as the average price of wheat, and at this rate all the new valuations are made, so that the different farms may be placed on an equal footing, and the whole estate be in a position for an uniform principle to be adopted hereafter in any further adjustment that may be necessary. The result of these valuations has been a slight reduction of rent, though in one or two instances there was an increase. The Duke then makes a discretionary abatement, corresponding to the difference between the real and assumed rate of prices, and this has been 12 per cent. on the corn farms, and 8 per cent. on the turnip and stock farms. His Grace likewise executes the requisite drainage on his estates at a charge of 5 per cent. on the outlay, supplying the pipes from his kilns at prime cost. He expends what is necessary for the improvement of farm buildings without any charge on the tenant. In an instance where a lease and corn rent was desired by the tenant, the

Duke acquiesced. He offered leases to any of his tenants who might desire them, but it is said that his offer was accepted by only two farmers, — so highly does his Grace stand in the confidence of the county.

The Duke of Portland has a considerable estate in the county, much of it very poor land, but his liberality to his tenantry is proverbial. His abatements here are said to be equal to 30 per cent., and this on a rental confessed to be moderate. He drains to any extent for 5 per cent.

Earl Grey has not made a general abatement, but he relieves any tenant who wishes to quit his farm; and in a case where a tenant occupied several very valuable farms on Lord Grey's estate he was relieved of one which he desired to quit, and permitted to retain the others, which were better worth the money. Lord Grey executes any drainage that is required at a charge of 6 per cent.

Lord Vernon drains for his tenants, paying for both tiles and labour, free of all charge.

The Marquis of Waterford gives 20 per cent. of abatement, and makes a large expenditure in farm buildings for his tenants.

These instances may be taken as illustrating the manner in which the landlords of Northumberland are meeting the present crisis; though we regret to say, that on some estates of great extent the proprietors are so embarrassed that they plead their inability to meet their tenants either in one way or another.

In the southern division of the county the practice, in letting farms, has generally been by the agent naming the rent and offering the farm to a tenant. On the turnip farms of the North this has been the exception, the rule being to advertise for tenders and encourage the utmost competition.

Some particulars as to the cultivation of individual farms will best convey an idea of the agricultural practice of the different districts of the county. On a turnip farm belonging to the Duke of Portland, 400 acres in extent, the tenant grows sixty acres of turnips, twenty-five acres of which are swedes. He has a considerable portion of old grass; and the rest of the farm has been hitherto managed in a five-course. The tenant contemplates extending this to a six or seven-course, by keeping his land three or four years in grass. On this farm, which is superior turnip land, and considered a model in its neighbourhood of liberal management on the part of the tenant, four cattle are fattened in winter; and about thirty-six are half-fed in winter, and finished on the best grass in summer. A stock of eighty Leicester ewes is kept, the produce of which, about 120 lambs in number, are fattened on the farm. The best are ready as soon as they are shorn, and last year were sold at 33s. each, besides their wool, which fetched 7s. more. The rest are put on young grass, and sent off as they become ready. Five pairs of horses in winter, and six in summer, carry on the work of the farm. Ten carts of strawyard dung and twelve bushels of bones per acre, are applied to the turnip crop; but this is reckoned an unusually liberal dose. The old grass land is occasionally cut for hay, and seldom receives any manure. The stock are all fed in open yards during winter.

The ordinary management of strong-land farms is the four-course, occasionally extended by the introduction of beans. Very few turnips are grown; and, as nothing else is tried on the fallow division, nearly the whole of it is bare fallowed. Twenty-four bushels of wheat may be reckoned the average produce from the best land. One farm which we visited had been increased 20 per cent. in its produce of wheat by pipe drainage; and the farmer declared himself satisfied that this was a more valuable boon to him than a temporary abatement of rent to the same amount. On this, as on very many farms in the district, a steam engine is employed for thrashing the corn crop; but it is not applied to crushing corn, cutting hay or straw, steaming or cooking food, or, in fact, to any other purpose than thrashing. No manure whatever is purchased, and no cake, corn, or other feeding stuff.

On the home farm at Howick, the Hon. Captain Grey has been for some years back endeavouring to introduce many improved practices which the experience of other counties has sanctioned. By thoroughly draining, subsoiling, and heavily manuring his land, he has greatly increased its produce both of corn and green crops. The corn is all drilled and horse-hoed. The root crops are taken from the ground, without injury to its surface, by the use of Crosskill's portable railway. The rails are found very easy to shift; and the work goes on with great expedition. The swedes are carried on the rails to the headland, where they are stored, till required, in long narrow heaps, thatched with straw. A handsome cattle lodge has been erected for the stall and box-feeding of cattle, but on a scale unnecessarily costly. The weight of the different kinds of food consumed by the animals is registered; and their relative progress is ascertained monthly by putting each animal on the weighing machine.

Experiments have been made to ascertain the respective merits of stall and box-feeding. A trial last season proved the advantage of house feeding in summer, as compared with grazing in the field. Two short horned cattle, two and a half years old, as nearly of the same quality and condition as possible, were weighed on June 14th, and each found to be 78 stone. The one was turned out on good pasture; the other was put into a loose box in the cattle lodge, where it received cut grass, with the addition of 2 lb. of oilcake daily. The two cattle, with others likewise experimented on, were weighed every month; and, on October 22nd, the box-fed animal was found to have gained twenty-six stones, and the pastured one only thirteen. The saving in the consumption of food far more than compensated the cost of oilcake and attendance, so that the increased gain of weight, besides the accumulation of valuable manure, formed a clear advantage in favour of the box-feeding and soiling system. The other cattle submitted to the same experiment showed the same result, though not in an equal proportion of increase. The monthly weighings showed that, while the

B B 4

pasture was fresh and juicy, and the weather warm, the cattle made most progress in the field.   After the 27th of August, the grass began to fail in quality ; and the pastured lot then fell back greatly, while the box-fed lot continued to increase and improve.

An experim ent was tried, two years in succession, to ascertain the advantage of putting ewes on rape at the end of September, which clearly proved an increase of nearly 50 per cent. in the produce of lambs as the result of this practice.   Small highland cattle, bought at the August Falkirk tryst, at from 2*l.* 16*s.* to 3*l.* each, are found a very payingclass of stock here.   They are turned out on rough pasture, where they feed during the winter, getting but a very small allowance of turnips in snow or frost, and therefore kept at a trifling cost; then placed during the summer on better pasture, and sold off fat in October at 10*l.* to 11*l.* each.

The farm of Beal, on the line of railway between Belford and Berwick, is considered to be one of the best managed strong land farms in Northumberland.   It contains 1050 acres, 270 of which are permanent pasture.   The rest of it is managed on the four-course system, varied by having a portion of the clover-break in drilled beans, which removes the clover on that part to a greater interval, and makes it a more certain plant.   Where the beans have been taken the land is left two years in pasture. 360 acres are annually in white crop — wheat, barley, and oats ; 60 acres in beans ; 120 acres in sown grasses ; while 180 acres are in bare fallow and turnips, 80 of which are swedes and 80 white turnips.   The average yield of wheat may be 30 bushels, of oats (Hopetown or Angus) 44 bushels, of barley 40 bushels, and of beans 30 bushels per acre.   The whole of the fallow is manured with dung, the swedes with 20 carts of dung and 2 cwt. of guano per acre ; and the beans are manured with dung, and cultivated in raised ridges, 27 inches apart, like potatoes, and repeatedly horse and hand-hoed, and kept clean.

The stock kept on this farm consists of 25 calves reared ; 60 cattle are fed in the yards and boxes during the winter, 40 of which receive full feeding of swedes, with 4 lb. of oilcake and 4 lb. of bean meal each, daily, for the last six weeks, and go out of the yards fat ; 20 are half-fed in winter on turnips and swedes, and are finished off on the best grass early in summer. By this management, Mr. Scott, the tenant, finds he can advantageously keep a larger number of cattle to break down his straw into manure than if he were fully to fatten a smaller number on the same quantity of turnips. When the cattle go to grass to be fed off they receive 4 lb. of oilcake each daily, and with this addition to their food the field can fatten off 20 cattle better than it can 14 without cake, while the land is yearly improving in consequence. 240 Leicester ewes are kept, and their produce, 320 hoggets, are sold fat off the farm annually. The rest of the grass land is pastured by a flying or shifting stock. The lease was entered into in 1839, and the rent of the farm is 1700l., tithe-free.

Mr. Scott has tile-drained 600 acres of the farm at his own expense, the landlord having allowed him one acre of clay to burn at his own risk, but with leave to sell to others as well as to supply himself. At that time, 1839 to 1846, tileries were less common than they are now, and Mr. Scott found tilemaking a very profitable business; so much so, indeed, that the expense of draining his farm was thereby materially lightened.

The cattle are principally kept in courts with sheds, one lot being fed singly in boxes. Mr. Scott contemplates feeding a portion of his stock in the yards during the summer on cut grass, for the purpose of more perfectly converting his great bulk of straw into good manure. At present the manure heaps have to be turned twice or three times before they can be sufficiently rotted. The farm is clean, and in high condition; but the arrangement of the buildings, which are old, is very defective, and must occasion considerable waste of labour to the tenant.

The best mode of drainage, and the propriety of levelling down the old crooked ridges on the strong lands throughout this county, have been frequently brought to our notice. The practical men all agree as to the advantage of having pipes or tiles with a sufficient orifice to insure a circulation of air, as well as to carry off water. Anything less than two inches they think inadequate to this purpose. With regard to the depth of drains, there is likewise considerable difference of opinion, the most experienced being agreed on the advantage of a depth, if possible, of not less than $3\frac{1}{2}$ feet. In many cases in this county, however, drains are still being put in with the old and expensive 3-inch horseshoe tiles and soles, at 2 feet in depth and 15 feet apart, just as if the experience gained during the last few years in all other parts of the country had never been heard of here. Every farmer we have spoken to on the subject strenuously disapproves the ploughing down of the old crooked ridges. These were made ages ago by our forefathers, for the purpose of effecting that which we now more perfectly manage by under-drainage. But the system has been so long continued that the subsoil, when cut across (as was shown to us by Mr. Dand, of Field-house), presents the same curvature as the surface; and, in cases where the attempt to plough down has been carried out, the result is said to have been, that the top of the old ridge, being completely bared of good soil, produces little or nothing, while the old furrow, by being rendered too strong, gives a "lodged" and badly-filled crop. In a case in which the opinion of practical men is so unanimous, we think it behoves others with less experience to be cautious. Nor, indeed, after a field has been drained, is there any practical inconvenience worth mentioning in the shape of these crooked ridges. Mr. Scott, of Beal, never ploughs his down, but he drills his turnip crop across them; he draws off his land for sowing his corn crops across them, for the convenience of correct seeding and reaping; and he ploughs his lea across them for the same convenience. In working his green crop land he again forms the old ridge, and maintains it in its ancient shape.

Labourers are everywhere fully employed and at good wages.

LETTER XLIII.

NORTHUMBERLAND, *continued*.

FARMING ON TILLSIDE — MR. THOMSON, OF PASTON — ROTATION AND PRODUCE OF CROPS — MANAGEMENT OF SHEEP — THEIR PRICES AND THAT OF WOOL — FEEDING OF CATTLE — RENT, RATES, AND LABOUR — LINSEED GIVEN TO HORSES WITH ADVANTAGE — THE FARM OF WARK — WANDON — ADVANTAGE OF MORE LIBERAL EXPENDITURE IN MANURE — RUINOUS CONSEQUENCES OF UNDUE ENCOURAGEMENT OF COMPETITION FOR FARMS — FARMERS' CAPITAL — REDUCTION OF RENT — CONDITION OF LABOURER — COTTAGE ACCOMMODATION REQUIRES IMPROVEMENT — CONTRAST BETWEEN THE VILLAGES OF FORD AND WARK — DISCREDITABLE STATE OF THE LATTER.

WOOLER COTTAGE, Jan. 1851.

IN our last letter we gave some description of the strong-land farming of Northumberland. The district from which we now write is celebrated for its turnip farms, and for having originated that system of husbandry now generally recognised as the Northumberland, or five-course, viz. (1.) oats, (2.) turnips, (3.) wheat or barley, (4.) grass, and (5.) grass. This system is particularly favourably to an enlargement of farms. It affords a very regular succession of crops, and, the land being a friable turnip soil, it is well suited to sheep husbandry. It can be conducted with a moderate capital, and, being simple in its details, it is easily understood. Many of the farms along the base of the Cheviot range of hills are well situated for holding a stock farm in conjunction with the arable, on which a breeding stock can be cheaply kept, and its produce fed on the richer land adjoining. This is found a very safe kind of holding, as the sheep can be managed with little more expense than the wages of a shepherd, and they are brought down to the low grounds during severe weather, and sent to the high land when it is found necessary to relieve the pastures for a brief season.

The farm of Mr. Thomson, of Paston, affords a good illustra-

tion of the Northumberland style of farming. It has been held
by the present tenant during a lease of twenty-one years, just
expired, and a lease for the same period has now been entered
into, exactly on the same terms as before, except that the landlord is to expend such sums as may be necessary for draining at
5 per cent., and that he is to make some small addition to the
buildings without charge.

In explanation of this we cannot do better than use the words
of the tenant himself: —

" I believe I stated to you that I had retaken my farm at the old
rent. Had I lived under a game preserving, grasping landlord, I
should not have done so in such times of depression, but have insisted
upon a considerable reduction of rent, or at once have quitted the
holding. My landlord is always ready to contribute his fair share
of any improvement required on the estate. If manufactures and
commerce flourish, agriculture must participate ; I therefore trust,
notwithstanding all that has been told us to the contrary, to hold my
head above water."

It is proper to mention that, when this farm was taken 21
years ago, the general opinion of the county was that the rent
was too high, and that the tenant could not long continue to
hold it without an abatement. The farm is 900 acres in extent,
500 acres of which are arable, and 400 acres hill and low
pasture. Part of the grass land is of rich old feeding quality,
but the most of it is green hill pasture. The arable is all dry
turnip soil, some of it rather steep, and creeping too much up
the hill-sides; the rest undulating and well sheltered. The
500 acres of arable land are divided into five equal portions as
nearly as may be, 100 acres each in extent, embracing several
fields as nearly contiguous as the necessity for having an equal
distribution of the different qualities of land admits. Contiguity facilitates operations by preventing the shifting of
implements to distant points of the farm, as we have seen it
necessary to do in Devonshire, where a man may have 10 or
more little turnip fields scattered here and there about his farm,

thus causing constant waste of time and labour in fetching the implements and horses backwards and forwards from field to field. 100 acres are in oats and beans, 100 in turnips, 100 in wheat and barley, 100 in sown grasses, and 100 in second year's grass. Ten acres of the lea are sown with beans broadcast, $2\frac{1}{2}$ bushels per acre, which are carefully hand-hoed and kept quite clean. They yield an average of 30 bushels an acre, and are consumed by the stock on the farm, the straw being found very useful as fodder for the horses. In the East Riding we found that the farmers consider bean straw injurious to the health of horses, on account of the earth which adheres to the root, and from which it is not easy entirely to detach it, but no injury of that kind is here experienced. The remainder of the lea, 90 acres in extent, is sown with sandy oats, three and a half bushels an acre broadcast, and yields on an average 48 bushels an acre. After the oat crop is removed in autumn, one-third of the land is dunged on the stubble, with 10 tons of farmyard manure an acre, which is then turned in with the winter furrow. The rest of the stubble is ploughed at the same time. In spring the ground which was dunged is prepared as early as possible for swedes, to be sown in the month of May, in raised ridges 27 inches apart, in which a mixture consisting of $1\frac{1}{2}$ cwt. of guano and two bushels of dissolved bones have been previously applied. The rest of the turnip land is then prepared, and, when ridged up, it receives 10 tons of farmyard manure and the same quantity of guano and bones. On that part of the farm which is too steep to be easily accessible, the turnips are sown last, and, all the dung on the farm being by this time applied to the other land, it receives 2 cwt. of guano and two bushels of dissolved bones, and no dung. The turnips on this part are all eaten off with sheep. One-third of the whole turnip crop is swedes, one-third yellow bullock, and one-third white globe. As soon as the earliest turnips are consumed or removed from the ground the land is ploughed, and, if the season admits, sown with wheat. The wheat is sown broadcast, at the rate of two up to three bushels an acre, the latter quantity being used as the season progresses. It is sown at any favourable time from the middle of October till the 1st of March. The average produce is 30 bushels an acre. After the 1st of March the turnip land is sown with barley as early as it can be got ready, at the rate of three bushels an acre, and 36 to 40 bushels are reckoned an average yield. There are usually 80 acres of wheat and 20 acres of barley on the farm, the whole of which is sown out with grass and clover seeds. Of the 100 acres of young grass, 25 are cut green for soiling and hay for the farm horses and fat cattle; the rest is pastured, and the following year it is all pastured with sheep and cattle.

The sheep stock consists of 400 Leicester ewes, which rear, on an average, 540 lambs. The ewe stock is kept good by draughting all that have thrice borne lambs, and replacing them by an equal number of selected ewe hoggets. The lambs are fed the first winter on turnips, and in spring the swedes are cut for them as soon as their front teeth begin to fail. After they are shorn, they are grazed till October, when they are sold. Mr. Thomson, however, now intends to alter his system, as the taste of the public for fat meat is quite changed, and he is persuaded he will be better paid by selling his hoggets as soon as shorn, and before they become so very fat as they have hitherto been when kept six months longer. The selling prices of the sheep stock of this farm for the last twelve years are as follows :—

	Draught Ewes.		18 Months old Sheep.		Wool, per stone of 24lb.	
	s.	d.	s.	d.	s.	d.
1838	32	0	37	0	38	6
1839	32	0	39	0	29	3
1840	33	0	42	0	29	3
1841	33	0	41	0	24	0
1842	30	0	32	0	23	6
1843	26	0	30	0	22	6
1844	31	0	37	0	28	0
1845	35	0	42	0	28	0

	Draught Ewes.		18 Months old Sheep.		Wool, per stone of 24lb.	
	s.	d.	s.	d.	s.	d.
1846 ...	36	0	44	0	25	6
1847 ...	37	0	38	0	22	6
1848 ...	36	6	38	0	20	0
1849 ...	31	0	34	6	22	0
1850 ...	31	0	32	6	26	0

Twenty-five short-horned calves are reared on the farm and sold off fat at three years old. The average price for the last twenty years has been 20*l.* a-head. They are fed during the last winter in sheds and courts, four or five in each. Before being put up they receive turnips on their pastures during the month of September. They are then put into their yards, and supplied with as many turnips as they can consume. About Christmas they begin to receive cake and meal in lieu of part of the turnips, 2 lb. of oilcake and 2 lb. of bean-meal being mixed together and given to each animal, in two equal portions, morning and afternoon. This quantity is gradually increased to 4 lb. of cake and 4 lb. of meal for a month or six weeks before the animal is fat.

The farm is managed with sixteen work-horses. There are seven ploughmen, two spadesmen, two shepherds, one byreman, one steward, besides women for barn work and turnip-hoeing. The farm is let tithe-free, and the rent is 1,100*l.* a year, payable in money, without reference to the price of corn. The county rates are 2*d.* and the poor rate 4*d.* per pound, but there is not an idle or unemployed man in the district. Day labourers' wages are from 9*s.* to 10*s.* a week. The ploughmen are paid in corn and produce, the particulars of which will be afterwards referred to.

In feeding the farm horses Mr. Thomson gives them half a pound of boiled linseed every evening, mixed with their feed of bruised oats, during the time that they receive straw for their fodder in winter, and he has found this to benefit them greatly both in health and condition.

A brief consideration of this system will show that it is a quiet, safe, and inexpensive one. The whole of the sales consist of the produce of the farm exclusively. All the mutton, wool, beef, and corn have been produced on it. The farmer does not divide any of his profits with the breeder or cattle dealer. Excepting a few tons of cake and a moderate expenditure on guano and bones, his payments are limited to his rent and the wages of his day labourers, for the principal part of his labour (that of the ploughmen, shepherds, and steward) is paid for chiefly in produce.

Along Tweedside are some of the finest turnip farms in the kingdom. The farm of Wark, on the south bank of the Tweed, opposite Coldstream, is 930 acres in extent, 800 of which are kept constantly in tillage, under the four-course rotation. The rest is in permanent pasture. The greater part of the farm consists of rich deep friable land, laid out in large level fields, and capable of producing any description of crop in the greatest perfection and abundance. It has been lately entered on, for a nineteen years' lease, by the present tenant, Mr. Dove, who is carrying on improvements of every kind in the most spirited manner. The deepest land receives nearly 9 tons an acre of lime at a cost of 4l. 4s., the effects of which are expected to last during the lease. Two hundred acres of turnips are grown on this farm annually, manured with eighteen cartloads of dung and 2 to 3 cwt. of guano per acre ; eighty acres of the turnips are swedes, eighty white globe, and forty yellow. New farm buildings have been erected, which cover 1¾ acres of ground, and give accommodation in stalls, boxes, yards and byres, to eighty cattle feeding fat, fifty half feeding, twenty-four horses, besides cows, pigs, and poultry. The barn, granaries, straw houses, cart sheds, and carpenter's shop, occupy one side of the building. The whole has been erected at a cost, exclusive of cartage, not exceeding 2000l. ; but everything, including tradesmen's wages, have this last season been unusually low. Thirty-six bushels of wheat and seventy-two bushels of oats are reckoned on as the average produce of this farm. Besides the cattle already mentioned, a very large stock of sheep are brought down to the farm to be "finished" from other farms held by Mr. Dove, where they are reared.   The rent is payable one-half in money and one-half according to the annual price of corn, as ascertained by the averages of the adjoining county of Roxburgh, in which the crop is chiefly sold.

Mr. Maddison, of Wandon, occupies a farm of 500 acres under the Duke of Northumberland.   This farm was revalued in 1847, when the rent was reduced about 10 per cent.   It is now regulated by the prices of corn.   Besides reducing the rent, the landlord expended 1,800*l.* in new buildings without any charge, and executes what drainage is required at 5 per cent.   The buildings are compactly and judiciously arranged, box-feeding having superseded sheds and yards, and a railway being laid down to convey the corn from the stackyard to the barn.   The farm is managed on a five-course; and the average produce of this and similar land may be stated at 30 bushels of wheat, 36 of barley, and 44 to 48 bushels of oats per acre.   Seven pairs of work horses are required in summer and six pairs in winter. Twelve men and ten women or boys are employed in winter, with twenty additional hands for turnip work in summer, and about thirty-five more for a month in harvest.   The young cattle receive 1 lb. of oilcake per day with their turnips during winter, to prevent "blackleg;" and the fattening cattle, besides their turnips, receive 3 lb. each of oilcake daily for three months before they are considered ready for the fat market.

These instances explain the mode of farming adopted on the turnip farms of Northumberland.   It remains very much the same as it has been during the last thirty years, guano being now partially substituted for bones, which were then more extensively used.   There is no general attempt made to increase in any material degree the annual yield of the green crops by larger application of manure.   Box-feeding has been in one or two instances tried with partial success, the most observant farmers affirming that they are satisfied of its advantages over

the common system of yards and sheds, both in the progress of
the animal, the saving of food, and the superior quality of the
manure. The sheep are allowed to gnaw the turnip on the
ground, though most farmers use the turnip-cutter in spring,
when the front teeth of the young sheep are shed, and they
become unable to bite the root. On land of inferior quality
the hoggets are not hastened on by the assistance of corn and
cake, though that would materially increase the returns of the
succeeding corn crops. Far too great a proportion of the
comparatively innutritious white turnip is grown, — two-thirds
generally to one-third of swedes, though the vast superiority of
the latter in its feeding qualities is acknowledged by all experienced feeders of stock. An expenditure of 20s. or 30s. an
acre on artificial manures, in addition to what is generally
applied to the green crops in this district, would greatly
increase the crop, enable the farmer to keep larger stocks of
cattle and sheep, add largely to his dungheaps, and raise
altogether the character and produce of his farm. Walking
through a turnip field with a farmer, we came to a thin knoll,
on which the crop was as good as on the richer land beside it;
and he accounted for that by saying that he had applied to the
knoll a double quantity of guano. He had faith that the increased application would double his crop on the worst part of
his land, and admitted, that, though he had not thought of it, a
still greater effect would have followed on the land which was
naturally of richer quality.

The evils attending the system of letting land by tender, and
encouraging competition to the utmost, are too instructively
exemplified in the northern division of this county, and especially in that fine tract of arable land extending for some miles
up and down the south side of the Tweed, near Cornhill. The
farms are all first class, both in regard to quality of soil and
extent. They are very desirably situated, and possess a soil
suitable for every crop, admirably adapted for turnip husbandry,
and at the same time deep and strong enough for wheat.

Seven of these first-class farms, all contiguous, and the very pick of the county, tell the following tale. The first, after having been held seven years, was given up, offered to the public by advertisement, and then relet at a reduction of about 20 per cent. The second, the tenant having become bankrupt, has been let to a new tenant at a reduction of rent. The third was given up by the tenant, and has been relet to another at a reduction of about 22 per cent. The fourth, the tenant having failed, was let to a new tenant at a reduction of 13 per cent. The fifth, the tenant having also failed, has been relet to a new man. The sixth has been relet at a reduction of 20 per cent. The seventh has been given up, and is now offered at a reduction of 20 per cent. These are melancholy facts, and show beyond all question the disastrous results to which competition, unduly encouraged by the landlord, must inevitably lead. Tenants were invited to add farm to farm, with the idea that a man holding one farm, on which he lived, could afford a higher rent for another on which the expense of housekeeping was saved. Men were thus induced to extend their holdings far beyond their capital; but so long as the landlord saw his rents increasing he found no fault with the system, and perhaps gave himself no trouble to inquire into its probable results. The bubble has burst at last, and he pays dearly for his neglect, in having his farms thrown upon his hands during a period of unprecedented depreciation. But the loss falls still more irretrievably on the unfortunate tenant, who, being compelled to vacate during a period of transition, sacrifices from 30 to 40 per cent. of his capital, by being forced to realise at any price. The rents of several of the farms now referred to vary between 1,400*l.* and 2,200*l.* a year. One farmer paid for his various farms 7,700*l.* a year, 6,000*l.* of which was to one proprietor.

The experience which has thus been gained during the last four or five years is beginning to produce its natural fruit. The landlords who bore most hardly on their tenants are now carefully shunned, and find it difficult to get a good tenant for their vacant farms. The agents who, with more wisdom and prudence, selected their tenants for their qualifications of skill and adequate capital, and then let their farms at a fair value, have still plenty of applications when they have a farm to let. Landlords now see the advantage of having resident tenants to look minutely into every expenditure, and turn all things to the best advantage; and they are pressing their business on those agents whose prudence and foresight have retained for their employers a solvent tenantry and a certain rental. The encouragement of reckless competition has deceived both landlord and tenant, leading the one to anticipate a prosperity and to maintain a style which it is difficult for him to abandon, and leaving the other, after a brief period of apparent success, in hopeless embarrassment.

Our information, from very competent sources, leads us to say that, on the whole, farming has been a most unprofitable business in this county for the last few years. The capital considered requisite to carry on a farm is from 4l. to 5l. an acre, 3l. in many cases being nearer the reality. But want of industry and attention are worse qualities even than deficient capital; and many who have started with most capital have been least successful, chiefly through want of application.

Many considerable farms have recently been relet, nearly in every case, at a reduction of from 12 to 15 per cent. At this reduction, which is now readily conceded, desirable farms are sought after with much more anxiety than they were last year, principally by tenants changing from estates where no encouragement or assistance is given by the landlord. Tradesmen and shopkeepers from the country towns, and other small capitalists, are also in many cases going into farming as a business. Strong clay land farms are here, as everywhere else, completely gone out of favour; and a deduction on such is both more required and more readily conceded.

The condition of the labourers in Northumberland, in so far

as regards wages and food, is much better than in most of the southern counties.   Their wages are paid chiefly in corn ; and they generally keep a cow.   A hind or ploughman receives for the year 42 bushels of oats, 24 bushels of barley, 12 bushels of beans or peas, 3 bushels of wheat,—all of the best quality produced on the farm.   He has a cow kept summer and winter ; one-sixth of an acre of potato ground, 12 lb. of wool, 4*l.* in cash, and a house and garden rent-free.   He is bound to keep a girl or strong boy to work on the farm, when required, at 1*d.* per hour.   The bailiff or head man receives from 2*l.* to 6*l.* more. When corn sold at high prices, prudent managers turned their produce wages to good account ; and there are several instances in which hinds have been enabled to take farms, and are now, by industry and intelligence, among the most successful and extensive farmers in the district.

But the state of the labourers' cottages throughout Northumberland is, in the majority of cases, most discreditable to the county.   It will hardly be believed that the labourer's cow and his pig are still lodged, in too many cases, under the same roof, and go out and in by the same door as himself and his family, the cow house being divided only by a slight partition wall from the single apartment which serves for kitchen, living and sleeping room, for all the inmates.   That great exertions are being made by many of the landlords of the district to remedy this state of matters, is true ; and by none more than the Duke of Northumberland, whose munificent outlay will soon provide comfortable lodging for every labourer on his great estates.

The village of Ford, built by the Marquis of Waterford, on a plan prepared by Mr. Stewart, of London, presents a great contrast to the general character of . Northumbrian villages.   The cottages are placed singly and in groups, with convenient outhouses and waterpipes, while at some distance apart is a general cowhouse and barn.   Each cottage contains two or four rooms, and, with a garden, is let, according to size, at 3*l.* to 4*l.* which includes the use of the cowhouse, and the privilege of turning a cow upon the adjoining common reserved for that purpose by the landlord.   At the upper part of the village is a reservoir of water for the supply of the cottages, and drinking-ponds for the cows, picturesquely laid out with shrubs and walks, and green banks where, on a summer evening, the labourer may stretch his weary limbs, and look down over as sweet a landscape as can be desired.   Let us describe another village—that at Wark Castle, on the banks of the " Silver Tweed."   From the top of the lofty mound which is now all that remains of that proud old border castle, the eye rests with delight on the rich and fertile vale through which the river winds in graceful sweeps, here shaded by groups of lofty trees, there gliding slowly past far-stretching holms which every returning harvest covers with golden corn.   Beside us is the village itself, the very picture of slovenliness and neglect. Wretched houses piled here and there without order—filth of every kind scattered about or heaped up against the walls— horses, cows, and pigs lodged under the same roof with their owners, and entering by the same door—in many cases a pig-stye beneath the only window of the dwelling—300 people, 60 horses, and 50 cows, besides hosts of pigs and poultry—such is the village of Wark, in Northumberland.   We have been in some of the most wretched villages in Ireland, betraying poverty far greater than this, but nothing more abject in filth and uncleanliness.

LETTER XLIV.

DERBYSHIRE.

MATLOCK, Feb. 1851.

ON a wet and stormy winter night we arrived by railway at
Matlock, unable to get a glimpse of the country, the only
outline indication of which was the glimmering of window-
lights — some beneath us, some at the same level, and some far
above us. The morning opened with a crisp light frost, and
presented a scene from the esplanade in front of Mr. Greaves's
hotel, contrasting pleasantly with the plains of the Vale of
York, which we had traversed on the preceding day. Deep in
the recess of a rocky valley lay the little village of Matlock
Bath, while dotted about the sunny face of the steep hill by
which it is shut in, at all various points of elevation, stood
villas of different architecture, prettily interspersed with trees
and little patches of green field or garden ground. In the
bottom of the valley flowed the swollen Derwent, brown with
winter floods, hemmed in on its opposite bank by lofty crags of
mountain limestone, every crevice of which and each bluff pro-
jecting eminence were clothed with wood. Entering the railway

carriage, we were soon whirled through a tunnel about a mile in length, whence we emerged on a more open country skirting the stream of the Derwent. Along the river the fields are chiefly in grass, forming the meadows which yield winter food for the stock of the dairy farmers.

The southern and northern parts of Derbyshire are very dissimilar ; the former being in many respects like the adjoining parts of Leicester and Stafford, — the latter celebrated for the beauty of its scenery, and its constant succession of hill and valley. The back-bone of England commences on the moors of North Derbyshire, whence a continuous range of mountain stretches northwards by Yorkshire and the northern counties to the Scottish border. The principal geological features of this district are the grit and mountain limestone, the river Derwent forming the boundary of each. On the grit the soil is earliest, and vegetation springs rapidly. On the limestone the land is richest, and its pasture stands out longest. The best feeding pastures, and those which yield the highest rent to the land-lord and the largest returns to the farmer, are found on the latter. The High Peak is a region of bleak high moors, inter-sected by deep valleys, where the native breed of white-faced moor sheep are the only stock that the severity of the climate admits. It is very subject to violent storms of wind and rain, which, with the high elevation of the country, render it cold and backward, and the vegetation more bulky than nutritious.

In the lower country, within the limits of profitable cultiva-tion, the land is still very hilly, but it is cultivated to the tops of the hills. Wheat, which is grown as high as 600 feet above the level of the sea, does not, on the whole, succeed well. It is generally thick chaffed, and does not yield in proportion to the bulk of straw. Oats are more common and much more to be depended on. They are grown successfully at an elevation of 900 feet. Nine-tenths of the county are in grass. It is a dairy and rearing district, the growth of corn being of quite inferior consideration. The farmers of the lower hills rear, from their dairy stocks, short-horned cattle, which are sold to the graziers of the low country to be fattened. The same practice is generally followed with Leicester sheep, which, after the first winter, are passed off to the richer clover of the low country farmer. In some places, however, the pastures are of very rich feeding quality, and for the most part they are sweet and healthy for stock.

Passing from a general description, we proceed to detail the management of the farm of Birchills, on the Duke of Devonshire's estate in the parish of Bakewell, and occupied by Mr. Furniss. It is 300 acres in extent, 100 acres of which are permanent pasture and meadow, and 200 acres tillage, the half of which is annually in crop. The course of cropping is not very clearly defined, the great object with Mr. Furniss, as with all tenants of highly manured farms, being to grow as heavy crops of the most valuable kinds as can be grown without the risk of lodging. The fields are laid out in divisions of twelve or thirteen acres each, the landlord paying the expense of new fences, and the tenant doing the team work. Drains, where requisite, are unskilfully made, only thirty inches in depth, and still laid with the expensive three-inch tiles and soles; all at the Duke's charge, except team work. On this outlay no interest has yet been charged. The soil is a fine friable loam, with a considerably undulating surface.

On breaking up from grass the first crop is usually oats, sown on one furrow, and yielding an excellent crop, seventy-two bushels to the acre, and weighing 42 lb. to the bushel. The next crop is swedes, for which the land receives the usual autumn and spring cultivation. In the end of May it is drawn into ridges, about twenty-eight inches apart, into which fifteen tons of well-rotted dung and sixteen bushels of bones are applied per acre, and the seed is then sown. This is uniformly a successful crop, the average for the present year, as weighed on the ground by impartial judges, being twenty-seven tons an acre. As the winters are generally severe, the swede crop is taken up before winter, and stored in little heaps in the field, containing about one ton and a half in each, covered with straw and nine inches of soil. In these the turnips remain protected from the weather, or the depredation of game, till wanted. But the expense of this operation seems very great, being not less than 15s. an acre for lifting and pitting, besides cartage. The turnips are followed by oats or barley, which are sown out with red clover or mixed seeds, part of which are mown and part pastured. The "seeds" are dressed with dung, the clover does not require it. The clover stubble, or second year's " seeds," as may be, is broken up, and, after being ploughed, is pressed, and sown broadcast with red lammas or Burwell wheat, eight to ten pecks to the acre. The yield varies from thirty to thirty-six bushels. Instead of swedes or turnips, rape is sometimes taken, sown in ridges at the same distance, and manured with bones. The rape is eaten off with sheep early in autumn, and the land is then sown with wheat.

Seventy head of cattle, young and old, are kept on this farm. They are all high bred short-horns. A milking stock of twelve cows is kept, which, besides rearing twenty calves (the requisite number of young calves being purchased in the neighbourhood to make up this quantity), yield, in cheese and butter, about 8l. each. The calves are fed for the first fortnight with six quarts a day of new milk (three quarts at a time); after that, with two gallons a day of skimmed milk, with which half a pound of boiled linseed is mixed. This is continued till the calves can help themselves to other food. The sales annually consist of twenty three-year olds, either fat oxen, or in-calf heifers, at an average price of 10l. to 12l. each.

The sheep stock comprises 110 high-bred Leicester ewes, the produce of which used to be sold at twenty months old, but, by an improved method of winter feeding, are now ready for the fat-market in little more than twelve months. In a sheltered situation not far from the farm buildings, and so placed as to admit of access on all sides to different pasture fields, a nicely contrived establishment for the winter feeding of sheep has been erected.   It comprises, in the centre, a house for turnips, with a loft over for hay and cake or corn, and accommodation for the shepherd in the lambing season.   Behind this are two yards, open in the centre, and shedded all round for shelter.   A rack for hay runs right round the shed, and under it a manger for the cut turnips and cake.   The yard is littered with straw or haulm. Each yard opens into a pasture field, to which the sheep have access for exercise.   As soon as the pastures fail in autumn the young sheep are put into these yards, and there receive the whole of their food.   They have cut swedes twice a day, 1 lb. of oats each, and ½ lb. of cake each, for the last ten weeks, besides hay in the rack.   Both fleece and carcase are improved by this management, food is economised, and the stock are less subject to casualties.   The year-old sheep sell at 30s. each, besides their wool, at present worth 6s. more.   The ewes are lambed in one of these yards.

Five horses and a riding nag do the horse-work of the farm. The manual labour is performed by six men and a boy, four of whom are boarded in the farm-house, and three are on weekly wages.   These average 10s. to 12s., without beer.   The in-door men have 10l. to 12l., besides their food, which, as it is very substantial fare, we may detail for the instruction of some of the large corn farmers in the Southern counties, whose poorly paid labourers must often go to bed on a supper of bread and water. For breakfast they have porridge, then bread and cheese.   They take with them to the field each man his pint of ale, and as much bread and cheese as he likes.   At one o'clock they have dinner, which is either bacon, beef, or mutton, and pudding, with small beer *ad libitum.*   At seven o'clock they have supper of milk porridge, then bread and cheese.   The men are stout and muscular, and work hard.   During harvest they have a quart of ale per day.

Since 1831 the stock kept on this farm has been much increased, whilst at the same time a larger extent of land is now under crop.   The yield per acre has also greatly increased.

The rent of the farm is 24s. an acre ; poor and all other rates
1s. 6d. per pound.

On the home-farm of Mr. Thornhill, of Stanton, near Bakewell, great improvements have been effected, and, as they
illustrate the advantage of such improvements, and show how
much may yet be done by well-directed enterprise, to increase
the produce of our fields and give employment to labour,
we shall describe them somewhat in detail.   The farm extends to 400 acres, 200 of which are grass and 200 arable.
Mr. Thornhill took it into his own hands in 1840.   The
farm then kept 16 cows, producing 2½ cwt. of cheese each.
About six young cattle and 50 to 60 sheep were sold off
the farm annually.   Four farm horses were employed in working it ; and, besides an annual produce of 60 quarters of oats,
there might be, once in three years or so, a field of five or
six acres of the best land in wheat, which, after a clean summer
fallow, yielded 27 bushels an acre.   Such was the whole
produce of the farm in stock and corn.   It now maintains a
regular stock of 43 milch cows and their produce, 30 of which
are sold fat every year at three years old.   Each cow, besides
rearing a calf, produces equal to 4 cwt. of new milk cheese.
200 sheep, old and young, are now kept on the farm, and
160l. worth of pigs were last year sold off it.   The average
yield of wheat is now 40 bushels an acre, and of oats 60
bushels.

The land lies on the gritstone, and is all on a considerable
slope, the lowest part being 620 feet above sea level, from which
it rises over the top of the hill to an elevation of 900 feet.   It
is well sheltered by plantations and good stone walls, and the
fields have been laid out in convenient enclosures.   The
soil is now dry and friable, and the field operations can be conducted without impediment.   To render it so a very large
expenditure has been incurred, the land having been full of
great blocks of stones, all of which have been removed, either
by being broken and placed in drains, or by being carried bodily

from the field, or by being broken to pieces, and then covered with trenched earth to a depth beyond the reach of the plough. This latter operation is at present being carried into effect on a corner of a field, for the purpose of making the fence straight. The ground is literally paved with huge blocks of gritstone, which are blown to pieces by gunpowder, or split by wedges, and then, after being spread along the face of a trench, are covered to a considerable depth by fine friable soil, got by the workmen in great abundance under the bed of the different massive blocks as they are removed. The cost of this operation is 50*l*. an acre, and can only be justified on the score of convenience in laying out the adjoining better land. But the reclamation of the whole farm has been an expensive operation, 200 acres of it having cost 15*l*. an acre for drainage, trenching, and fences.

The arable land is managed on the four-course system, with this peculiarity, that on the upper land oats are the only corn crop taken, and on the lower and richer land wheat only. On the upper land the turnips and clover are both eaten on the land, the sheep getting also cake or corn. On the lower land the turnips are drawn for consumption in the stalls, and the clover is cut for soiling or for hay. The general style of management is as follows: — (1.) The "seeds," which are a mixture of 14 lb. of red clover and two pecks of Italian rye-grass per acre, are watered with liquid manure from the tank in April. The first cut is made into hay, and the ground is then watered a second time with excellent effect. The second cut is given to the horses, and to the cows when the grass on the pastures begins to fail, in August, at which time the gritstone land gives way, and the cows fall off in produce a half-cwt. of cheese as compared with those fed on limestone land. The cut grass more than counterbalances this natural defect of the soil, the increase of produce in consequence of this additional food being from a half-cwt. to one cwt. of cheese each. The whole of this land is ploughed up for wheat in October, the worst of it being first dressed with ten tons of farmyard dung per acre. The land is then sown with (2.) wheat, eight to ten pecks of Spalding's Prolific being drilled across it, in rows of seven to eight inches apart. The wheat-crop is never hoed. Last year the average yield was forty-eight bushels an acre. When the crop has been harvested, the stubble is gone over by men with forks, who fork out all the twich. This, after being exposed to the weather, is gathered into heaps and mixed with lime. The land is then ploughed and prepared in spring for (3.) swedes, mangold, and yellow bullock turnips. The swedes are sown in the end of May, twenty tons of dung being previously spread in the ridges. The crop averages twenty tons. It is in all cases drawn in autumn, and pitted. The other green crops are treated in the same way. On the most distant and elevated fields sixteen bushels of bones and one cwt. of guano per acre are used without dung, which cannot be conveniently taken so far; but the crop is there consumed on the field by sheep, the turnips having been previously taken up and pitted in little heaps, to preserve them from frost or other injury. The turnips are taken out of the little pits as required, and given, cut, to the sheep in troughs, with $\frac{1}{2}$ lb. to 1 lb. of cake each daily. The green crop is followed by wheat on the best land, by oats on the inferior land.

The cattle being all fed in stalls, and the buildings spouted to carry off rain-water, a large quantity of liquid manure is collected in an underground tank, which is found most valuable as an application to young grass.

The dairy produce chiefly consists of cheese, which weigh from 27 lb. to 30 lb. each. They are coloured, and salted by being placed in brine in a trough for two days. The calves are fed for the first fortnight on four quarts of new milk a day each, for the second fortnight on six quarts, and after that on scalded whey and 1 lb. of oilcake, steeped over-night in boiling water and hay tea.

The accounts on this farm are kept minutely and accurately, and for last year they show a charge, in addition to the old rent, of 7 per cent. interest on expenditure on buildings, 5 per cent. on other permanent improvements, 10 per cent. on implements, 10 per cent. on live stock,— amounting altogether to a charge of 885*l*. against the farm for rent and interest of capital. After deducting an abatement of 10 per cent. on the rent for "present prices," and adding the usual expenses of cultivation, the produce of the farm in stock and crop last year leaves a balance over to the credit of the farm. Mr. Thornhill has, therefore, the satisfaction of having furnished remunerative employment to a large extent by his enterprise, besides ameliorating the face of the country and engaging himself in an occupation most useful to the neighbourhood, and which not only does not interfere with, but adds zest and interest to the other occupations of a resident landlord.

On the farm of Ashford, occupied by the Hon. Mr. Cavendish, M.P., and situated 600 feet above the sea, it has been found necessary, on account of the rankness of the crops, to adopt, a six-course, with two successive corn crops, as follows:— (1) clover, (2) oats, (3) wheat, (4) turnips, or mangold, (5) potatoes, (6) winter barley. The winter barley is eaten down in April or May by sheep, affording, at that time, very useful feed, and is then left for a crop, which last year proved a heavy one. The sheep, which are Shropshire Downs, are fed during the winter in sheds and yards; the cattle are wintered in stalls.

There is a privilege, in this part of the country, enjoyed by the public, which very much interferes with the economy of an arable farm. In the mineral districts, and on " King's field,"— that is where the Sovereign is lord of the manor,—in the Duchy of Lancaster, any one may enter where he likes, or whatever crop may be in the particular field, and dig for ore without paying damages to the owner or farmer of the land. He has a right to keep the pit open for a certain time, and can extend that time by occasional workings. There are some restrictions connected with the exercise of this privilege, but not in any degree commensurate with the injury done to the surface where the pits are opened in valuable tillage or grazing land.

There is no general custom of compensation to outgoing tenants for manures or management; but on the Duke of Devonshire's estate a special agreement has been introduced, by which tenants receive the following payments: — for labour and manure on fallows the last year of lease; for lime, its value as for two years on ploughed land, for seven on pasture; for purchased manure, as for two years; for inch bones, four years; for bone-dust, three years on tillage land, and double that time on grass land; the price of " seeds; " the expense of paring and burning for the turnip crop; for drains, as for seven years; fences, seven years; and anything further that, in the discretion of the Duke's agents, the tenant may have a just claim to.   On a farm of 268 acres, the payment made by an incoming to an outgoing tenant, under this agreement, was summed up thus: — "Amount of tillages, including the above items, 568*l*. 1*s*. 10*d*.," or rather over 42*s*. an acre.

In Bakewell there are no poor but the frame stocking knitters, who were established in their trade before power-looms were invented.   They still continue to work at a business to which they were brought up, although it scarcely now affords them maintenance.

LETTER XLV.

DERBYSHIRE — *continued.*

CHATSWORTH — EDENSOR — DUKE OF DEVONSHIRE'S MODE OF LETTING PART OF HIS PARK TO THE LABOURERS AND TENANTS — DUKE OF RUTLAND'S REVALUATION — CAUSE OF RENTS NOT FALLING WITH PRICE OF WHEAT — GAME DAMAGES — BUILDINGS ERECTED BY LANDLORD — RENT OF LAND — BARBAROUS SUCCESSION OF CROPS — ECONOMICAL MODE OF FEEDING DAIRY STOCK — CULTIVATION OF FIELD CABBAGE AT DRIFFIELD — SUCCESSFUL TOP DRESSING FOR POOR PASTURE — DRAINS TOO SHALLOW — WAGES, RATES, AND RENT — DERBYSHIRE AND NORTHUMBERLAND COMPARED. — RUTLAND — LAND NEAR STAMFORD, WHICH ARTHUR YOUNG COMPLAINED OF AS UNENCLOSED, REMAINS SO STILL — PRICES IN 1778 AND 1851 — BURLEIGH — DESPONDENCE OF FARMERS — DESTRUCTIVE EFFECTS OF GAME — IMPROVEMENT FORCED BY LOW PRICES — REVALUATION MAY BE UNJUST TO A GOOD FARMER.

DERBY, Feb. 1851.

A DESCRIPTION of Chatsworth, the residence of the Duke of Devonshire, does not fall within our province, though no one ought to pass through that part of Derbyshire without spending a few hours among its varied beauties, both natural and artificial. It is a privilege which all are alike capable of enjoying; and it is a trait in the Duke's character worthy of being mentioned, that he takes a particular pleasure in witnessing the gratified and happy countenances of the wondering artisans and their families, who are brought up in crowds from the " black " country in Staffordshire by the excursion-trains, and are permitted to walk through and inspect his superb apartments and ornate grounds.

But the village of Edensor which the Duke has erected within the park for the accommodation of his labourers, and the arrangements he has made for their comfort, may be briefly described. It comprises the parish church; a commodious and elegant school, and a considerable number of cottages, standing singly or in groups, and all disposed in such a way as to produce the most

D D

pleasing diversities of effect. They are constructed substantially of white freestone, with variegated roofs, and interspersed with pretty green slopes and shrubs; their pointed gables, Italian towers, and snug picturesque little porches show that here the labourer has both a comfortable and an elegant home. The park itself is partly devoted to their comfort, the best of it being reserved for the cows of the cottagers and labourers on the estate. The rates paid by the labourers for joisting a cow are from 50s. to 55s., which are very moderate, and must add much to the comfort of a labourer's fireside.

Another part of the park, about 300 acres in extent, is joisted to the tenants, who are thereby enabled to ease their farms of young stock in summer, and to reserve part of their grass for hay. The rate charged to the farmers for year-olds is 25s.; for two-year-olds, 35s.; for young horses, 50s. each; and for a mare and foal, 5l. We are persuaded that this is a plan which might be advantageously adopted on many large estates, and which would afford, on moderate terms, very useful keep to the neighbouring tenantry, and possibly with more direct advantage to the proprietor than he, on the average, secures from speculating in the grazing of cattle on his own account.

The principal proprietors in this part of the county are the Dukes of Devonshire and Rutland. On neither estate have there been any reductions of rent, but both are believed to be moderately let. On the Duke of Rutland's estate it has been thought proper, on account of the complaints of the tenants, to order a revaluation. That has not yet been completed; but, so far as it has gone, the farms are now revalued at the same rent, with the average of wheat under 40s., as they bore eighteen years ago, when the average was 65s. This is thought a wonderful result, and only to be accounted for by supposing that at that time the farms were greatly undervalued. The solution, in our opinion, may be found, without casting any imputation on the skill or good faith of the former valuators, by considering that in a district where nine tenths of the land are in grass, the price of wheat cannot truly regulate the value of the soil to the farmer. His main dependence is on stock; and the price of meat of every kind, and of dairy produce, wool, and vegetables, is as good as it was then, while the demand for these articles is constantly increasing. Not a farm on this extensive division of the Duke of Rutland's estates has been given up. If by any chance one becomes vacant, there are many competitors for it, the Duke's character as a landlord standing very high with his tenantry. But not only are there no farms vacant, but we were told, on most competent authority, that at the rent audits there had not been more than one defaulter for the last five years, and at the last audit only one complaint of the times, among a tenantry numbering 1100, including village and cottage as well as farm tenants.

The farms are generally small, being from 50 to 100 acres in extent. Where drainage is required, the Duke pays half the expenses, the tenant charging all team work as part of his share. The stock being chiefly for dairy purposes and fed in stalls, the liquid flowing from them is collected in tanks made at the landlord's expense. In the neighbourhood of Rowsley the tenantry used to suffer very serious loss by the strictness with which game was preserved; but the hares and rabbits have been greatly reduced, and the winged game only is now preserved so strictly. The damages paid by the Duke to the tenants last year for the destruction of their crops by game were 600*l*. His Grace makes no charge for repairs or additions to farm buildings, looking upon this outlay as a landlord's investment, which is as requisite to enable the tenant to pay his rent as the possession of the land itself. There are few changes of tenancies, but the farms are periodically revalued. One case, of rather an instructive character, may be mentioned, where a man of slovenly habits fell into arrear with his rent, and got behind in every way. He was warned that he must either improve his habits or leave the farm. He did improve, and has now paid off all arrears, and has his farm better stocked than ever it was before.

The rent of the best grass land may be stated at 2*l.* per acre; arable, 15*s.* to 30*s.* — all tithe free.

The eastern district of North Derbyshire is more of an arable country. Great improvements have been and are being made by Mr. Arkwright, of Sutton, on his extensive estates, which we regretted that we had not an opportunity of visiting and inspecting.

In the southern division of the county, near Duffield, the land is chiefly in grass, the best of which lets from 40*s.* to 60*s.* an acre. In preparation for hay, it is top-dressed with dung. The hedges are neatly trimmed, but the fields are small, and encumbered with numerous hedge-row trees. The corn is sown broadcast. Some miles further to the north, the land, while under crop, is not uncommonly cropped in this barbarous manner — (1) oats, (2) wheat, (3) oats, (4) fallow, (5) wheat, (6) "seeds," (7) wheat. Good green crop land is fallowed when it becomes too foul to bear a crop. It is then limed, and cropped again. Many farms have no turnips whatever; and the accommodation for stock is generally defective. There are cases, however, of much better management.

The following particulars of a farm in the parish or district of Shottle may convey an idea of the system pursued. The farm consists of 130 acres, between 30 and 40 of which are under tillage. A portion is under meadow and old pasture, and the rest in pasture in rotation with arable. Two successive crops of oats are taken when the land is broken up from grass; then fallow, limed and dunged; then wheat sown out with "seeds," which remain in grass two years more. The oats are said to average 32 bushels an acre — certainly a very moderate crop, considering the quality of the land. There are twenty-two dairy cows kept on this farm, — short-horns or crosses. They are housed during the winter, and get very few turnips, but are kept in fresh condition by the following mixture of food, in addition to their fodder of oat-straw : — The refuse of the oat-straw in the cribs and any damaged or inferior hay are cut into chaff, over which is poured half a peck of ground linseed, which has been previously steeped twenty-four hours in cold water. The mixture, which is damped a little, begins to heat in twenty-four hours, and is then given to the cows once a day with a handful of bruised oats, but the half-peck of linseed mixture serves the whole stock three days. When the cows calve, their daily allowance is doubled, with a few turnips besides. This feeding is said to keep the dairy cows in good healthy condition when not giving milk, and it certainly is not costly. The dung on this farm is carefully managed, and mixed with absorbent earth, and the whole liquid is collected in a tank, whence it is pumped over the dungheap during the winter, and taken out to the meadow in spring.

On the land farmed by Mr. Bell Crompton, of Duffield Hall, a stock of Ayrshire dairy cows has been successfully introduced. Mr. Crompton finds them excellent "doers," and more profitable than any other stock on his land, considering the quantity and quality of food they consume. The produce of these cows by a short-horned bull are very fine animals. He grows the large drum-head cabbage for feeding his cows, the young plants being transplanted into the field in June, at one yard apart every way, and manured with seven tons of dung per acre. The plants are laid in every third furrow, a forkfull of dung being placed on each plant, which is then covered up by the next furrow. They are taken up when most convenient in winter, the good cabbages being carted off to a plot of ground near the feeding houses, and there placed top downwards, each plant on the ground, where they remain fresh till wanted. The bad plants are used at once by the pigs and young stock, but none by the milch cows, as they would affect the taste of the milk injuriously.

We may mention a top-dressing which has been used here with great success on poor pasture. It consisted of 2 cwt. of rape dust, 3 cwt. of superphosphate, and 1 cwt. of salt per acre, mixed together and applied early in spring. Three cwt. of guano tried beside it caused a more rapid growth of coarse grass, but the former raised the thickest and most nutritious herbage, especially so of clover. Mr. Crompton finds it a good plan to mix his new-mown hay, when only one day cut, in layers with oat or wheat straw. The juice and flavour of the hay make the straw palatable to the stock, and the mixture is eaten eagerly by the milch cows and young cattle in winter. The liquid manure is here also carefully collected in tanks, and used with much advantage as a top-dressing on grass.

There is still a great deal to be done by drainage in improving the moister part of this district. The drains which are made are too shallow, 2 to 2½ feet being the general depth.

Labourers' wages are from 10s. to 12s. a week; cottage rents, with gardens, from 3d. to 1s. 6d. per week; poor rates, 7½d. in the pound; and the rent of land about 28s. an acre.

Before ending the description of Derbyshire we may mention the general impression made on us, and the contrast afforded with the county we had previously visited — Northumberland. The situation and soil of the two counties are certainly very different, but not more so than the state of agriculture. The rate of rent, wages, and taxes of all kinds, in Derbyshire is higher than in Northumberland. The farms are better cultivated, and the farmers infinitely more prosperous and contented. In Derbyshire the land is chiefly in grass, carefully managed, and the small proportion of ploughed land receives minute attention. The farms are small comparatively, being from 100 to 300 acres, and the farmers superintend their own business. They are not encouraged by their landlords to add farm to farm without being provided with adequate capital. They depend for their returns more on the produce of the dairy, breeding, and sheep stock, than on corn. The low country of Northumberland, again, is chiefly under the plough, most of it undrained, the small farms held by men of insufficient capital, the large ones by men who had capital, but who have been tempted to dissipate it over far too great an extent of land; the price of corn has failed them, and they have little stock to fall back upon. They have overploughed and entangled themselves with large undrained farms, the returns from which will not pay the expenses of cultivation. Derbyshire is a pleasant, picturesque county, in which landlords, tenants, and labourers seem mutually content, where the pastures are well managed, the ploughed lands neatly cultivated, and the stock suitable to the soil and carefully tended.

Passing through Leicestershire, we traversed the small county of Rutland, which seemed undulating and well wooded. The grass land management appeared to be very good; the turnips inferior. Close round Stamford much of the land is still uninclosed, and held in little patches by farmers whose fields are intermixed with each other. Eighty years ago Arthur Young described it exactly as it is at present, adding then that "it is melancholy to think that, in an age wherein the benefits of inclosing are so well understood, such tracts should remain in such a comparatively unprofitable state." And yet so they remain to this day!

The following is a table of the prices in this neighbourhood in

	1770.	1851.
Beef, per lb., ...	3d. ...	5½d.
Mutton, per lb.	3d. ...	5d.
Butter, per lb.	6d. ...	1s.
Pork, per lb.	... ...	5d.
Milk, per quart...	...	2d.
Bread, per lb.	2d. ...	1¼d.
Wheat, per quarter ...	41s. 4d. ...	40s.
Labourers' wages per week	6s. ...	10s.
Women at weeding corn and haymaking		8d. to 1s. a day
Boys who can plough		5s. a week without food.
Cottage rents	{ 20s. with an acre of land	30s. on great estates, with 1 rood of land; 80s. in open villages, with small garden.
Average produce of wheat on good sandy loam	} 20 bushels...	28 bushels.
Rent per acre of farms	5s. to 7s. ...	20s. to 30s.

At Stamford we passed into Northamptonshire, obtaining a glimpse of the Marquis of Exeter's finely wooded park and mansion of Burleigh. This magnificent place, founded by Queen Elizabeth's lord treasurer, Cecil, with its grand old trees and noble park, is just the place a foreigner should see to give him an idea of the wealth of our English nobility.

The tenants on this estate are represented as being in the most hopeless state of despondence on account of the present low prices of agricultural produce; and, as they were complaining vehemently, the Marquis offered to have the farms of any who desired it revalued. Only one on this great estate accepted the offer. There have been no farms of any consequence yet given up, and for those which do come into the market there are plenty of offerers, though men of capital are become chary, and will only look at very desirable farms. The estate is said to be low rented. Small farmers, of whom there are many, are suffering most severely, as they have not saved anything in good times to fall back upon now. Some of them are, indeed, greatly reduced; and we heard of one who had applied to his parish for relief. Others have sold everything off their farms, and some, we were told, had not even seed corn left with which to sow their fields.

In a fine country, with a gently undulating surface and a soil dry and easy of culture, laid into large fields moderately rented, one is surprised to hear that there is so much complaint and so much real suffering among the poorer class of farmers. It is only in part accounted for by the devastation of game, which on this and some other noblemen's estates in North Northamptonshire, is still most strictly preserved. On the 24th of January last, seven guns, as we were told, on the Marquis's estate, killed 430 head of game, — a most immoderate quantity at such a late period of the season. The fields are all stuck about with bushes to prevent the poachers netting; and the farmers feel most severely the losses they sustain in order that their landlord and his friends may not be deprived of their sport.

The strict preservation of game on this and some other estates in the northern parts of the county, was described to us in the bitterest terms, as " completely eating up the tenant farmer, and against which no man can farm or live upon a farm." It is " the last ounce that breaks the camel's back ;" and men who might have made a manful struggle against blighted crops and low prices, are overborne by a burden which they feel to be needlessly inflicted, and of which they dare not openly complain.

In consequence of the distress among the small farmers, many of the labourers would have been thrown out of employment had work not been found for them by the Marquis in stubbing and clearing woodland, which will thus be reclaimed for cultivation. The improvement is expected to be amply remunerative in the end; and it is one of the unlooked-for results of free trade, which are to be met with in every part of the country, that a landlord is compelled by circumstances, various in kind, to improve the neglected portions of his estate, and which, without such impelling cause, might have long lain unproductive. Every such improvement is not merely an addition to the arable land of the kingdom, but it becomes also an increased source of employment to the labourer.

The offer of a revaluation, which is made by many landlords to their tenants, may be declined by the tenants, and yet be no proof that they complain unjustly. On every large estate there are tenants of various degrees of enterprise and skill; and one farm, of the same soil naturally with another, may be doubled in its productive qualities by the superior industry and skill of its occupier. The want of these qualities may have actually reduced the natural fertility of the other farm. Now, these farms may have been originally valued to their respective occupiers at the same rent, and a revaluation now would increase the rent of the one, and diminish that of the other. In the one case, the landlord would obtain a benefit from the skill and capital of the tenant over and above the intrinsic value

of his land ; in the other, he would be deprived of its fair value
on account of the mismanagement of the tenant.   The tenant
of industry and skill, who had employed his capital to the
advantage of his neighbourhood, would be actually fined for
his enterprise ; while the indolent or incapable man, who had
benefited neither himself nor others, would obtain a premium
for his misconduct or negligence.

We may mention an offer which was made by one landlord
in this district to his tenants, in order to meet the difficulties
of the time.   He proposed that an outlay of 30 per cent. on
the rental should be expended in cake, manure, and any other
beneficial object the tenant preferred, on condition that this
outlay should be borne equally by landlord and tenant.   It
was equivalent to the offer of a reduction of 15 per cent. in the
rent, — with this important difference, that that reduction was
to be made the basis of future fertility.   An outlay of 30 per
cent. could not fail to be attended with the best results, inasmuch
as the crops would be greatly increased and the groundwork be
laid for solid prosperity.

LETTER XLVI.

NORTHAMPTONSHIRE.

LIGHT LAND FARMING AT WITTERING — MANAGEMENT OF STOCK — MODE OF FEEDING HORSES — WAGES — FARMING CAPITAL — VERY DEFICIENT — FARMING THEREFORE INFERIOR — GOOD MANAGEMENT AT WANSFORD — WAGES — DIFFERENCE BETWEEN RENTS OF COTTAGES UNDER THE DUKE OF BEDFORD AND THE SMALL OWNERS IN VILLAGES — GEOLOGICAL CHARACTER OF THE COUNTY — GRAZING FARMS — PROPORTIONS OF ARABLE AND GRASS LAND — GOOD FARMING THE EXCEPTION — LANDLORDS EMBARASSED, AND MANY EMPLOY UNSKILFUL AGENTS — INEQUALITY OF RENTS — TOO GREAT EAGERNESS ON PART OF TENANTS TO TAKE FARMS BEYOND THEIR CAPITAL — COURSE OF CROPS ON LIGHT AND HEAVY LAND — MUSTARD PLOUGHED IN GREEN AS A PREPARATION FOR BARLEY — MR. SHAW'S FARM AT COTTON END — USE OF SALT AS MANURE — WAGES — CHANGES OF TENANTS — GENERAL LOWERING OF RENT.

NORTHAMPTON, Feb. 1851.

THE farm of Mr. Sharpley, of Wittering, a few miles south of Stamford, presents us with the details of good agricultural management in North Northamptonshire. It contains 630 acres, 480 of which are arable and 150 in grass. The arable is managed in a four-course rotation of 120 acres in each division.

Beginning with (1) wheat after seeds, the land is ploughed and pressed, and sown from the middle of October to the middle of November with three bushels an acre of Spalding's red wheat, scattered broadcast. It is rolled in spring with Crosskill's clod-crusher. The average produce is 28 bushels an acre. The stubble is then ploughed in autumn in preparation for (2) turnips, and in the following spring it is worked twice with Finlayson's harrow, which not only takes out the twitch better than the plough, but is less expensive, and keeps the moisture in the land, which the turning over by the plough in dry spring weather entirely dries up. When the land is thus sufficiently cleaned and prepared, the turnip seed is drilled on the flat, in rows 16 inches apart, from the middle of June till the middle of July, with 10 bushels of bone-dust and 70 bushels of ashes per acre.   The ashes are burnt in spring from bottoms of hedge banks, road sides, and any waste corners, and are found a most valuable adjunct to the bones.   There are no swedes grown on this farm, as they are believed to exhaust the land, are found more difficult to grow, and are not considered better than common turnips for a breeding stock of ewes.   There are two varieties of turnips cultivated — the white, which are meant to serve till Christmas; and the green top white, which are to carry the stock on till the beginning of April.   The crop produced by this management is equal to the keep of eight sheep per acre, for twenty weeks.   It is consumed on the ground by the whole of the flock, in two divisions, — the hoggets first, and the ewes following.   Such sheep as are being fed fat go loose before all.   No cake is given, nor are any turnips cut.   As soon as each piece is eaten, the ground is ploughed about three inches deep, and prepared for (3) barley, which is sown broadcast as soon after the middle of March as possible.   The seed is " scuffled" in with long-tined strong harrows. This crop yields 40 bushels an acre.   It is followed by (4) seeds, 100 acres of which are sown with 12 lb. of white clover, 2 lb. of trefoil, and half a peck of rye-grass per acre, which is grazed with sheep only, except that for a few weeks at first the young cattle are also admitted to it.   The remainder of this division, 20 acres in extent, is sown with 14 lb. of red clover and half a peck of rye-grass per acre, to be mown for the horses and for hay. The red clover piece is changed at each return of the course, and its recurrence on the same ground is in that way postponed for several rotations.   Forty acres of the grass land are mown for hay also, the remainder being grazed.

The whole of the dung from the yards is carted out during the winter, and laid in large heaps in each of the fields of " seeds," where, after being well rotted, it is applied before ploughing for wheat in September.   If it could be got rotted in time, Mr. Sharpley would greatly prefer applying it on the seeds in spring, that they might receive the first benefit, as the additional feed eaten on the ground would equally prepare it for wheat. But this is thought to require a year's dung in advance, which would necessitate the use of artificial manure for a year on a scale more expensive than the farmers here are yet accustomed to.

The stock on this farm consists of 400 Lincolnshire ewes, which lamb in March, and rear about 400 lambs. The wedders are sold at one year old (last year at 31s. each), the ewes are kept for stock. Nine hundred sheep altogether are wintered. Four hundred ewes and lambs and 100 young ewes are kept during the summer.

Besides a few milch cows, there are every year about 18 or 20 heifers which have a calf and rear it. Twenty-six calves are thus reared altogether. The steers are kept till three years old, and sold in March or April to go to the rich grazing lands to feed. The average price last year was 13l. The heifers, after rearing a calf, come into the straw-yard to be wintered, and they are sold in spring with the steers. During winter, 80 to 90 head of cattle altogether are kept in yards on this farm. They never receive a turnip, as Mr. Sharpley thinks it a waste of labour to draw home the turnips and take back the manure : besides that, he finds the turnips to pay better by sheep feeding, and as the cattle are kept only in a rearing state, they can be carried on very well, and not expensively, with straw and cake. They are managed in this fashion : — During summer they are grazed on the grass lands, and in winter are put into separate straw-yards with sheds. The yearlings get hay or clover chaff and 1 lb. of cake each daily. The two-year-olds get barley straw and 2 lb. of cake each daily, besides barley and wheat chaff, of which they are very fond. The three-year olds get two fodderings of straw, one of clover hay, and 3 lb. to 4 lb. of cake daily. They have all an abundant supply of water in their yards, and look fresh and thriving.

When the cattle leave their yards for the grass, their places are supplied by the work horses, twelve in number, which are then taken from their stables, and during the summer receive

their food in the cattle yards.   The whole of the straw on the
farm is thus made into good dung.   In the winter the horses
are put into a stable in which there are no division stalls ; but
if any horse is inclined to be vicious, a bar of wood is hung
up between him and his neighbours.   The winter food of the
horses consists of two parts oat-straw with the corn, and one
part clover hay, cut together, and given in the manger, as
much as they can eat without waste.   The quantity of oats thus
consumed by each horse in the day, besides the straw and hay,
may be about 12½ lb.   During the summer they receive green
clover and oats.   The work of the farm horses is very light, as
the land is easy of tillage.

Eight men and two boys are regularly employed on the
farm, the men receiving 10s. a week, and the boys who can
plough, 5s.

This farm was entered to on the 25th of March, and the
first half-year's rent is payable on the 1st of January there-
after, the second on the 1st of July.   The whole implements
were bought two years ago, quite new, and, with the live stock,
cost the tenant 3,500l.   Besides that sum, he had to pay for
labour before getting any of his crop turned into money.   His
invested capital altogether amounted to 6l. an acre; but there is
here no draining, building, or permanent improvements which
the tenant has either to execute himself or to aid in doing; he
has just to stock and work an easy light land farm.   It is
commonly thought, in this part of the country, that an arable
farm requires less capital than a grass farm, and many men
without adequate capital enter to arable land with the intention of trusting to cropping entirely.   But an arable farm, if
fully stocked and *fully farmed*, cannot be carried on without a
good capital.   Here there is said to be a great deficiency among
the farmers in that important matter, many having taken to
arable farming with the idea that ploughing and sowing, with
seed and labour, were the only requisites.   The low range of
prices is compelling greater attention to business, and, as we
were significantly told, the fox hunting farmers are becoming

a gradually diminishing body.   Industry, capital, and skill may,
it is conceded, still carry a man through with difficulty; a
deficiency in all these qualities must be fatal to him.

The only tenant right or compensation in this part of the
country, is that an outgoing tenant is allowed the whole of his
" bones " bill and the half of his " cake " bill for the last year of
his occupation ; but, in general practice, little or no artificial
food or manure is purchased.   The dung of the few poorly
straw-fed cattle is used to raise the turnip crop.   That is eaten
on the ground by sheep, and insures a fair crop of barley.   But
there is no progress here, no addition to the powers of the soil
to compensate for their continued exhaustion, and consequently
there can be no increasing averages to make good the defi-
ciencies of price.

In the neighbourhood of Wansford we come on the Duke of
Bedford's estate, where the well-managed farm of Mr. Perceval
at once arrests the notice of the traveller by its neatly trimmed
fences, well kept roads and cleanly cultivated fields.   The
swedes are a fine crop.   They are taken up in the beginning of
winter, and stored in little heaps covered with earth on the
field where they grow.   When the sheep come over the ground,
a few hurdles are placed round each heap and the turnip cutter
inside, and the cut swedes are then served out to the sheep in
boxes.   The crop is by this means protected from injury by
game, it is kept juicy till wanted, and it neither exhausts itself
nor " draws " the ground by shooting up a seed stem in a
mild winter or at the beginning of spring.   Winter tares and
rye are sown in autumn, to be cut for the horses in spring and
summer ; and these are followed by white turnips, which are
eaten on the ground by sheep.   The next farm, that of Mr.
Leeds, of Stebbington, is also very neatly farmed.

On this portion of his estate, the Duke of Bedford is draining
all the heavy and wet fields of such of his tenants as are unable
to do so themselves.   In these cases he supplies and carts the
tiles, makes the drains, and finishes the whole free of all charge to

the tenant. The drains are made 4 feet deep, and 33 feet apart, and are found very efficient on strong land.

All the labourers are employed; and the general rate of wages is 9*s.* a week. No beer is given to the labourers. The Duke's cottages are let at very low rents; but others in the village are extremely high,—as much as 3*s.* 6*d.* a week,—with much less accommodation and a more limited extent of garden ground than the more fortunate tenants under the Duke, who pay 1*s.* to 1*s.* 6*d.* a week.

Nearly the whole county of Northampton lies on the lower oolite formation. The southern division is celebrated for its grazing qualities. For ten miles round the town of Northampton, one-third to one-fourth of the land is in grass. Grazing, which was formerly the most profitable occupation of the Northamptonshire farmer, is now interfered with by the mode of feeding adopted on the arable farms, which, by the aid of artificial food, corn, and cake, turn out more fat stock than the purely grazing farms. Some of the farms in this division are wholly arable; but the most common proportions into which farms are divided are two-thirds arable and one-third old grass. In winter the cattle are generally turned out on the grass lands during the day, and seldom receive any cake in the yards, as after such treatment they are found to fall off when turned out in summer to be fattened on the pastures, the cake not being then continued. The appearance of the country generally is well wooded and picturesque, undulating, with a fine friable red soil, admirably suited for green crops, corn, and grass.

Though there are many excellent farmers in the county, and much improvement has taken place in its agriculture, good farming is still the exception. For this there are several causes. In regard to the landlords, in the first place, many of them have no interest in their farms beyond the annual rent they receive, know nothing of the management of land themselves, and do not employ an agent who does. Some employ men of low standing with a small salary, and in a dependent position, butlers, gardeners, and sometimes gamekeepers, performing the functions of land-agent. Lawyers are employed by some; but they merely receive the rents. The duties of a competent agent, embracing an inspection of the farms, a general intelligent supervision of the property, with that confidential communication with the landlord as to the measures best adapted to promote the interests of both landlord and tenant, and the suggestion of such improvements as may be made at the least cost for the benefit of both, cannot, of course, by such agency be contemplated. Many of the landlords are straitened for capital, having their land heavily mortgaged or burdened with annuities, and who would yet rather embarrass themselves more by spending money in adding to their acres, than by improving those they have, though their tenants, from deficient buildings and want of drainage, are incapacitated from doing justice to their farms. In many cases the arable land is much injured by superfluous fences and hedge-row timber, the injurious quantity of which may be seen right and left from the railway between Blisworth and Rugby. The inequality of rents is also the cause of some districts and estates being better farmed than others. Many estates are let and have been rented for years at 20 and 25 per cent. higher than others. These are carefully eschewed by the best tenants; and any good farmer with capital, who may have the misfortune to be placed on such rack-rented estates, is constantly looking out for a vacant farm under a more liberal landlord, where he may expend his capital with security.

A great obstacle to good farming is the system adopted by some landlords, and those not the least popular among the tenants, of letting their farms at low rents, with the understanding that all improvements are to be made by the tenants. A good tenant keeps things in good order, and very possibly improves his farm; a bad tenant, most likely, deteriorates it. In the course of years a stranger is sent to make a new valuation of the farms; and he, of course, fixes the highest rent on the good farmer,

E E

whose spirit of improvement is thereby effectually curbed for the future.

On the part of the tenants the obstacles to good farming are those too common to their class in other counties as well as Northamptonshire— a headlong running after more land than they have capital to manage, and the employment of insufficient labour to work their farms.

On the light soils the four-course system of cropping is practised by some farmers, especially for the purpose of cleaning their farms, but the most general course on the red land is a six-course, thus — (1) turnips, (2) barley, (3) clover mown, (4) clover grazed, (5) wheat, (6) barley.   Though some farms are profitably managed under this course, they are never quite clean. If the land is not perfectly clean when laid down to clover, the two years' grass allows the root weeds to gain strength and strike deeper, and with two corn crops after the breaking-up of the clover leys the land gets very foul.   Where the land is left only one year in clover, and if well farmed and the wheat stubbles cleaned before sowing the barley, the crops are generally very good, and the land tolerably clean.

Where the *substratum*, instead of being red sandstone or sand, is of a clayey nature, the eight-course is successfully practised; viz., turnips, barley, clover, wheat, turnips, barley, beans, wheat. In some parts of the county the clay lands are very ill-farmed and imperfectly drained.   On the eastern side, however, there is some good clay farming where the land has been well drained, and the following mode of management is adopted : — Half the fallow is sown with vetches, the other half is a naked fallow well worked through the summer, and as soon as the vetches on the first half are folded off with sheep, the land is ploughed up and the whole worked together.   It is kept as rough as possible, and soon after harvest, before the land gets wet, it is manured.   In that state it is left till the spring, when, without again ploughing, the barley is drilled as early in the season as the land is dry.   Under this management heavier crops of barley are got on very strong clays than on the best turnip soils, especially in a dry season.   The barley is followed by clover, which is mown, the clover ley is sown with wheat, and the wheat is followed by beans.   Some farmers take the beans after the clover, and then wheat, which they find to succeed best after beans, and less subject to grub and wireworm than after clover.

On some very good and well-managed land within a mile or two of Northampton, on the farm of Mr. West, of Dallington, we saw a fine crop of swedes which had been drilled on the flat, the dung having been previously ploughed in.   When ridged over freshly-applied dung, the crop, though bulkier, is said to be more apt to decay.   The course adopted here is seeds, wheat, barley, turnips, barley.   As soon as the wheat stubble is ploughed the land is sown with mustard, which is ploughed in green, as a preparation for barley, with much success.

Mr. Shaw, of Cotton-End, near Northampton, adopts the four-course.   His "seeds," which are very early and fine, are chiefly Italian rye grass, which is grazed the first year, dunged in autumn, and after yielding two months' keep to the sheep the following spring, it is to be ploughed with a skim coulter plough, and the land planted with potatoes.   The swedes are all taken up in November, and stored in heaps on the field covered with earth, whence they are used as required, and given to the sheep cut, in boxes, with 1 lb. of cake each, for the last ten weeks before being sold fat.   The wheat land after being sown in autumn is dressed with 7 cwt. of salt. per acre, which is found to have a very beneficial effect in destroying all small weeds, and in strengthening and brightening the straw. It renders loose land firm by glazing over the surface, and for that reason probably would not be a suitable application on strong or wet land.   Being in the immediate neighbourhood of Northampton, a town of 20,000 to 30,000 inhabitants, Mr. Shaw kept a dairy stock on this farm, which he found very profitable, as milk sells at 2d. per quart, and there is a constant demand for it; but his stock was so much injured by pleuro-pneumonia that he has for the present discontinued a dairy. The rent of this farm, 300 acres in extent, is 45s. an acre, tithe free; the land is of superior quality, and the situation very advantageous.

The regular labourers on the farm are hired by the year, the best receiving 12s. a week and a house rent free, others 11s., and the lowest 9s. The average rate of wages for the county is 9s. a week: there are very few labourers unemployed, and scarcely any able-bodied in the workhouses. As a class, it is said, they were never better off, and yet there have been more incendiary fires than in any former winter.

The farmers generally are very desponding, and there can be little doubt than many of them have been losing money during the last two years. Those of small capital originally may be unable to recover the shock, as that can only be done by greater exertions on the part of both landlord and tenant, and the means of the latter are already gone. It is anticipated that there will be many changes of tenants — those who have not capital and industry being obliged to give up, and others who are quitting one landlord to go to another under whom they expect better conditions. The landlord who gives least encouragement and assistance to his tenants in this crisis will suffer most severely at last, as all good tenants will go to good landlords, and the careless and indifferent must content themselves with just such as they can get. The best farms of the best landlords will probably maintain their value; the inferior farms and those of cold clay must fall very considerably. In the latter there will probably be a new basis of valuation altogether, more in accordance with their relative value, which hitherto has been rated too high. The greater expense of cultivation on clay as compared with stock land will now bear much more heavily on the balance left for rent than formerly, when the value of the produce was relatively high. On the whole, there can be little doubt that the first effect of all these changes will be a lowering of rent to a greater or less extent throughout this county.

LETTER XLVII.

NORTHAMPTONSHIRE — *continued.*

FARMING AT OVERSTONE — DRAINAGE — USE OF BROKEN STONES ABOVE PIPES RECOMMENDED BY MR. BEASLEY — TRENCHING — ITS COST — REMOVAL AND RENEWAL OF FENCES — SUBDIVISION OF FARM — COURSE OF CROPS — TURNIPS, BARLEY, SEEDS, WHEAT, MANAGEMENT OF EACH DESCRIBED — USE OF SALT AS MANURE TO WHEAT — MR. BEASLEY'S REASONS FOR NOT SOWING THIN — ARRANGEMENT OF THE BUILDINGS — FEEDING OF CATTLE AND SHEEP — THEIR PRODUCE — INCREASE OF STOCK KEPT ON THE FARM — COST OF LABOUR — MATERIALS FOR MAKING MANURE — LENGTHENED PERIOD WHICH MR. BEASLEY'S HIGH BRED STOCK TAKE TO ARRIVE AT MATURITY—ESTATES OF LORD SPENCER, MR. LLOYD AND LORD OVERSTONE— THEIR ARRANGEMENTS GIVE THEM THE ADVANTAGE OF A CHOICE OF GOOD TENANTS — TENURE — IMPROVED COTTAGES — SIR C. KNIGHTLEY'S ESTATE.

NORTHAMPTON, Feb. 1851.

As a breeder of short-horned cattle and new Leicester sheep the name of Mr. John Beasley, of Chapel Brampton, is well known beyond the county he resides in. A detailed description of the management of his farm at Overstone will, we are confident, prove generally useful and instructive. It contains altogether about 727 acres, 420 of which are arable, and 300 pasture. Three-fourths of the whole are a "convertible" soil, a good red loam upon a substratum of red sandstone, which is in parts very near the surface. Some portions have a considerable admixture of sand, and the soil there is consequently weaker. The remainder of the farm is a strong soil lying on very stiff clay.

The whole of this portion of the farm, the clay-land, has been underdrained. The drains are made in straight lines, seven yards apart, without any regard to the old high-backed crooked lands. They are dug to a depth of three to four feet, a pipe-tile laid at the bottom, and over it broken stones about nine inches in depth. The drains being very narrow at the bottom, few stones are required, a cart-load sufficing for four chains in length, and, as the stones are got on the farm from the red land,

the cost is not much increased. The whole expense amounts to
4*l*. 10*s*. an acre. When the land is drained it is ploughed, harrowed, scuffled, and worked across the high-backed crooked
ridges, which are thus gradually levelled, and there is said to be
no perceptible difference between the crops on the ridges and
furrows. If the stones on the tiles answer no other purpose,
they assist the drainage in the first two or three years; for the
water does not find its way very quickly to a depth of three or
four feet on strong clay soils, which, perhaps, have never before
been moved more than four inches. By degrees the clay soil
will crack to a considerable depth, when the air has been secured
an entrance, and that is materially assisted by the subsequent
processes of deep ploughing and subsoiling. If the draining is
made more efficient, by the addition of the stones, for the first
two or three years, the expense will be repaid, and the drain is
not so liable to accident by the breaking of a tile, or the stopping up of one. Where the stones can be cheaply got, this
practice may be advisable; but the advantages it possesses are
not sufficient to justify any considerable outlay, as it has been
abundantly proved that tile-drains at this depth and distance
will, if properly constructed, effect perfect drainage. The
drainage is carried off by a brook, in which an increased fall has
been obtained by making it deeper, wider, and straight in its
course, and which has at the same time greatly improved the
drainage of the adjoining land. This land is now perfectly dry,
and can be worked at almost any season of the year. One field
was trenched or dug with spades to the depth of 14 inches. It
was first manured, the labourers digging in the manure, and
picking out any twitch or weeds as they went on. It was not
touched again until the end of April, when it was drilled
on the flat with mangold wurzel, and produced a good crop
where anything approaching to a good crop of vegetables had
never before been seen. The digging cost 2*l*. per acre, and supplied work at a time when it was very scarce. Some of the
men, by working hard, earned 10*s*. a week.

Over the whole of the arable land the old fences have been taken up and new ones planted, and the fields made of the same size, 21 acres each. Five of these fields, or 105 acres, are in the same description of crop every year; three adjoining fields on one side of the farm and two on the other. The same description of work is thus always being carried on at the same place The hedges are kept very low and neatly trimmed, occupying the smallest portion of land. There are no open ditches, tile-drains supplying their place where necessary. The fields are all square, being the same width at both ends, except where a public road interferes, when the unequal side is put next to the road, so that the field is ploughed square up to a small portion of the last part. So perfectly square are these fields that the ridges for the turnips, as well as the drills for the corn, are frequently commenced in the middle of the field and finished on each side up to the hedges, the last row running in a perfectly straight line with the hedge. The old and bad trees have all been grubbed up, the best oaks — and they are very fine — having been left in the open fields with excellent effect.

The whole of the farm is managed on the four-course rotation — turnips, barley, clover, wheat, 105 acres of each. On the heavy land, now that it is drained, white turnips are grown, to be eaten off early, and mangold and cabbage answer very well The general management begins with the wheat stubbles, which are ploughed early in the autumn, six inches deep, with the common Scotch iron plough with two horses abreast. About one-third, or as much as can be got through in a season, is subsoiled with Grey's subsoil-plough eight inches under the first furrow, making in all fourteen inches. The horses attached to the common plough walk on the unploughed land so as not to trample on the furrow which has been subsoiled. The land is left in this state through the winter. In spring, when it is sufficiently dry, a scuffler is drawn across the furrows, which, where the land has been subsoiled, will work to the same depth it has been ploughed. The land is then rolled and harrowed, and the twitch

brought to the surface and picked off. The whole force of the
farm is applied to one field; and when the whole of the twitch,
brought to the surface, has been removed, the field is left for a
time.   It is again ploughed, then scuffled, harrowed, and picked
as before; and this is repeated until it is perfectly clean.   The
practice of autumn cleaning adopted by Mr. Outhwaite, of Bainesse, and described in a former letter, might, we have no doubt,
be introduced here with great advantage, as such repeated turning over and exposure of a dry soil in the hot sunny weather of
April or May must sometimes render a plant precarious.   When
the turnip-sowing commences, the land is ploughed into ridges
25 inches apart, and 20 loads of good rotten farmyard dung is
placed in the ridges and covered up.   The seed (2 lb. per acre)
is then drilled on them with Hornsby's drill, with concave
rollers, made in the shape of an hour-glass, and which give a
good finish to the work.   Part of the dung, which is made in the
autumn and early part of winter, is carted out upon the cleanest
wheat stubbles, and immediately spread and ploughed in.   This
saves a great deal of labour at the turnip-sowing time, and expe-
dites the work.

The dung is never removed from the yards except to be
applied directly to the land.   In the yards it is trampled very
firmly by an unusual number of cattle, the buildings are all
spouted, and drains are laid from all the yards and feeding-
houses to the liquid-tanks.   The manure heaps are carefully
levelled on the top every day, and, if too dry, in the spring of
the year the liquid from the tanks is thrown over them.   When
turnip-sowing commences, the top of the manure heap is laid
aside, and the rest carted away and at once put into the ground.
The top of the heaps and the spring-made manure are thrown
up and turned over to cause fermentation, so as to be sufficiently
decomposed for application to the turnip crop.   No artificial
manure is used for turnips,— linseed cake, beans, and barley
being consumed by fattening cattle to a large extent, and the
farmyard dung being thus all of good quality.

About 20 acres of the wheat stubbles are sown in the autumn with vetches, a slight dressing of dung being first applied. They are drilled at the rate of three bushels to the acre, and the young plants are watered with liquid manure in the winter and spring. A small portion of these vetches are cut for the horses, the remainder are fed off early with sheep, which are kept in folds, the vetches being mown and put into cribs. The land is then ploughed, and cleaned for turnips.

About two-thirds of the green crop land are sown with swedes, the remainder with white turnips, mangold, cabbage, and potatoes. The first swedes are sown about the last week in May, and the whole are completed by the beginning of July. Earlier sowing is found to be attended with much greater risk of mildew in autumn. The crops average upwards of 20 tons an acre, and, when all eaten on the land, are found equal to the keep of 20 sheep an acre, for 20 weeks. They are hand-hoed three times at a cost of 8s. per acre, and horse-hoed four times at a cost of 4s., by which perfect cleanliness is attained, there not being a weed or a particle of twitch to be seen in the autumn. In November and the beginning of December the swedes are pulled up, cleaned, thrown into conical heaps on the field, and covered with soil, a light coating of stubble being previously laid on. When the turnips are stacked without any straw between them and the soil, they are found to be very dirty in wet weather, and in that state they purge the sheep. This operation costs 6s. per acre. Every third heap is carted off and consumed in the fold-yards by cattle; the rest are eaten by sheep on the ground, the turnips being cut with Gardner's turnip-cutter, and given in troughs. The whole of the sheep have chopped hay or clover with their turnips, and the fattening sheep cake or corn also. The hoggets and ewes have neither, except some weak ones; but the sheep which have cake or corn exchange pens regularly with those which have only turnips and hay, that the land may be equally manured.

In preparation for barley, the land, as the turnips are consumed, is ploughed five inches deep. Barley-sowing commences in the first week of March, and is finished about the 5th of April. Three bushels and a half of chevalier barley are drilled to the acre, and the average produce for the last eight years has been 45 bushels 1 peck; the quality good, weighing 55 lb. per bushel, and fetching the highest market price.

The grass seeds sown with the barley consist of the following mixture: — Two-fifths or three-fifths with 10 lb. of red clover, 3 lb. of white clover, 3 lb. of trefoil, 1 peck of Italian rye grass, and half a peck of common rye grass per acre. The remainder is sown with 10 lb. of white clover, and the same quantity of trefoil and rye grass as above. Beans and peas have been tried, in the place of clover, to produce a more varied course; but the crops were light, the land being too dry for them, and the wheat was much lighter than when sown upon the clover ley, solidity of soil being considered of the first importance to wheat on this kind of land. Two of the five clover fields are mown for hay, and three depastured with sheep and young calves. The sheep are folded during the night.

For wheat, the ploughing of the clover ley begins about the 10th of September, and the wheat is generally all in by the 20th of October. The first sown is drilled at eight inches apart with $2\frac{1}{2}$ bushels, the last with 3 bushels an acre of Valpin's red Spalding wheat, which is the only sort now grown on this farm, having been found most productive and of good quality, weighing 62 lb. per bushel. The average crop for the six years preceding 1849 has been $34\frac{1}{2}$ bushels an acre. The crops of 1849 and 1850 are not yet thrashed, but are estimated at much more. Immediately after the wheat is sown, the land is pressed with Crosskill's clod-crusher, and, if the weather admits, it is again pressed in the same way in spring. Eight cwt. of salt to the acre is sown upon the wheat, 4 cwt. in autumn, and 4 cwt. in spring. This is found to give solidity to the land, while it checks the weeds, prevents mildew, blight, and rust, and improves the quality and increases the produce of the crop. On the clay soil portions of the farm the application of salt has been discontinued, as it was found to keep the land too damp and sad, and to give the wheat a starved and unhealthy appearance. The wheat is always hoed between the rows, but Mr. Beasley is of opinion that if the land could be kept perfectly clean without hoeing it would be better, as the hoeing, by cutting the small fibres, has a tendency to let the wheat fall.

We must make a short digression, to explain Mr. Beasley's reason for sowing so thickly on land in every way so well prepared and in such high condition. He does it because, in his opinion, corn ought not to be encouraged to tiller. If the plants are sufficiently thick in spring, they at once send up the stalk ; but if the roots are thin, they send out lateral shoots, which strike in the earth and produce new plants. The first plant is weakened by having to produce auxiliary plants, and the plants of the second growth do not come to maturity so early as the original or parent plant. The quality of the crop is thus injured, as there are always more light and defective corns in a thin-sown than in a thick-sown crop ; besides that there is less seed to meet the contingencies of wireworm, grub, or very severe weather.

The accommodation of the stock and crop is provided for in a set of farm buildings, which have been erected, at a moderate cost, out of old materials, with the aid of larch timber, and stone, procured on the estate. The buildings are on a large scale, but compact, and in the centre of the Overstone farm. They include a house for a steam-engine, which drives thrashing machinery, millstones, saw-mill, and turnip-cutter. The waste steam can be used for steaming food. The thrashing and dressing of the corn, including coals and oil, costs 1s. 3d. per quarter. The feeding-houses are 15 feet wide, with a manger, rack, and water-trough at the head of the cattle. The cattle are tied by the neck in pairs, in stalls 8 feet wide. They are well littered and kept perfectly clean. The water is supplied by a pipe from the well in the yard, and when one trough is full it supplies the next, until all are full. The young stock are kept loose in yards, with shelter sheds, and the in-calf heifers and cows are kept in the yards where the manure from the feeding-houses is emptied, which they compress by treading.

The stock is of the improved short-horn breed, bred with much care for many years, chiefly from the stock of the late Earl Spencer, and crossed with bulls from Sir Charles Knightley and other eminent breeders. About 35 cows and heifers are kept for breeding, 40 calves being reared every year, a few of the best that can be got being bought to make up this number. The calves begin to fall in February, and continue till Midsummer. About six of these are sold for bulls by the time they are a year old. For the first fortnight the calves have new milk, for another fortnight half new and half skim ; afterwards skim milk, mixed with linseed porridge. They are turned out into the young clover very early, returning to open sheds at first for the night, where they receive bruised oats, or cake, as soon as they will eat, and until they are able to gather a living for themselves by grazing. The first winter the calves are kept in four paddocks, in each of which there is an open shed, in which they are fed with turnips and hay, and the youngest with 2 lb. of cake a day in addition. In spring they are turned out to grass with the ewes and lambs, and remain on the pastures till Christmas, when they are brought into the fold-yard to straw and turnips. They are kept in the same way for another year, and, when nearly three years old, they are placed in the feeding pastures, which are not very rich, and in autumn on the aftermaths. In November they are tied up in their stalls in the feeding-houses, when, after a short time, they are placed upon full feeding. They are then fed four times a day, and their daily supply consists of

	s.	d.
7 lb. of linseed cake, at 3½ farthings per lb. ...	0	6
1 gallon of beans ground into meal, at 32s. per quarter, including grinding	0	6
1 bushel of swedes, at 3d. (10s. per ton) ..., ...	0	3
8 lb. of hay, at 3l. 10s. per ton	0	3
	1	6

or 10s. 6d. per week for the last eight or ten weeks. In the middle of February they and the fat sheep are sold by auction on the farm. The average price last year was 22l. 2s. 2d., including some old cows, which scarcely made 5d. per lb. At such a price it is very doubtful whether this mode of rearing and feeding is profitable. Earlier maturity, we are convinced, would pay better; and we have often seen cattle of inferior breeding, and on no better land, made fat in half the time, — certainly not the same weight, but fetching greatly more than half the money.

We now come to the management of the sheep stock, which are of the new Leicester breed, bred from the best flocks since the days of Bakewell. The ewes, 350 in number, rear about the same number of lambs, but being for the most part bad nurses, the lambs are consequently small, and are taken early from their dams and put upon clover or good pasture until November, when they are placed upon turnips during the winter, as already described. In the spring the ewe hoggets are put into a store pasture, and the wethers are grazed upon vetches and clover. In the autumn the draught ewes and theaves, and the whole of the wethers, are put to turnips, when they receive a pint of beans or a pound of linseed cake per day also, whichever is to be had cheapest. As many shearling wethers are bought in summer as, besides those bred on the farm, are required to consume the turnips. They are all treated alike, and are sold fat in February, by auction. Last year the average price was 46s. 7d. These sheep yield 6 lb. of fine long wool.

The whole stock on the farm in February 1838, was 77 cattle, 525 sheep, and 25 pigs. In February 1849, there were 184 cattle, 879 sheep, and 33 pigs. The farm was then 520 acres in extent. Two hundred acres have since been added, and in February, 1851, there were 202 cattle, 1017 sheep, and 70 pigs. Mr. Beasley intends to increase the sheep stock to 1300, and to diminish the number of cattle in the same proportion.

The labour of the farm costs 19s. 6d. an acre for the whole, or 28s. an acre for the arable, and 7s. for the pasture. The labourers are receiving 9s., 10s., 11s., and 12s. per week, according to their ability, character, and the time they have worked upon the farm. They have all been reduced 1s. per week since last year. Much of the work is done as task-work. With the exception of the strong land, the farm is light and easy to manage, and the arrangement of the fields and buildings greatly facilitates and economises labour.

A leading object on this farm has been to make as much good manure as possible. To effect this, a very large stock is kept, all of which are well fed, and a considerable quantity of artificial food is consumed. Where the relative values of different kinds of food do not greatly differ, feeding cattle will generally thrive best upon a variety. One-third of all the turnips grown upon the farm are consumed in the stalls and yards by cattle. These, with 60 acres of meadow hay, 40 acres of clover, and the straw from all the corn crops, make up the materials for the manufacture of manure. The quantity made has gradually increased, and the crops are likewise increasing. The condition of the farm is aided by the sheep being partly fed with artificial food. It has been now brought to a point of cleanliness and condition that the corn crops scarcely admit of increase. If the barley crop is made much more luxuriant, the straw will be more productive than the corn, and the quality will be apt to deteriorate. It therefore becomes a question whether the four-course should be continued, or whether, as we think, the farm has now reached the point at which successive corn crops might be occasionally taken with advantage.

The experienced reader cannot fail to remark the lengthened period which this very high bred stock takes to arrive at maturity. This is somewhat unusual, and appears to us the most vulnerable point in Mr. Beasley's management. The chief excellence of short-horns consists in their earlier maturity than other breeds, for which we are willing to sacrifice in some degree the quality of the meat; but if they are kept till four years old, this advantage is lost, and we might as well feed West Highlanders or Welsh runts, as these would get fat at that age, and be of much primer quality. The high bred Leicester sheep, too, have the failing of being bad nurses, and not prolific. They do not appear to have any countervailing advantage, as with the same feeding and at the same age any of our good crosses would give as much money. It would thus appear that merely for feeding purposes it is unnecessary to spend money on very highly bred stock, as Mr. Beasley, with the best short-horn and the purest Leicester blood in England, gets neither earlier maturity nor greater weights than many farmers with stock of very inferior breeding.

We make no apology for occupying so much space with a detailed description of Mr. Beasley's farm. It comprises within itself an instructive little treatise on agriculture, affording much matter for reflection, and many points of comparison to the skilful practical farmer.

The estates of Earl Spencer, Mr. Lloyd, and Lord Overstone are managed by Mr. Beasley. For the last two years there has not been a farthing of arrears on the whole of these extensive estates, comprehending tenants from 1,000l. a year to the humble cottager, and including 600 of the latter class. This is attributed to the farms being moderately let, and to the erection by the landlord of suitable buildings for lodging the cattle and saving their manure, and to drainage. It is not that the farms are let lower by the acre than other estates, but that they are let truly as farms, fitted by the landlord with those accommodations by which a tenant is enabled to farm successfully. This liberality of the landlord is fully appreciated by the tenantry, and gives the agent an immense advantage in the selection of tenants when a farm becomes vacant. He has the choice of the best men; and there can be no doubt that an estate can be most effectually and economically improved through good tenants. The farms are not advertised, and never let by tender; they are examined by the agent, who fixes the rent, and selects his tenant. In valuing a farm, Mr. Beasley assumes that all adequate accommodation will be provided by the landlord. No percentage, therefore, is charged on any outlay by the landlord, either for buildings or drainage. The land is valued at its intrinsic or natural worth, with such ameliorations as the landlord ought to make at his exclusive cost; and thus the rent of good and bad farmers is raised alike. If the bad is thereby compelled to quit, so much the better. The more common practice of valuing land as it stands, without regard to the landlord's outlays or the tenant's improvements, increases the rent of the good farmer in consequence of his own exertions, and lowers that of the negligent one as a reward for his neglect. Instances have often occurred where farms of precisely similar character and rent have been revalued, and one that had been well farmed was raised 10s. an acre, while the other, which had been badly farmed, was lowered 10s. an acre, the landlord in both cases having dealt equally by both tenants in doing nothing for either, but leaving each to follow his own plans. An abatement of 10 per cent. has been made on the estates under Mr. Beasley's management for the present year, more as a mark of sympathy on account of the deficient crop of last year than as a permanent readjustment. The time for that is not yet come.

The farms are all held from year to year; and there is no desire on the part of the tenants for leases. The security under such landlords is felt by the tenants as quite sufficient; and yet there have been many instances where a change of owner has completely altered the confidence formerly subsisting between tenant and landlord. On Earl Spencer's estate, however, good landlords are believed by the tenantry to be hereditary. Some of the farms on the estate have been held by the same family for 300 years ; and the average period during which all the farms on this estate have been held by the same families, exceeds ninety years. Nor is there any written agreement or other document to bind either landlord or tenant. The rent is entered in the rental ; and the tenant pays it punctually as a matter of course. Crop books are kept for every farm ; and the agent visits every field once a year. He interferes with the tenant's management as little as possible, and chiefly in the way of advice. On each of these estates large sums have been expended on buildings, farmyards, and in better arranging the farms. Draining-tiles are given almost without limit.

But the landlord's expenditure is not confined to the requirements of his farms, the comfort of the labourers on these estates has met with an equal share of attention. On Lord Spencer's estate, within a short period, seventy-four new and substantial cottages have been erected, in groups of two, three, and five, with a pump and kitchen common to five cottages, fitted up with oven, copper, ironing-board, &c. To each cottage is attached a rood of land, a pigstye, wood barn, &c. They are let by the week at a yearly rent of 3*l.* 10*s.*, including land. The average rent of cottages on these estates is under 2*l.* There are also many garden tenants, who have a rood of good land (in all cases near their homes), and for which they pay 10*s.*, the landlord paying rates. Besides building new cottages, Lord Spencer has put into order an immense number of old ones, and is still continuing to build, but on a less expensive plan.

The tenants of bad land on Sir Charles Knightley's estate, besides getting their farms drained free of charge, have received equal to 20 per cent. of abatement. In the neighbourhood of Weedon, and to the south of it, two-thirds of the land

is in grass of prime feeding quality; the other third is cultivated in a six-course, thus : seeds, wheat, beans, wheat, turnips,
barley.   There being so large a proportion of the land in grass,
a heavy stock is kept on the different farms, fed in winter with
cake; and thus a great quantity of manure is made, by which
the arable land is kept in high condition, and yields abundant
crops.

Land of prime feeding quality in this part of the county is
let at 2*l.* per acre, tithe free; and the rates are from 3*s.* to
4*s.* an acre.   It is not uncommon to see five fat and powerful
horses yoked in line in a plough, turning over a barley-seed
furrow not more than four inches deep.   Amid so much complaint of distress, it is wonderful that such a heedless waste of
power is continued.

LETTER XLVIII.

BEDFORDSHIRE.

WOBURN—DUKE OF BEDFORD'S MODE OF LETTING FARMS—DRAINAGE, FENCES,
AND BUILDINGS MADE AT LANDLORD'S EXPENSE — COMFORTABLE ACCOM-
MODATION OF LABOURERS—THEIR COTTAGES HELD DIRECTLY FROM THE DUKE
—COST OF ERECTING COTTAGES—RENT—SCHOOLS FOR LABOURERS' CHILDREN
— WORKSHOPS AT WOBURN — EXPERIMENTS IN FEEDING CATTLE — COST OF
FARM BUILDINGS — PRINCIPLES WHICH REGULATE THE CONNECTION BE-
TWEEN LANDLORD, TENANT, AND LABOURER ON THIS ESTATE — WORTHY OF
GENERAL IMITATION.

WOBURN, Feb., 1851.

PASSING from Northampton to Bedfordshire, we proceeded to the Park Farm at Woburn, the seat of the Duke of Bedford. It was certainly with no feeling of idle curiosity that we endeavoured to acquaint ourselves with the relations subsisting between the head of the house of Russell and his numerous tenantry and dependents. A nobleman of the highest rank, the owner of one of the largest landed estates in the kingdom, all situated in purely agricultural districts, and deriving no direct aid from the neighbourhood of any of our hives of manufacturing industry, and yet the possessor of a name identified with the progress of all our liberal institutions,—it could not fail to be instructive to learn how this large property was administered.

The farms are never advertised, or let by tender. When a farm becomes vacant, it naturally forms the subject of conversation at the market table, and parties wishing to take it make application. The farm is then valued by the local agent, a practical man, who estimates it as in perfect order in so far as the landlord's improvements are concerned. Anything that is requisite to be done, either in regard to drainage, fences, or buildings, is done by the landlord as a matter of course. The Duke then selects his tenant from the various applicants, and offers the farm to him at the rent fixed by the agent. It is generally accepted at once, and by a picked man. All the tenants have the option of, and are encouraged to, take leases subject to fluctuation in the price of corn. One half of the tenantry accepted leases of various duration—twelve, sixteen, and some twenty years. Those who prefer a fixed rent have shorter leases—seven or eight years, and then a readjustment of rent, according to prices. The rental of the estate at present is rather more than in 1834 and 1835, but a very large outlay has been made in improvements to maintain it. In some cases these improvements are equivalent to a reduction of 12 to 15 per cent. There is no system of general temporary abatements. If a complaint is made, the case is at once considered on its own merits, and, if requisite, the rent is readjusted. At the end of every lease a readjustment takes place. A farm taken in 1843 at a fixed rent then calculated, with prices at 56s. as the basis, is now being converted by adding the value of such improvements as have since been made by the proprietor, and then charging the rent on the basis of 40s. for the quarter of wheat. The corn rent is in some cases all corn, in others part corn and part money, varying with the character of the land, and the proportion in which its produce is dependent on the prices of corn. The basis for present (Spring) lettings is 40s. for the quarter of wheat, regulated afterwards by taking the average of the whole country for four years, each year taking off one year and adding another. Game is not preserved, and hedgerow timber injurious to the tenant is at once felled and removed.

A system of husbandry is prescribed to the tenantry, from which they are not permitted to deviate except by consent of the agent. On light land that system is the four-course; on strong land the same, with the substitution of beans in lieu of a portion of the clover, and such extent of dead fallow as may be necessary. On new land, much of which has been broken up

in consequence of the Tithe Commutation Act, two white crops
are allowed at the commencement. The land is generally breast
ploughed, burned, and sown with cole-seed. This destroys wireworm. Oats are then taken, followed by wheat, then beans,
then wheat. The breaking up of inferior pasture has been a
great boon to the farmers, as they have had heavy crops from it
at little expense, and strong land carries good green crops after
first being broken up. The introduction of winter beans into the
rotation has been of immense benefit to the light land farmer, by
enabling him to alter his crops. They require to be planted in
September, if possible, and hence the difficulty of getting them
sufficiently early into the ground in the northern counties, and
consequently their greater uncertainty there.

The comfortable accommodation and welfare of the labourers,
is a consideration with the Duke of Bedford not less important
than equitable arrangements with his tenantry. Cottages are
built in numbers sufficient to suit the wants of the different
farms, with a due proportion for the mechanics also necessary.
The cottages are situated near the farms on which their occupants are to be engaged. They are held directly from the
Duke, from week to week, so that both the labourer and the
farmer are kept in some degree of check. Thus an ill-con-
ducted labourer can be promptly dismissed from the estate,
while a trifling jealousy or pique on the part of the farmer is
not necessarily acquiesced in by the landlord. All the cottages
have two rooms on the ground floor, and two or three sleeping
apartments up stairs. They are fitted with kitchen range, and
copper, — and one fireplace up stairs, — outbuildings for wood,
ashes, and other conveniences, — and an oven common to each
block of cottages.

The cottages are built in a substantial manner, of various
designs, the situation being so chosen as, if possible, to combine
the advantages of a genial airy exposure with a plentiful supply
of water. Ornament is employed, but not further than is in
accordance with the character and objects of the buildings.

While needless expense is thus avoided, the cottages are substantially constructed, so that they may not be subject to
frequent repair.   The use of hollow brick will, it is expected,
not only cheapen the cost of construction, but add materially to
the dryness of the walls and to the healthy ventilation of the
house.   Cottages built of hollow brick, with wall 9 inches
thick, cost 90*l.* to 100*l.* each.

Field allotments, from an eighth to a quarter of an acre, are
provided close to each cottage, and in the case of villages, as near
at hand as they can be conveniently had.   The rent is charged
at rates varying from 20*s.* to 40*s.* an acre, inclusive of rates.
The rent of cottages varies from 1*s.* to 1*s.* 6*d.* a week, according
to accommodation, and is paid half-yearly with great regularity.
It is believed to give a return of nearly 3 per cent. on the
outlay, exclusive of the value of the site.

But the education of the labourers' children is not forgotten
while their bodily comfort is so amply cared for.   Schools are
being built at the Duke's expense, in central villages, for the
accommodation of two or three adjoining parishes, for the more
advanced scholars ; and, in most parishes, infant schools are
established, at which the youngest children receive a little
instruction in the immediate neigbourhood of their own homes.
To all of these the Duke subscribes, and the children pay, so
that the schools are partly self-supporting, and the independence of the parents is not compromised.

On an estate of such magnitude as that of the Duke of
Bedford, where the duties as well as the rights of property are
so fully recognised, there being constantly new sets of farm
buildings and cottages in progress, it has been found necessary
to erect a complete set of workshops for the construction of
every article required on the estate.   In the yard at the Park
farm appropriated to this purpose, 100 workmen are constantly
employed, chiefly skilled mechanics, under the superintendence
of a resident engineer.   This is conducted with all the method
of a private speculation, the workmen attending throughout the

year, from 6 a.m. till half-past 5 p.m., with intervals of half an
hour for breakfast and one hour for dinner.   The premises are
lighted, when necessary, with gas, and an equal temperature is
maintained by steam pipes in the different workshops.   These
comprise a wood yard, with sawing sheds for cutting up into all
requisite sizes either foreign or home timber, the refuse of
which is split into faggots for the use of the Abbey.   Next, a
foundry for all manner of castings; then a smithy; then an extensive carpenter's shop; then a plumber, glazier, and painter's
several apartments.   A 25-horse power steam engine saws
the wood, blows the smithy fires, gives motion to the lathes in
the carpenter's shop, and to planing and other machines, — while
the waste steam from the boiler dries the sawn wood in the
drying shed, warms the workshops, and heats an oven where
the men may cook their dinners.   Every kind of work is done
on the premises, and fitted and put together before being
sent out.   The windows, doors, and stairs of farm buildings and
cottages, being made of certain dimensions and of certain
uniform sizes, are constructed in sets more economically and
substantially than they could be by country tradesmen.   During winter the different articles are prepared in-doors, and in
summer the carpenters and other workmen are sent to put them
up where they are required.   Not the least interesting department of this establishment is that where troughs for water,
slabs with the Ducal crest or cypher, and other ornamental
parts of architecture, are formed of concrete, possessing all the
hardness and durability of stone.

Adjoining these buildings are the extensive farm premises of
the Duke's home farm.   Here another powerful steam engine
gives motion to every variety of machinery used in working up
the crop on the farm.   Many interesting experiments in the
feeding and management of cattle are here being carried on
the *data* and results being carefully registered for the instruction of the agricultural public and the Duke's own tenantry.
Comparative trials are being made of the respective advantages

of box and stall feeding, of the advantages or otherwise of feeding with corn and linseed, as against oilcake, and of the effects of certain chemical applications in fixing the ammonia in the manure of the box-fed cattle.   All the cattle in the feeding houses were in the primest condition, so that a spectator could form no opinion as to the merits of the different modes of feeding; but it may be remarked that the box-fed cattle were all under one roof, not exposed with an open side to the air, as is frequently the case, but in every way as warm as those in the stalls.   The quality of the dung from the box-fed cattle was said to have proved itself far superior to that from the stall-fed, but on more minutely inquiring into this, we found that the dung of the stall-fed cattle had been thrown into an open yard and mixed with that of the *lean* cattle, and in this state tried against the box-fed cattle manure taken directly from the boxes.   Such an experiment proved nothing, and it just shows how guardedly we must watch every particular of detail before accepting conclusions as fully proved.

The fattening cattle are being fed, one part with 5 lb. of barley, beans, and linseed, and the other with 5 lb. of oilcake to each animal, boiled and poured over 14 lb. of cut clover hay and 45 lb. of cut swedes in layers, in large boxes, which are covered up and left for twenty-four hours, and the mixture is then given in three feeds.   The cattle get no other food, and no water. The milch cows, when they calve, receive cut hay, and 1 lb. of oilcake daily.   On this they do extremely well till the grass is ready, better than on mangold, and swedes are never given as they taste in the milk.   The year-olds receive cut hay, with 1 lb. of meal sprinkled over it, and 1 peck of cut swedes daily.   A very fine herd of Hereford stock is kept, and a first rate cross, for quality of meat, is got from an Ayrshire cow by a Hereford bull.   We must not omit mention of the pig department, with its ample and unusually elegant feeding-house, and the various contrivances for cooking, and for conveying the food to the animals without disturbing them.

The liquid from the different cattle houses and yards is conveyed to a covered tank, over which a wooden house is erected, where ashes, night soil, wood ashes, and other dry refuse are stored, and also the solid droppings from the feeding stalls. The liquid is pumped over the ashes and the whole turned and mixed together to dry, in which state the mixture is drilled in as manure with the turnip seed.

The farm buildings throughout the estate are many of them very extensive and new, but we cannot say that they appeared to us to be designed with that regard to economy and arrangement which would render them models for other estates. They comprise extensive barn accommodation, stables, feeding stalls, and large open yards, with sheds for young cattle. A farm let at 25*s.* an acre, or 400 acres for 500*l.*, costs five years' rent for all new outbuildings, including dwelling-house for the farmer. A farm at 600*l.* will cost somewhat less in proportion, and one at 400*l.* considerably more, so that farms of not less than 500 acres are found the most economical division for an estate. A set of farm buildings is at present being erected for a small farm in which the whole stock and manure are to be under cover.

We have already referred to the business-like arrangements which the Duke makes with his tenantry. The connection subsisting between them is of an intelligent character, inasmuch as a tenant receives his farm in fitting order for the employment of his capital, neither cramped with insufficient accommodation for his stock, nor wasting his means in undrained land. His crops are not destroyed by game, nor injured by hedgerow timber. He has the option of a lease and a corn rent. With these advantages his rent is moderately charged, but proper opportunities are taken for a readjustment, by which the landlord receives his fair share of the increased returns, partly the result of his own expenditure, partly arising from the general progress of agriculture, the increase of population, and the accumulating wealth of the country. Tenants remain long on the estate, but a change is made, without hesitation, when believed to be necessary.

" To improve the dwellings of the labouring class, and afford them the means of greater cleanliness, health, and comfort in their own homes, to extend education, and thus raise the social and moral habits of those most valuable members of the community, are among the first duties, and ought to be among the truest pleasures, of every landlord." Such are the words of the present Duke of Bedford, and truly is he carrying them into practice. Recognising in their fullest extent the responsibilities of his high position, he rests himself not on the possession of great wealth or the pride of ancestry, but in the performance of those duties which secure the confidence of his tenantry, and engage the affectionate respect of the labourers. If we should venture to say to other landlords, " Go, and do thou likewise," we may be met with the reply, that they have not equal means at their disposal. Yet the same circumstances which limit or extend their property, limit also or extend the claims on their justice; and great though the expenditure of the Duke may be, it is governed by that prudent foresight and adherence to economical principles which, while it provides for a fair return from the investment, at the same time draws forth the intelligent energies of those who share in the prosperity thereby created.

LETTER XLIX.

BEDFORDSHIRE — *continued.*

FARMING AT LIDLINGTON — MANAGEMENT OF THE VARIOUS CROPS — VALUE OF WINTER BEANS — CLAY LAND — MANGOLD — COMPARATIVE VALUE OF LEICESTER AND SOUTHDOWN BREEDS OF SHEEP — MANAGEMENT OF SHEEP — OF OXEN — COOKED FOOD ECONOMICAL — FEEDING OF HORSES — FARM BUILDINGS AND MACHINERY — ECONOMY OF STEAM POWER — COST OF THRASHING — CONSUMPTION OF COALS — COMPARISON BETWEEN SHORT HORNS AND DEVONS — PROFIT ON GALLOWAYS — GRASS LAND INJURED BY DRAINAGE — DIFFERENT MANAGEMENT AND EXPENSE OF SAND AND CHALK SOILS — LAND AS IT BECOMES RICHER IS BETTER FITTED FOR WHEAT THAN BARLEY — RELATIVE VALUE OF THE TWO CROPS — ADVANTAGE OF LEAVING A TENANT FREE TO ADAPT HIS MANAGEMENT TO CHANGES OF CIRCUMSTANCE — COMPENSATION FOR REDUCE DPRICES — RENT — RATES AND WAGES — CARDINGTON TO SOUTHHILL — STRAW FOR PLAITING AT DUNSTABLE.

BEDFORD, Feb., 1851.

FROM Woburn, towards Bedford, by Lidlington, we pass through a country of various geological character, including the greensand and the clays of the middle and upper oolite formations. On the former the surface is much undulated, rising into dry rounded ridges of considerable elevation, the soil sometimes a blowing sand, sometimes a fine loam, and generally good land for turnip crops and sheep feeding. From these the road drops suddenly, and by steep descents, into the level plains of the clay districts stretching towards the Ouse and its tributaries, where the mode of husbandry is regulated by the heavy character of the land. At a point where the two districts of country join, one sees in immediate contrast the modifications of management dictated by experience, and at Lidlington we were fortunate in meeting with a large farmer whose occupation includes both the heavy soil of the plain, and the light turnip soil hills of the greensand.

Mr. Thomas, of Lidlington, a tenant of the Duké of Bedford, occupies a farm of 740 acres in extent, 240 of which are in

grass, and 500 under the plough. Of the latter, 240 acres are
light land, and 260 acres clay or heavy land. The light land is
farmed on the four course, thus:—

1. Turnips—in preparation for which the land is ploughed
and crossploughed before winter, stirred in spring with Finlayson's harrow, and after being cleaned it is ploughed the last
furrow. The seed is then drilled on the flat, at 2 feet intervals,
with 3 cwt. to 4 cwt. of superphosphate, and, where the clover
leys have not been dunged, with dung also. Two-thirds of the
crop is consumed on the ground by sheep. These are stored
very cheaply by throwing six rows together from each side to
the centre, and covering them by a bout of the plough. When
wanted, they are taken up and given to the sheep, cut, in
troughs. Two-thirds of the turnip crop are swedes. White
turnips are grown after autumn-sown tares, as soon as these have
been taken off the ground, for early feed.

2. Barley.—For this crop a good tilth is absolutely indispensable, and must regulate both the time of sowing and the
quantity of seed. The sooner it can be got in after the new
year the better, provided the state of the soil and the weather
admit. With a fine tilth 2 bushels an acre are sufficient for
seed, and 3 bushels the most under any circumstances. It
is generally drilled, though, when it can be done, it succeeds
better by being sown broadcast and scuffled in. One-half of
this division is sown out with "seeds," the other, after harvest,
with winter beans. The half of the "seeds" is sown with 14 lb.
red clover and 6 lb. trefoil per acre; the other half with 14 lb.
white clover and 6 lb. rib grass. No rye grass is sown, as it has
been found to injure the succeeding wheat crop greatly.

3. "Seeds" and winter beans, half of this division in each
alternately. As much of the "seeds" is dunged in September
and October as there is dung for. For winter beans the stubble
is dunged before being ploughed: they are sown in September
and first ten days of October, and are a very hardy and prolific
variety. In spring they stool or tiller in a remarkable manner,

so that it is necessary to be very cautious about the quantity of seed sown. It should never exceed 1½ bushels per acre, and the rows should not be less than 24 inches apart. The yield is good, 50 bushels an acre on sandy as well as light land, being got quite successfully and regularly. The introduction of this crop into the four course rotation on light lands is considered one of the most important improvements of the alternate system of husbandry, both by postponing, and thereby rendering more certain, the clover crop, and by the intrinsic value of the bean crop itself. Cow grass, which used to be sown instead of red clover, comes so much later in spring, that, having no compensating value, it has been discontinued.

4. Wheat. — Spalding's red wheat is the variety used, at the rate of 2 bushels an acre on the clover leys. It is drilled by the light Bedfordshire drill with wonderful accuracy, on hill sides so steep that it is difficult even to walk on them. In spring the wheat is generally hand-hoed. Mr. Thomas expressed a strong opinion against very thin sowing, into which he had been several times beguiled to his loss.

The course of management on the clay land is prescribed by the terms of lease, and is a six course, as follows: — 1. Fallow, with cole-seed or tares dunged; the bare fallow limed. 2. Wheat sown on fallows with 1½ bushel of seed, and top dressed with 2 cwt. of guano in spring. 3. Clover. 4. Oats, the stubble dunged for 5, Beans. 6. Wheat. Mr. Thomas has little doubt that wheat might be taken every other year with advantage, but he is restricted as above.

Deep loam recently broken up from pasture is found best for the growth of the mangold, which does not succeed on the sand land. It is drilled in at 2 feet intervals in the last week of April, or first week of May, with 4 cwt. of superphospate per acre, and if the land has long been in cultivation, with dung ploughed in besides. The yellow globe is the variety most approved.

A flock of 500 ewes is kept on the farm — one-half Leicesters, and one-half Downs — both breeds being kept pure and distinct.

As Mr. Thomas has followed this practice for nearly twenty years he has had an ample opportunity of studying their comparative excellence.   But so difficult does he find it to judge between them that he is still undecided as to which has proved the most advantageous flock.   The male produce of the Leicester ewes is sold of at fourteen months old, averaging 20lb. per quarter, and producing about 7lb. of wool.   The Downs, with similar management, do not become fit for the butcher until eighteen months old, when they are also sold, averaging, as the others, about 20lb. per quarter, and clipping about 4lb. of wool.   So far this shows an average in favour of the Leicester, but a 10 stone (of 8lb.) Leicester is now making about 36s. and a Down of the same weight 40s.   The fleece is also worth a trifle more per lb.   From 250 Leicester ewes there are rarely more than 250 lambs raised, while the 250 Downs often rear 350 lambs. The casualties, too, attending the Down flock are much less than those which attach to the Leicester.   Parturition is much easier, and they rarely prove barren.   From the fact that upon rich grass land the Down sheep often becomes poor, while in the same field the Leicester becomes extremely fat, it would seem that each breed has its proper locality; and it may be that farms possessing both descriptions of soil would be best stocked by crossing the two breeds.   The best shearlings which this county produces are a cross between the Down ewe and the Costwold ram, but it does not do to breed from this cross, the produce in such trials having proved very bad.

The shepherd is paid 1s. a head for every lamb alive and well on the 1st of June, over and above the number of ewes put to the ram at the previous Michaelmas.   The lambs are wintered in three flocks, two of which are fattening tegs, and one, ewe tegs, to be drawn into the flock next year.   The old ewes which are culled out are immediately put on the best keep and are sold as soon as fat, or as stores, but they are not put to turnips, as they eat enormously and seldom pay for their food. No vegetable will fatten an old ewe faster than cole-seed.—(We noticed a practice in Northamptonshire of nipping off the front teeth of the old ewes and then turning them into the swedes to eat the tops, on which, with half a pint of lentils daily, they fatten quickly. The sheep appear to suffer little pain from this operation, and the succulent leaves of the swede are thus consumed while in their most nutritious state, the bulb being left untouched.) In spring — April or May — Mr. Thomas purchases all the tegs that are required to run on the old pastures among the fattening oxen. Those of his own tegs which do not go away fat from turnips are put on the best " seeds," and forced as much as possible on corn and cake. As they go off, their places on the " seeds " are filled up by the purchased tegs from the grazing pastures. Thus, towards Midsummer, the grass fields are relieved of sheep, and the proper succession of mutton is kept up for Smithfield. The Leicester tegs have cake and corn during the winter, $\frac{3}{4}$ lb. of beans, and, towards the end of the turnip season, $\frac{1}{2}$ lb. of oilcake also. The Downs, in addition to cut swedes, receive clover chaff. There are no subdivision fences on the sand land division of this farm, and the sheep are therefore folded at all seasons, being placed in summer on a new piece of clover daily, and shifted back during the night on the piece they had the day before.

In the management of oxen the system followed by Mr. Thomas is to purchase every spring or summer 80 good two-year-old Devons, or short-horns, or three-year-old-Scots; to run them upon inferior sward that summer; winter them well; graze them on the best land the following year; and get them all off fat from grass as early in autumn as possible. These details of the management of sheep and cattle are founded on long experience and capable advice.

In the winter-feeding of cattle Mr. Thomas finds the following mixture an excellent and economical substitute for turnips : — for fifteen beasts, 5 lb. per head of bruised lentils and offal wheat and barley are mixed up in equal portions, and put into 80 gallons of water, which is then boiled by steam for half-an-hour ;

it is then taken out by buckets and poured over layers of chaff, half hay and half oat straw; this stands for twenty-four hours in a close box, when it is served out to the cattle once a day. Besides this, they have straw fodder in the evening, but no turnips or other food, and are kept in very fresh, good condition. The cooking and steaming apparatus made by Stanley of Peterborough is used on this farm, and at the Park farm at Woburn, and other places we visited, and is very highly approved of.

The farm horses receive each a bushel of oats and a bushel of beans (split) per week throughout the year. The stables open into a yard with a covered shed, into which the horses are turned loose every night. There they receive lucerne or tares in summer, and during the winter they have, with their corn, in the stable, cut chaff, half hay and half straw, and 14 lb. of hay each at night under the shed. They are very rarely turned out to pasture in the fields.

The farm buildings, which have been erected within the last few years, are extensive and commodious. They include feeding houses for fattening cattle in stalls, which seems the mode generally preferred in this county, as having been found to make earlier maturity with the same expenditure of food. Mr. Thomas, at his own cost, has erected a very complete suite of barn machinery, comprising a 6-horse steam engine; thrashing mill with hummelers, shakers, and dressing machinery; French burr stones for grinding; linseed crusher; and hay and straw cutter. This last machine is by Ferrabee of Stroud, and is here preferred to that of Cornes, which has been laid aside. The difficulty of sharpening the latter is the only objection to it; but as that must be frequently done on a large occupation where a great quantity of chaff is cut, the time lost in unscrewing the knives and taking them off to be sharpened is a serious objection, when brought into comparison with the ingenious mode in which Ferrabee's machine is sharpened. By merely reversing the motion and applying a stick of emery to the face of the knives, they are sharpened in a few minutes and with very little delay. The cost of the engine and machinery altogether has been about 500*l.*, and will, no doubt, be considered by many very extravagant. Mr. Thomas finds it an excellent investment, as it has effected a saving to him of 200*l.* a year in labour, which, released from barn work, is now more advantageously employed in other departments of his farm. One ton of coals, costing 15*s.*, keeps the engine in constant work for three days, thrashing and dressing 200 bushels of wheat per day, at a cost, including labour and every expense, of 8*d.* per quarter. The same work, Mr. Thomas calculates, would cost by horse-power 2*s.* 8*d.* per quarter, and by hand, 3*s.* 4*d.* to 4*s.* The whole crop on this farm, last year, was thrashed by the use of 26 tons of coals.

The advantage to the grazier of feeding the most profitable breed of cattle is a matter of much importance. He must endeavour not only to stock his land with good cattle, but with that particular description which, with the same consumption of food, will leave him the largest profit. The following experiment between short-horns and Devons was tried by Mr. Thomas, to satisfy himself as to the respective merits of these two favourite breeds:—

SHORT-HORNS.

Purchased at Darlington, Easter Monday fair, 1842,—
41 2-year-old steers, at £10 18 0 each.

1843, same fair,—
45 ditto, at   -   - £12 13 6
These two lots of beasts were kept till the following Christmas twelvemonths, and made respectively 19*l.* and 21*l.* 11*s.*, leaving a difference respectively of 8*l.* 2*s.* 8*d.* and 8*l.* 17*s.* 6*d.*, or an average increase of 8*l.* 10*s.*

DEVONS.

Purchased at Aylesbury, May 27. 1842,—
50 2-year-old Devons, at £8 0 each.

1843, same market,—
45 ditto at -   -   - £8 10
These two lots were kept until the following September twelvemonths, and made, from grass, respectively, 16*l.* 14*s.* 2*d.* and 19*l.* 12*s.*, leaving a difference respectively of 8*l.* 14*s.* 2*d.* and 11*l.* 2*s.*, or an average increase of 9*l.* 18*s.*

Both lots were fed the first winter on turnips and straw, and grazed precisely alike; but the Devons, with three months' less keep, left the largest increase in value, and, considering also the smaller amount of capital employed in their purchase, proved themselves a good deal more profitable on Mr. Thomas's farm than the short-horns.

Mr. Thomas finds Galloway Scots a pleasant and generally profitable stock. His average advance for a year's keep for this breed of cattle, bought at Barnet on the 4th of September, and sold in the end of August, during the last fourteen years, has been 7*l.* a head. They are wintered on turnips and straw, and kept in summer on the best grass land.

While speaking of the grass lands, we may mention that on strong land here, which was apt to poach in wet weather, but produced no rushes, tile drainage, at 18 feet distances and 3 feet depth, *proved injurious.* The finer grasses disappeared and gave place to hassocks of coarse bent, which the stock leave untouched. These Mr. Thomas proposes to get rid of by laying cake on them which the cattle will eat, and at the same time, perhaps, the coarse grass also. The best grazing lands of South Leicestershire are said to have been much injured in the same way by an injudicious application of the modern system of tile drainage.

In the grazing of sandy land the clover should be allowed to grow ankle deep, otherwise when ploughed for wheat it turns up like a desert. In the chalk country, again, it is found most advantageous to eat the clover very bare during summer, and let it get up at last before being ploughed. On account of the greater readiness with which the sand land runs to weed, and partly, also, from an inferiority in the labourers, Mr. Thomas, who some years ago farmed extensively in Hertfordshire, found a difference in the expense of labour in favour of a chalk and flint soil in that county of 10*s.* an acre, the chalk land of Hertfordshire costing him 18*s.*, and the sand land of Bedfordshire 28*s.* an acre for labour, with the same system of manage-

ment.   We are inclined to attribute some portion of this
increased charge to the gradual and certain, though perhaps
imperceptible, increase of employment caused by the progress
of agricultural improvement, not the least important benefit
of which is that every addition to the annual average of our
crops increases the amount of labour necessary to manage and
manufacture them.

Another consequence of improved farming was here brought
under our notice, which further illustrates a principle of much
importance.   That the produce per acre of barley does not
increase under high farming in the same proportion as the straw;
that, in fact, the crop runs to straw to the injury of the corn,
while, on the contrary, the wheat crop increases in yield with
higher cultivation.   During the last fourteen years the wheat
crop on this farm has averaged 35 bushels an acre, and the
barley 42½ — the former progressing, the latter stationary.   The
respective average realised values of the two crops during the
same period has been, of

Wheat per acre - - - - -	£13	3	6
Of barley per acre - - -	8	6	0
Difference in favour of wheat per acre -	£4	17	6

The experience of Mr. Thomas coincides with that of many
other eminent farmers whom we have visited, in this, that, on as
much of his turnip lan l as can be cleared in good season, nothing
but the terms of his agreement prevent him from realising this
difference of value, as he has no doubt that, with good farming,
wheat might be taken, on soils suitable for its cultivation, every
alternate year.   This we state, not as any matter of complaint
against his landlord, to whom he has not applied for such a
change, but to show to landlords and agents how important it is
that they should be careful not to bind their tenants down to
rules of management which the progress of agricultural knowledge has rendered obsolete, and which, besides being the cause
of serious loss to the tenant, are positively injurious to the land-

lord by preventing the full development of the capabilities of his land.   The science of agriculture is progressive; every year is adding new facts to our knowledge, and opening up new sources of fertility for our farms.   A system founded on the principle of making a farm self-supporting, might be a very pro- per one when the price of purchased manure and the expense of transporting it were greater than the value of the additional produce created by it.   But the discovery of guano, the manu- facture of artificial manures, the facilities of transport by railway, and the vast increase of population, are completely changing these relative values; and the landlord or the tenant who remains blind to these changes, and fights against them, must have a losing game in competing with his neighbour who has the wisdom and prudence to turn them to his advantage.

The circumstances to which Mr. Thomas looks to compensate him in some degree for the fall in the price of corn are, a partial corn rent, the saving he has effected by the use of the steam engine, which he calculates at 200*l.* a year, the partial substitution of wheat for barley in the rotation, and the growth of potatoes for the London market.   For this crop the sand land of the farm is well adapted, and its situation, close to a railway station about 50 miles from London, gives it great facility of transport.

The town of Bedford is surrounded by a fine, low, rich country.   Towards Cardington the fields are open and the soil of various quality.   Strong, deep, friable land, suitable for the production of all kinds of crops, is let at 36*s.* an acre; very good, sharp land, level and easy of culture, at from 20*s.* to 30*s.*; and level clay land as low as 13*s.*, all tithe free.   The rates altogether are about 3*s.* a pound additional.   The county rates are increasing, on account of the police and other expenses, over which the ratepayers have no control.   Nor do they expect to be much benefitted by Mr. Milner Gibson's bill, if he adopts the recommendation of Sir George Grey to restrict the choice of the boards of guardians, who are to represent them at the county finan-

cial boards, to such only as are magistrates, the practical effect of
which would be, that the ratepaying farmers would have no real
representation, as a *bonâ fide* farmer, however otherwise eligible,
is seldom or never placed on the commission of the peace.

In this district the labourers' wages vary from 8*s.*, the
lowest, to 9*s.* and 10*s.* a week.   The four-course system of cultivation is beginning to be changed, by taking oats after wheat
in order to postpone the recurrence of turnips for a season.
Winter beans, as a change from clover, are also being adopted
in this part of the county.   40 bushels of barley and 25 of red
wheat are reckoned good average crops.   No potatoes are
cultivated for sale, and very little dairy produce.   A little cake
is purchased for the stock, but no artificial manure whatever to
be applied directly to the crops.   On a farm of 400 acres the
number of cattle wintered were 4 feeding fat, and 30 young
cattle, getting 3 lb. of cake, and some meal and millers' offal
daily.   The pure Leicester sheep ceasing to breed, are now
being crossed with great advantage by Southdown rams.   Land
not of the strongest character is very commonly fallowed during
the summer, once in the course, and sown with barley in spring.
The horses, which are strong and good, are sometimes yoked
three in line, but generally two abreast.   The crop is altogether
thrashed by hand ; and, though the mowing of wheat is gaining
ground, reaping is more common, followed by the mowing of
the stubble.

On the estate of Mr. Whitbread, who, next to the Duke
of Bedford, is the largest landowner in the county, the farm
buildings are generally very old fashioned — wood and thatch.
Hedgerow timber is being cut down, but there has been
no reduction nor abatement of rents, as the farms are understood to be moderately let.   There are no leases on this estate,
but the tenants are not disturbed nor changed, and if a farm
is given up there are numerous competitors for it.   In the
neighbourhood of Southhill, where there is a tract of very fine
country, we found spring wheat after turnips being substituted

G G 3

to a great extent for barley, on account of the low price of the latter, and partly, also, on account of the value of the wheat straw, for which there is a demand for plaiting at Dunstable. The wheat is sown thick and broadcast, to improve the quality of the straw for this purpose, and as suitable straw brings from 6*l.* to 8*l.* a ton, it is as valuable by the acre as corn.   In some districts of the county this business is followed with much advantage.

LETTER L.

HERTFORDSHIRE. — MIDDLESEX.

DESCRIPTION OF THE COUNTY — RENT AND RATES — COURSE OF CROPS — SIZE OF FARMS — WAGES — FARMING AT LAWRENCE END — TOP DRESSING FOR GRASS LAND — PRACTICE OF "INOCULATING" GRASS LAND — PROFITS ON FAT CATTLE, AND MANAGEMENT — OF SHEEP — EXPENDITURE IN MANURE, FOOD, AND ON LABOUR — ROTHAMSTEAD PARK — MR. LAWES' VALUABLE EXPERIMENTS — SUCCESSIVE WHEAT CROPS — AMMONIA THE ESSENTIAL REQUISITE OF MANURE FOR WHEAT — THE ADVANTAGES OF ITS APPLICATION LIMITED BY CLIMATE — NEW PLAN OF FARMING CLAY SOILS SUGGESTED — VALUE OF OTHER MANURES — SHED FEEDING OF SHEEP — PLAN OF DOING SO ON THE TURNIP FIELD — FARM HORSES — MIDDLESEX GROWTH OF ITALIAN RYE GRASS AT MR. DICKENSON'S FARM — VALUE OF LIQUID MANURE — EXTRAORDINARY PRODUCE — WANT OF DRAINAGE NEAR LONDON.

HEMEL HEMPSTEAD, Feb. 1851.

Entering Hertfordshire near Hitchin, we passed through a picturesque country much resembling the quick succession of hill and valley in South Devon, with the same winding narrow lanes, shut in on either side by close and lofty hedgerows. Nearly the whole county lies upon chalk, at greater or less depth beneath the surface. On the Northern side the chalk substratum chiefly influences the character of the soil, clay prevails on the South and Essex borders, and rich, sandy loams are found in the valleys along which the various rivers and streams flow through the county.

Taking a line from Hitchin to Hemel Hempstead, the average rent for a large district is 25s. an acre, tithe free, all rates being covered by 3s. an acre at the utmost. The common system of cultivation is the five-course — viz., turnips or fallow, barley, clover, wheat, oats. Naked fallows are very much adopted. The average extent of farms is 200 acres. Small farmers holding from 50 to 100 acres are not doing well.

Some may manage just to keep going, but no more, while many of them must go out of the business.   As an example of the common mode of husbandry in the district, we found that on a farm of rather more than 200 acres, the whole stock was six cows and 100 ewes.   The calves are fattened, and the dairy produce sold as butter.   The lambs are either fattened, or sold as stores at 16s. or 18s. each.   About twenty pigs are fattened. Six horses work the farm, three in line in a plough.   Part of the straw is sold at about 1l. a load, and soot purchased in lieu of it.   The wheat crop yields twenty-two bushels an acre, oats twenty-four, and barley thirty-two.   No cake, and very little, if any, artificial manure is purchased.   Labourers are paid 9s. a-week.   No land is given up by the tenants, and farms continue to let readily.

The management adopted by Mr. Oakley, of Lawrence End, is of a very different character.   He holds in his own occupation two farms of about 400 acres each ; 140 acres of the home farm being grass.   He tries to have wheat every second year, instead of alternating with barley, as he finds it to pay better.   His average crops of wheat are now thirty-five bushels an acre, having nearly doubled them within seven years, and his barley forty bushels an acre.   He very rarely takes oats in succession to wheat, and only then when the land is in high condition.   Begining with the

1. Turnip crop.— On such part of the land after barley as the clover has not taken well, white turnips are sown in April, with 3 cwt. of Lawes' superphosphate, to be eaten off in August and September by sheep with corn, and to be followed by wheat drilled in October.   The main crop of swedes is sown in May, the land having been previously manured with twelve tons of dung, ploughed in, if possible, in February.   About the middle of May the seed is drilled on the flat with 2 cwt. of Lawes' superphosphate, mixed with eight bushels of ashes to the acre.   Green top turnips are sown in the middle of June, to be eaten by the " couples" and ewe tegs.   In the end of June, a few acres of green top swedes are sown for feeding at the last

of the season, and that land (then too late for a corn crop) is sown with white turnips or rape, to be followed in autumn by wheat, which is found a good system.

2. Turnips are followed by wheat or barley, chiefly wheat, which is sown at the rate of two bushels an acre. The varieties sown are the golden goody — a red wheat, very productive, but thin looking on the ground in winter — and red straw white wheat. In spring a spring variety is sown till February, but not later; after that barley is sown, or if the land is much trodden and cloddy, it is sown with oats.

3. Seeds.— A mixture of 16 lb. red clover, and 5 lb. of trefoil, is sown for mowing, once in eight years; 16 lb. of trefoil alone, for grazing with fattening sheep. The couples are first allowed to run over the trefoil without eating it down close, and it is then regularly penned off with sheep, fed also on corn. It is ploughed in October, and drilled with

4. Wheat. — The wheat is not hoed in spring, but any thistles or other large weeds are picked out.

The grass land that is mowed for hay is dressed with 2 cwt. of guano and twelve or fourteen bushels of ashes every year. The increased produce of hay repays the manure, and the after-grass is very superior. The ashes are made by burning hedge sides and waste places.

The system called "inoculating" grass land, is practised here very successfully. The object is to obtain at an early period the natural grasses of an old pasture on newly laid out land, and is managed thus : — A small plough is passed along an old pasture, from which it throws out about three inches of turf and leaves a little more, returning again with another strip of turf, of the same breadth, until the requisite quantity is obtained. A corn drill is then passed over the ground to be inoculated, the coulters of which mark it off in rows at eight inches apart. The sod is then cut into little pieces and laid down in the rows, each piece about four inches apart, by men who then tread it into the ground. This must be done in damp weather in September or October. In spring the ground is rolled and a little Dutch clover is sown, after which the whole is allowed to seed itself, and stock is put on in autumn.   By this process a fine pasture is rapidly formed, and on that portion where the strips of turf had been cut out, the ground soon covers itself from the adjoining rows of grass.

In the grazing of stock Mr. Oakley buys part Scots and part Herefords, and changes twice a year, fattening off in each case. In summer all the cattle receive half a peck each of split Egyptian beans daily, given in boxes on the pastures or in an adjoining yard.   In winter they receive clover hay chaff, as much as they will eat, a bushel of cut swedes daily, and 10 lb. of foreign oilcake each — that being the cheapest at present. This management not only greatly enriches the pastures and improves the quality of the foldyard manure, but enables Mr. Oakley to realize about 14*l.* for the year's keep of each of his cattle, or 7*l.* a-head for each of the two lots.   But his stock are of the primest quality and sent direct to a west-end butcher, who gives so much a stone (sending down the weight of the animal when killed), and thus saving the salesman's commission and the risks of Smithfield.

A stock of 300 Leicester ewes is kept, which are crossed with a Cotswold ram.   The produce have more wool and less fat than the pure Leicester.   They rear, on an average, a lamb to an ewe.   The wethers are fattened off at a year old, bringing, between mutton and wool, about 40*s.* ; having constantly received corn summer and winter, on the seeds and turnips. When the lambs are weaned, in June, the stock ewes are put on inferior pasture.   The culled ewes are placed on " the seeds," and receive corn to fatten them with all expedition, then on white turnips in August, and are ready for sale in September or October.   They are sold at four-years old.   Never more than 100 of any age are kept in one pen at a time.

About 500*l.* are expended annually on this farm in purchased food, beans, and oilcake, and about half that sum in artificial manures, guano, and Lawes' superphosphate.   Labour costs

25*s.* an acre, and 5*s.* an acre more is expended in stubbing hedges, and other improvements not strictly farm work.

From Lawrence End we proceeded to Rothamstead Park, the residence of Mr. Lawes, whose papers on agricultural chemistry in the Journal of the Royal Agricultural Society have thrown so much light on the *rationale* of farming. Other writers are obliged to depend on experiments over the details of which they have little control, and for the certainty of which they must rely on the good faith of their informant; but Mr. Lawes carries into practice under his own eye the researches which his scientific attainments enable him to originate, and superintends himself the accuracy of results on which he ventures to found conclusions of importance. Any one conversant with the exactitude required in agricultural experiments, and the facility with which attendants may make inadvertent blunders, or ready assertions on matters which, the evidences being removed, cannot be disproved, will at once estimate the value of such scientific superintendence as that of Mr. Lawes. Laying down for himself a certain line of inquiry, he tests it under many various circumstances, noting everything as he proceeds, waiting patiently, week by week, month by month, year by year, till his cumulative facts have laid a foundation so broad and strong that his results may safely be relied on as a certainty. With patience and perseverance, practice so combined with science may in one man's lifetime do much for agriculture. The facts which are slowly accumulating will in due time be given to the world, and while there cannot be a doubt that they will extend the fame of their author, they will settle many disputed points in agriculture on which the practical farmer now vainly asks for a trustworthy guide.

On some very important points Mr. Lawes has arrived at results in our opinion most interesting and instructive. No branch of British agriculture has presented greater difficulties in its consideration in connection with the prospect of a considerably lower average of the price of corn than the management of clay soils. Increased breadths of green crop and grass, with a larger head of stock, may help to compensate the farmer of that description of convertible soil which is appropriately termed in some counties " stock " land ; but this remedy is not available, in anything like the same extent, to the farmer of stiff clay. In the common course of husbandry, green crops are got on such soils with so much expenditure of labour, that, apart from other considerations, they scarcely repay the expense. For fear of injuring the soil in autumn, they must either be taken off the ground before they have arrived at maturity, or left on it for dry frosty weather, when the season for sowing wheat advantageously on such soils is past. The most valuable crop in the rotation is thus superseded, or imperfectly managed. In the latter case the land becomes foul, each year adds to the mischief, and the disappointed farmer has no remedy at last but a long fallow. He has been disgusted with green crops, a wheat crop once in four years with present prices does not pay, and he reluctantly comes to the conclusion that if no other than the ordinary alternate system can be devised, the cultivation of stiff clays must be abandoned.

To this point, for the last ten years, Mr. Lawes has directed a series of experiments. On a soil of heavy loam, on which sheep cannot be fed on turnips, four, five, and six feet above the chalk, and therefore uninfluenced by it, except in so far as it is thereby naturally drained, ten crops of wheat have been taken in succession, one portion always without any manure whatever, and the rest with a variety of manures the effects of which have been carefully observed. The seed is of the red cluster variety, drilled uniformly in rows at eight inches apart, and two bushels to the acre, hand-hoed twice in spring, and kept perfectly free from weeds. When the crop is removed the land is scarified with Bentall's skimmer, all weeds are removed, it is ploughed once, and the seed for the next crop is then drilled in. During the ten years the land, in a natural state, without manure, has produced a uniform average of

sixteen bushels of wheat an acre, with 100 lb. of straw per
bushel of wheat, the actual quantity varying with the change
of seasons between fourteen and twenty bushels.   The repetition of the crop has made no diminution or change in the
uniformity of the average, and the conclusion seems to be
established that if the land is kept clean and worked at proper
seasons, it is impossible to exhaust this soil below the power of
producing sixteen bushels of wheat every year.

But this natural produce may be doubled by the application
of certain manures.   Of these Mr. Lawes's experiments lead
him to conclude that ammonia is the essential requisite.   His
conclusions are almost uniform, that no organic matter affects
the produce of wheat except in so far as it yields ammonia, and
that the whole of the organic matter of the corn crop is taken
from the atmosphere by the medium of ammonia.   There is
a constant loss of ammonia going on by expiration, so that
a larger quantity must be supplied than is contained in the
crop.   For practical purposes, 5 lb. of ammonia is found to
produce a bushel of wheat, and the cheapest form of ammonia
at present being Peruvian guano, 1 cwt. of that substance may
be calculated to give four bushels of wheat.   The natural
produce of sixteen bushels an acre may, therefore, be doubled
by the application of 4 cwt. of Peruvian guano.

To this, however, there is a limit — climate.   Ammonia
gives growth, but it depends on climate whether that produce is straw or corn.   In a wet, cold summer, a heavy application of ammonia produces an undue development of the
circulating condition of the plant, the crop is laid, and the
farmer's hopes disappointed.   Seven of corn to ten of straw
is usually the most productive crop, five to ten seldom yields
well.   The prudent farmer will, therefore, regulate his application of ammonia with a reference to the average character
of the climate in which his farm is situated.

Straw, the refuse of the corn crops, Mr. Lawes considers to
be of no value as an application for wheat, except for its

ammonia, which can be more cheaply obtained otherwise.   But
the turnip converts straw into food, and in this it is amazingly
aided by phosphates; hence one great benefit of growing
turnips on a farm.   On heavy soils turnips are out of place, as
already shown in the ordinary system of farming; but with
the view of converting the refuse of the corn crops into food
they may be very usefully grown.

The practical conclusion at which we arrive is this, that in
the cultivation of a clay land farm, of similar quality of soil to
that of Mr. Lawes, there is no other restriction necessary than
to keep the land clean.   That while it is very possible to reduce the land by weeds, it is impossible to *exhaust* it (to a certain
point it may be *reduced*) by cleanly cultivated corn crops.
That it is an ascertained fact that wheat may be taken on soils
of this description, (provided they are manured) year after year
with no other limit than the necessity for cleaning the land, and
that may best be accomplished by an occasional green crop —
turnip or mangold, as best suits — at great intervals, the straw
being brought to the most rotten state, and applied in the
greatest possible quantity to insure a good crop, which will
clean the land well.   If these conclusions are satisfactorily
proved, the present mode of cultivating heavy clays may be
greatly changed, and the owners and occupiers of such soils be
better compensated in their cultivation than they have of late
had reason to anticipate.

While below a certain point, Mr. Lawes has ascertained that
his soil cannot be exhausted by continued crops of wheat, he
has found that green crops, without manure, run out entirely,
and that, consequently, they exhaust the land of that which is
necessary to their growth to a far greater degree that the
wheat crop.   On light soils the turnip comes beautifully into
the four-course rotation.   It converts into food the refuse of
the corn crops, aided, as already mentioned, by the application
of superphosphate, which in the warmer parts of England, is a
certain manure for turnips.   The consumption of the turnip on

the ground supplies the ammonia needed by the corn, and the alternate crops thus work beneficially for each other.

Superphosphate of lime or dissolved bones, while a specific for the turnip crop in the warmer and drier parts of Britain, is not relatively so valuable in the North and West, where there is sufficient moisture for guano. In such a climate, guano is probably the cheaper manure for the turnip crop. For corn, Mr. Lawes considers superphosphate valueless, as also bone manure, except in so far as it supplies 5 per cent. of ammonia ; though to us this appears doubtful, as we have grown excellent wheat crops with no other application than bone manure.

Beans Mr. Lawes finds to be best manured by alkalies. But if land does not contain these alkalies, naturally, it will not pay, at present prices, to supply them artificially. It is, therefore, better to grow wheat, as guano cannot be relied on as an application to beans, while it is certain in its effect on wheat. All cereals probably follow the same law.

Besides the elucidation of such general principles, Mr. Lawes applies himself to specific points, especially in the feeding of stock, and the influence of different kinds of food, both as regards the animal and the quality of its manure. The results he has arrived at, will no doubt in due time be given to the public, and we shall only mention one or two points more connected with this interesting place.

All the turnips are taken off the land by Crosskill's railway, which is found to prevent completely the injury which carting in damp weather does to heavy land. The rails are used also for removing manure from the feeding houses to the compost heap. Wooden sheds with open raftered bottoms, large enough to contain twelve sheep, are being made, with wheels, to push forward on rails over the turnip crop, for the purpose of consuming it on this heavy land under cover. The ground will in this way be saved from being trodden, the crop will be consumed without the expense of drawing it off or bringing back the manure, and the sheep will have the benefit of shelter and a dry bed.

The shed-feeding of sheep is practised to a considerable extent, and successfully. They are accommodated in a long shed, open on one side for light and air. Along this open side the mangers are placed for their food, which consists of 1 lb. of oilcake and 1 lb. of hay chaff, with cut roots. The sheep stand on an open platform of rafters, about three feet from the ground, the manure falling through the interstices. The liquid drains off to a tank. The solid is mixed with earth, and dibbled in with mangold. Two pounds a-week of live weight, on tegs, is deemed a good average increase under ordinary feeding.

The farm horses are fed entirely on Egyptian beans and bran, 7 lb. of beans and 7 lb. of bran a day, mixed with cut hay, for each horse. This food is considered equal in its effects to the same money's worth of oats, and the manure contains double the amount of ammonia. Oats are, therefore, never given to the horses.

Before leaving the subject of Mr. Lawes's farm, we think it necessary to guard our readers against any misapprehensions of our meaning. However conclusive his experiments may be on his own farm, we must not forget that they refer to a given soil and climate. Other wheat soils may not be naturally so rich as to enable the farmer to dispense with every manure for his wheat crop except ammonia. But one important service he has rendered by his experiments, in demolishing the notion so long prevalent among land agents, that there was something so peculiarly exhaustive about a wheat crop that it could only be taken from the same land at considerable intervals and after the interposition of less valuable cereals,—a notion which, now that the farmer has the command of foreign manures, has done more to prevent the due development of the capabilities of clay soils, in a climate suitable for wheat, than can be easily estimated.

From Hertfordshire we proceeded to Willesden, in Middlesex,

to visit the farm of Mr. Dickenson, of Curzon Street, on account of his celebrity as a grower of Italian ryegrass. This farm, which is the property of All Souls' College, is about 100 acres in extent, very heavy clay, let at 3*l.* an acre, the tenant paying also tithe and rates, about 15*s.* an acre. It is in the vicinity of London, being only three or four miles west of Regent's Park. Sixty acres are in meadow, and forty in tillage, all of which is under Italian ryegrass.

The rapid growth of this grass, and the immense yield of forage which it gives under proper management, made us anxious to examine that of Mr. Dickenson. He sows four bushels of seed per acre, never later than September. The crop becomes thin the second year, and the ground is then ploughed up, well cleaned, and again sown down with Italian ryegrass. Seven, eight, and even nine cuttings are got in the course of the year, the land being dressed after each cutting with 3,300 gallons of liquid manure per acre. In the latter part of the season, owing to the wet character of the clay subsoil, this dressing cannot be applied without injury to the surface by the carts, and therefore nitrate of soda is substituted in warm weather, and guano in cold. The demand for seed is so great that two crops are sometimes taken in a season (though that is not recommended in ordinary circumstances), the first crop yielding from four to seven quarters an acre, the second about three quarters. In order to ascertain how much hay could be got from Italian ryegrass, the whole year's crop of one field was made into hay, and the produce of four cuttings on a field of 20 acres amounted to 130 loads of 18 cwt. each, or nearly six tons of hay per acre. After each cutting the land was dressed with nitrate of soda at the rate of 1½ to 2 cwt. per acre.

The fresh cuttings during winter, which are of course comparatively of light weight, are used by Mr. Dickenson for his cows, ewes, and early lambs, and sick horses.

The other meadows are manured with horse-dung (from Mr. Dickenson's very extensive horse establishment in Curzon

Street), which is spread thinly over the ground and left there till
the rains have washed the manure from the straw.   The
manure sinks into and enriches the ground, and the dry straw
is raked up and sent back for further use in the stables.

The land in the neighbourhood of Willesden and towards
London is an extremely heavy clay, very wet, and undrained.
Mr. Dickenson's farm is partially drained, about 20 inches deep,
by a mole plough.   The tenants do not seem generally anxious
to have their land thoroughly underdrained, and yet it is so wet
that they cannot put stock on it after October.   During an
open winter the fields have a pleasant green appearance, looking
richer than they really are, their summer produce being only
from one and a half to two loads of hay per acre.   The land is
chiefly used for growing hay for the London market, which is
made early in June, and the aftermath grazed with sheep and
cattle till October.   Surely if any land in the world would pay
for drainage this would, within three or four miles of the
metropolis, and yet so wet that stock cannot be put upon it
without injury after October.

LETTER LI.

CAMBRIDGE—HUNTINGDON.

CAMBRIDGESHIRE. — INCENDIARY FIRES — INDUCING, WITH OTHER CIRCUM-
STANCES, GREAT DESPONDENCE AMONG FARMERS — NO EXPENSE TOO GREAT
TO PUT THIS DOWN—LOW RATE OF WAGES—MR. JONAS WEBB'S CELEBRATED
FLOCK OF SOUTH DOWNS — HUNTINGDONSHIRE — RENT AND RATES —
FARMING AT WOODHURST, A RATIONAL SYSTEM OF MANAGING CLAY LAND
— FEEDING OF SHEEP AND CATTLE — NECESSITY FOR DRAINAGE ON THE
CLAY LANDS — FARMERS' PROSPECTS — THEY DEMAND REDUCTION OF RENTS
—LABOURERS' WAGES — WANT OF SYMPATHY BETWEEN LANDLORDS AND
FARMERS, AND BETWEEN FARMERS AND LABOURERS.

HUNTINGDON, Feb. 1851.

PASSING through the southern division of Cambridgeshire, we found the land a light turnip soil, laid out in large enclosures and managed generally in the four-course rotation. The buildings are chiefly wood and thatch, antique and inconvenient; and, from the combustible nature of their structure, both very tempting and very subject to the fire of the incendiary.

These incendiary fires are said to be of almost nightly occurrence in this and the adjoining part of Huntingdonshire. Many of the farmers live in constant apprehension of them, and, with their families, are kept in a state of nervous excitement which we had not expected to find in any English county. The corn ricks are built in different parts of the fields, seldom contiguous, so that if one should be fired the rest may have some chance to escape. One farmer had his buildings three times burned, and the Insurance Companies now decline to insure him. The culprits generally escape detection, — the mischief may be so quickly done, and without any trace by which to discover the doer.

In any district of England in which we have yet been, we have not heard the farmers speak in a tone of greater discouragement than here. Their wheat crop, last year, was of inferior quality, the price unusually low, and, to add to this, their live stock and crop are continually exposed to the match of the prowling incendiary. Such a state of matters is unendurable, and not a little discreditable to the police arrangements of the district. To get rid of so great an evil we should consider the rent of the county well expended in setting a watch on every corn rick in it, if no less effectual means of prevention can be adopted. To say that, in a district within fifty miles of London, property is so insecure and even life in some degree of hazard, is to tell us of a country in a semi-barbarous state. A man might as well expose his life to the risk of a shot from a Tipperary assassin, as live, like a Cambridgeshire farmer, in constant apprehension of incendiarism.

Whatever the cause, the evil itself must be put down. We were assured that no considerate or kindly treatment of his labourers on the part of an individual farmer was any protection to him. Fires break out indiscriminately among all, — the kindest and most large hearted as often as the most selfish and narrow minded. A few bad fellows in a district are believed to do all the mischief, and bring discredit on the whole rural population. The fact of its existence argues discontent among the labouring class, for which the low rate of wages may in some degree account, 7s. to 8s. a week being the current rate. Cottage rents are from 2l. to as much, in some parishes, as 4l. or 5l., so that a labourer on 7s. a week has little to spare for the necessaries of life after paying his landlord 1s. 6d. or 2s. out of it. Labourers are fairly employed.

The agricultural management of Cambridgeshire is so ably and fully described by Mr. Jonas of Ickleton, in the seventh volume of the Royal Agricultural Society's Journal, that we do not think it necessary to enter upon it here in detail. The manure applied to the green crops appeared scanty, and the crops light, while, in the in-door work of the farm, the flail still holds its place, and the economical aids of machinery have not been generally adopted, partly from the mistaken belief that labour would thereby be thrown out of employment.

At Babraham we had the pleasure of inspecting Mr. Jonas Webb's celebrated flock of South Downs. For symmetry and hardy constitution this flock has proved itself unequalled in England; and as the greatest attention is bestowed by Mr. Webb in developing and preserving this hardiness of constitution, as well as the qualities of weight and early maturity, his rams are sought for from all parts of the country. Mr. Webb is likewise turning his attention to the breeding of short-horns, some very superior animals of which are at present in his possession, though we have little doubt he will find this business attended with more risk and less profit than he most deservedly reaps from his breed of beautiful South Downs.

From Cambridge we passed, by St. Ives, into Huntingdonshire, which is a thinly-wooded county, and on its northern side principally a fen district, nearly one-fourth of the whole county being of that character. On the rivers Ouse and Nene are extensive meadows, subject to injury from floods, but very useful, when held in conjunction with an arable farm in the interior, for supplying the farmer with hay for his stock. Excepting the fens, which we described in a former letter, the soil of the greater part of the county is either a strong deep clay, more or less mixed with loam, or a deep loamy gravel, lying on the middle oolite formation. The rent varies according to the quality and situation. Near Godmanchester, prime deep land, the greater part of which is not too strong for turnips, and which produces all other crops of fine quality, is let at 30s. an acre tithe free, the rates being 3s. to 4s. an acre. The average rent of good strong land favourably situated in other parts of the county, may be stated at 25s. tithe free, and the rates from 1s. to 3s. a pound on the rack rent.

The farm of Mr. Ekyne, of Woodhurst, is a favourable specimen of strong-land farming in Huntingdonshire. It extends to 200 acres, 30 of which are in grass, the rest under cultivation. It is all strong land lying on a heavy retentive clay, and has been drained by placing a 3-feet drain in every old furrow. The drain is dug 2 feet through the soil, and 1 foot into the strong gault, and is laid with $1\frac{1}{2}$-inch pipes, which cost at the kiln 13*s.* per 1000.

The land is managed in a six-course, thus : — (1) fallow, which, after being wrought and cleaned, receives 12 bushels of bone manure per acre, and is sown in July with mustard, a small portion being reserved for tares. In autumn the mustard is ploughed into the ground, and the land drilled with (2) red wheat, sown at the rate of 2 bushels an acre, in rows 8 inches apart. The wheat is once hand-hoed in spring, at a cost of 2*s.* to 3*s.* This crop yields from 32 to 36 bushels an acre. After the wheat is removed the stubble is dunged in November, and then ploughed in. As early in spring as the weather will admit, the next crop, (3) beans, are drilled on the winter furrow in 22-inch rows, and 3 bushels of seed per acre. During the season these are generally twice hand-hoed, besides weeding the rows, at a cost of 6*s.* 6*d.* to 9*s.* per acre, by piece work. They are not horse-hoed. The average produce is 28 bushels an acre. The bean stubble is ploughed twice before winter ; and in spring the land is stirred across with a large harrow, and drilled with the fourth crop, (4) barley. With the barley 18 lb. of red clover seed are sown. The average crop of barley is 50 bushels an acre. (5) The " seeds " are manured with 8 to 10 tons of dung, and are part mown for hay and forage, but chiefly consumed on the ground by sheep, which receive also half a pint each of pease or beans, daily. About the middle of September the land thus enriched is ploughed, and after lying exposed to the action of the weather for about a month, the surface is harrowed and drilled with the last crop of the course, (6) wheat.

Between 40 and 50 cattle are fed fat during the winter in stalls, each animal receiving, daily, 9 lb. of the best oil cake, and $3\frac{1}{2}$ lb. of bean and barley meal mixed, with cut hay *ad libitum.* The produce of 40 acres of hay is brought to this farm every year from a meadow in the fen country. The feeding cattle have no turnips or other green food. Nine horses are required to work the farm, three and sometimes four in a plough, the ploughing being now done two inches deeper in consequence of drainage than formerly, and thus requiring more power.

A little consideration will show that this is a very rational method of dealing with strong land. It is first thoroughly cleaned by a spring and summer fallow, then manured with bones to ensure a good crop of mustard, which again is ploughed down to enrich the land for the main crop, wheat. The land, being by this preparation very clean, requires only one light hand-hoeing in spring. The stubble is then dunged, and the land ploughed in preparation for beans, which are sown in spring on the winter furrow rendered friable by the natural action of the weather. The beans are kept very clean by repeated hand-hoeing; and, when they are removed, the ground is twice ploughed at the time of year when it is in a dry state, and is thus both cleaned and pulverised, the growth of root-weeds being checked before they have time to gather strength. In spring the land is merely stirred with a large heavy harrow, in preparing for barley, the seed of which is thus sown in a suitable bed for it, the pulverulent state of the soil being likewise very favourable for the small seeds of the clover which are sown at the same time. Well rotted manure is, for the second time in the course, laid on the young " seeds" after the barley crop has been removed, to encourage a heavy crop for eating off with sheep, which, receiving corn also, are both profitably fed and greatly enrich the land for the following wheat crop. No turnips are grown on the farm, the soil being considered unsuitable for that crop; but their place is supplied by the hay which is brought from the fen country; and the cattle being likewise fed on cake and corn, the whole of the straw is made into good dung. The result of this management is, that the land is kept very clean and in high condition; and the yield of the different crops is very satisfactory.

But it is not possible to carry out this system with equal success on land which has not been drained; and unfortunately that
is still the case with by far the larger portion of the strong lands
of Huntingdonshire.   Let any unprejudiced man examine the
farm we have described and one of the undrained farms in the
same district, both possessing a soil naturally of the same
quality; and let him compare the yield of the two farms in
corn and stock, and he will then be able to appreciate the loss
sustained by the latter from want of drainage, and the impossibility of the tenant of such land competing with the other on
equal terms in the corn market.   If landlords would give this
matter the grave consideration it deserves, they would see the
absolute necessity of carrying out this improvement without a
day's delay.

Near Godmanchester and in the neighbourhood of Huntingdon, the land is managed chiefly in the four-course rotation.
The farmers, though not complaining quite so loudly as those of
Cambridgeshire, declare their inability to go on with the present
rents and prices.   One-third of them, it is said, must give up
the business if prices do not improve; and the rest, who feel
that their only remedy, supposing low prices to continue, is in
increased production, declare that they will not lay out their
capital, unless the landlords reduce their rents 25 per cent.

Labourers' wages in this county are from 8s. to 9s. a week;
and very few, except idle men whom nobody cares to have, are
out of employment.   In some cases the farmers are employing
less than their usual number of people, on account of its having
been said last year, in the House of Commons, that there was
no agricultural distress, as labourers were everywhere fully
employed; and to prevent such an argument being used again,
they resolved to send some to the Union.   That a farmer should
think himself compelled to resort to such a mode of proof only
shows how little real sympathy exists in the district between
landlord, tenant, and labourer.   Cottage rents are from 2l. 10s.
to 4l. and 5l.

LETTER LII.

CONCLUSIONS.

Results of this inquiry as compared with those of Arthur Young in 1770. — Increase of rent, produce, wages, and prices — no increase in price of corn, while stock and its produce have doubled in value — the rent of land capriciously fixed — evils of under-letting and over-letting — difficult to hit the just mean — three remarkable examples of the various modes of letting — value of land not now so much dependent on proximity to the metropolis as in 1770 — shown (by a comparative table) to be chiefly influenced by the kind of produce it yields — and by the size of farms — the corn counties being lower rented than the mixed husbandry and grazing counties — the kind of produce in greatest demand — examples showing the tendency to an increasing consumption of articles the produce of grass and green crops — results of this on the value of land — the direction in which agricultural enterprise will in future be most remunerative — this change will not diminish the supply of food, — or of labour.

Dec. 1851.

Having now traversed thirty-two of the forty counties of England, it is time that our mission should draw to a close; the many facts already collected forming a sufficient basis for an accurate estimate of the present condition of agriculture in this country. Since Arthur Young's tours in 1770, there has been no similar inquiry; the Agricultural Reports of Counties collected by the Board of Agriculture, and those at present in the course of publication by the Royal Agricultural Society, being the work of separate individuals — full of instructive information, but wanting that link of combination and comparison which is obtained from the single point of view whence one mind surveys, in succession, the various modes of husbandry practised throughout England.

An interval of eighty years affords ample room for denoting with precision the progress of agriculture. Young's " Tours " conclude with very specific data showing the actual state of rents, produce, prices, and wages in 1770, in the twenty-six counties which he then examined. The information on which his data are based, seems to have been the same as ours—personal inquiry from the most trustworthy sources. As regards rent and produce, it is obvious that, unless the same farms had been spoken of, exactness of comparison is impossible. The figures which we give are therefore not offered as perfectly correct, but as the nearest approximation to correctness in our power. Until Government shall take up the important question of agricultural statistics, we must be content with such broad results as it is in the power of individual inquiry to elicit, conscious though we may be of the comparatively limited data from which we are obliged to generalise.

Table showing the Rent of Cultivated Land per Acre, the Produce of Wheat in bushels, the Price of Provisions, the Wages of the agricultural Labourer, and the Rent of Cottages, in 1770 and 1850, in Twenty-six of the English Counties.

	Rent of Cultivated Land per Acre.		Produce of Wheat per Acre.		Price of Provisions. Bread.		Price of Provisions. Meat.		Price of Provisions. Butter.		Labourers' Wages per Week.		Cottage Rents.	
	1770.	1850.	1770.	1850.	1770.	1850.	1770.	1850.	1770.	1850.	1770.	1850.	1770.	1850.
	s. d.	s. d.			d.	d.	d.	d.	d.	s. d.	s. d.	s. d.	s. d.	s.
Northumberland	12 6	20 0	18	30	¾	1¼	2½	5	5	10	6 0	11 0	20 0	60
Cumberland . .	7 6	25 0	23	27	¾	1¼	2½	5	5½	10	6 6	13 0	20 0	55
Durham . . .	21 0	17 0	25	16	1	1¼	3¼	5	6	1 0	6 6	11 0	20 0	
North Riding .	12 6	29 0	21	20	1	1¼	3	5	5½	1 0	6 6	11 0	25 0	
East Riding . .	8 0	22 6	25	30	1	1¼	3¼	5	5	1 0	7 6	12 0	27 6	
West Riding . .	16 6	40 0	20	30	1¼	1¼	3½	5	6½	1 0	6 0	14 0	30 0	
Lancashire . .	22 6	42 0	26	28	1	1¼	3	5	6½	11	6 6	13 6	27 6	
Cheshire . . .	16 0	30 0	25	28	1	1¼	3	5	6	11	—	—	27 6	80
Nottingham . .	13 0	32 0	31	32	1¼	1¼	3½	5	6	1 0	9 0	10 0	25 0	
Lincoln	10 0	30 0	21	26	1½	1¼	3¼	5	6	1 0	7 0	10 0	20 0	80
Stafford . . .	17 6	30 0	23	28	1	1¼	3½	5	6½	1 0	6 4	9 6	40 0	70
Warwick . .	17 6	32 6	28	30	1½	1¼	3	5	6	1 0	8 0	8 6	35 0	
Northampton .	7 0	30 0	23	28	1¾	1¼	3½	5	6	1 0	6 6	9 0	20 0	80
Huntingdon . .	10 0	26 6	18	32	1¾	1¼	3¾	5	6	1 0	7 5	8 6	30 0	70
Norfolk	11 6	25 6	24	32	1¾	1¼	3¾	5	6	1 0	8 0	8 6	40 0	80
Suffolk	13 6	24 0	24	32	1½	1¼	3¾	5	7¼	1 0	7 11	7 0	40 0	80
Bedford	12 0	25 0	19	25	1¾	1¼	3½	5	7	1 0	7 3	9 0	40 0	
Hertford . . .	12 0	22 6	24	22	2	1¼	3¾	5	7	1 2	7 6	9 0	42 6	
Essex	13 6	26 0	24	28	1¼	1¼	4	5	8½	1 0	7 9	8 0	40 0	80
Buckingham . .	10 0	26 0	25	25	1¼	1¼	4	5	7	1 2	8 0	8 6	42 0	42
Oxford	19 6	30 0	26	25	1¼	1¼	3½	5	7	1 2	7 0	9 0	45 0	
Berks	19 6	30 0	28	30	1¼	1¼	3¾	5	7	1 2	7 6	7 6	50 0	55
Surrey	15 9	18 6	—	22	1½	1¼	4¼	5	8½	1 1	9 0	9 6	55 0	100
Sussex	10 9	19 0	22	22	1¾	1¼	4	5	7	1 0	8 6	10 6	60 0	100
Hampshire . .	12 0	25 0	20	30	1½	1¼	3	5	6¾	1 0	8 0	9 0	40 0	100
Wilts	12 0	25 0	20	26	1½	1¼	2¾	5	6	1 0	7 0	7 0	30 0	60
Dorset	10 9	20 0	20	21	1½	1¼	2¾	5	6	1 0	6 9	7 6	30 0	
Gloucester . . .	10 6	28 0	20	23	1½	1¼	3	5	6¼	11	6 9	7 0	30 0	
Averages . . .	13 4	26 10	23	26¾	1⅓	1¼	3¼	5	6	1 0	7 3	9 7	34 0 8d. per week.	74 6 1s. 5d. per week.

In twenty-six counties the average rent of
arable land, in 1770, appears from Young's   *s.   d.*
returns to have been   -   -   - 13   4 an acre.
For the same counties our returns in 1850–51
give an average of   -   -   - 26 10   „

Increase of rent in eighty years   -   - 13   6 or 100 per cent.

In 1770 the average produce of wheat was - 23   an acre.
Bushels.

In 1770 the average produce of wheat was - 23   an acre.
In 1850–51 in the same counties it was   - 26¾   „

Increased produce of wheat per acre   - 3¾ or 14 per cent.

s.   d.
In 1770 the labourers' wages averaged   - 7   3 a week.
In 1850–51, in the same counties they averaged 9   7   „

Increase in wages of agricultural labourers - 2   4 or 34 per cent.

	Bread.	Butter.	Meat.	
In 1770 the price of provisions was	1½d.	0s. 6d.	3¼d.	per lb.
In 1850–51 it was - -	1¼	1 0	5	„

s.   d.
In 1770 the price of wool was   -   - 0   5½ per lb.
In 1850–51 it was   -   -   - 1   0   „

In 1770 the rent of labourers' cottages *s.   d.*
in sixteen counties averaged   -   - 36   0 a year.
In 1850–51, in the same counties   -   - 74   6   „

It thus appears that, in a period of 80 years, the average rent
of arable land has risen 100 per cent., the average produce of
wheat per acre has increased 14 per cent., the labourers' wages
34 per cent., and his cottage rent 100 per cent. ; while the price
of bread, the great staple of the food of the English labourer, is
about the same as it was in 1770.   The price of butter has increased 100 per cent., meat about 70 per cent., and wool upwards of 100 per cent.

The increase of 14 per cent. on the average yield of wheat
per acre, does not indicate the total increased produce.   The
extent of land in cultivation in 1770 was, without doubt,

much less than it is now; and the produce given then was the average of a higher quality of land, the best having of course been earliest taken into cultivation. The increase of acreable corn produce has therefore been obtained by better farming, notwithstanding the contrary influence arising from the employment of inferior soils. The increased breadth now under wheat, with the higher average produce, bear, however, no proportion to the increase of rent in the same period; and the price of wheat now is much the same as it was then. We must therefore look to the returns from stock to explain this discrepancy.

While wheat has not increased in price, butter, meat, and wool have nearly doubled in value. The quantity produced has also greatly increased, the same land now carrying larger cows, cattle which arrive at earlier maturity, and of greater size, and sheep of better weight and quality, and yielding more wool. On dairy farms, and on such as are adapted for the rearing and feeding of stock, especially of sheep stock, the value of the annual produce has kept pace with the increase of rent. With the corn farms the case is very different. In former times the strong clay lands were looked upon as the true wheat soils of the country. They paid the highest rent, the heaviest tithe, and employed the greatest number of labourers. But modern improvements have entirely changed their position. The extension of green crops, and the feeding of stock, have so raised the productive quality of the light lands, that they now produce corn at less cost than the clays, with the further important advantage, that the stock maintained on them yields a large profit besides. In all parts of the country, accordingly, we have found the farmers of strong clays suffering the most severely under the recent depression of prices.

The rent of land is defined by Mr. M'Culloch to be "the result of the unequal returns of the capital successively employed in agriculture." But in practice we have found rent to be a very capricious thing, often more regulated by the

character of the landlord or his agent, and the custom of the
neighbourhood, than by the value of the soil or the commodities
it produces.   There is not a county in England where this is
not exemplified.   On one estate we shall find land let at 20s.
per acre, and on the next, farms of the same quality and with
the same facilities of conveyance, let at 30s.   With farmers of
equal skill and enterprise this difference of rent remains in
the pocket of the fortunate tenant who holds under an easy
landlord.   But exertion is generally the child of necessity, and
the man who must pay 30s. is obliged to be industrious, while
his neighbour may be indolent, and, in that case, the difference
of rent is lost to all, because indolence leads to diminished production.   The active and industrious man employs more labour
to raise an increased produce, that he may be enabled to pay
his higher rent.

Whilst however we deprecate the under-letting of land, as
injurious to the landlord, and frequently in its consequences to
the public, we must guard against an error of the opposite
character, which is much more hurtful,—the over-letting of land.
When the rent of land is raised to such a point that the profits
of the farmer's industry are absorbed by it, he loses the motive
for exertion, and, if a man of capital, he carries it on the first
opportunity to a farm where it can be more profitably employed.
In other cases he may struggle on, in hope of success, till his
capital is so seriously diminished that he has little to withdraw,
his farm all the while rapidly deteriorating in cultivation and
produce, till it is at length abandoned to the landlord.

It may be very difficult to hit the happy mean between those
extremes, and, if there was no extraneous element to influence
the result, that mean would probably be best regulated by supply and demand.   But the preference over other creditors given
to the landlord, by the law of distraint, is sometimes used to
encourage competition between men of capital and skill and
men who have little of either, and the rent may thus be unfairly
raised.   Competition in the open market, therefore, is not always,

in the present state of the law, the fair measure of the value of
land to the tenant of capital.

Three remarkable examples of the different results produced
by the mode of letting land have been detailed in these letters.
In Oxfordshire we have found a great landlord, so injudiciously
requiring an increase of rent, that his best tenants have left, and
a large portion of the estate is being abandoned to him.   We
have described the relations between another nobleman and his
tenants in Durham, where neither the rent is raised nor the
tenants changed, where the bulk of the land is confessed to be
underlet, and yet the tenants are not prosperous nor satisfied.
In these two cases we have the opposite extremes, producing in
the one case ruin and diminished produce, and in the other,
indolence and discontent.   How nearly alike is the result of
conduct dictated by principles so different.

The third example is shown in Bedfordshire, where another
landlord, fully recognising his duty, puts his farms into a proper
state as regards the permanent improvements necessary for their
profitable occupation, and then lets them on lease to selected
tenants, at a fair rent, estimated by his agent, who is practically
acquainted with the value of land.   Undue competition is thus
discouraged, while at the same time the landlord participates,
by gradually improving rents, in the increasing wealth of the
country.   The tenants have their capital left free for the cultivation of the land, in the full benefit of which they have the
option of being secured by lease.   A living sympathetic interest
is thus maintained between landlord and tenant, the result of
which is seen in improved farming, increased employment, and
a cordial understanding between a wise and considerate landlord and an intelligent and independent tenantry.   On many
other estates mentioned in these Letters, the same principles
are followed by similar results.

The influence of proximity to large populations in enhancing
the rent of land, varies in different parts of the country.   The
lowest rented counties in England are Surrey, Sussex, and

Durham, two of which may be said to be in the vicinity of the
metropolis, and the third has a large and well employed native
population. The highest rented counties are Lancashire and the
West Riding, many parts of which are continuous villages, and both
contain a large proportion of grass land. In 1770, distance from
the metropolis seems to have in a great measure regulated the
rent, which begins, according to Young, at 19s. 6d. in Berkshire,
and gradually falls to 7s. 6d. in Cumberland. But the means of
communication in his time are described by him as "execrable."
" Let me most seriously caution all travellers," he says, " who
may accidentally purpose to travel this terrible country, to avoid
it.   They will here meet with ruts which I actually measured
four feet deep, and floating with mud, only from a wet summer.
I would advise no one to journey further north than Newcastle-
under-Lyne.   Until better management is produced I would
advise all travellers to consider this country as sea, and as soon
think of driving into the ocean as venturing into such detest-
able roads." Matters are changed now. We have railways
traversing every part of the country, steam vessels sailing from
almost every port, and generally good roads of accommodation
between every village and market town.

   Rent, in so far as regulated by external circumstances, we shall
find depends now on other influences than proximity to or distance from the metropolis. To illustrate this, among other points,
we have prepared the following table, which divides the country
into two sections from north to south, with reference to climate,
the one embracing the eastern and south coast or corn side of
the island; the other the midland and western counties, where
the system of husbandry is more a mixture of corn, stock, and
dairy farming.

Table

Table* showing the Average Rent of Cultivated Land, the Produce of
Wheat in bushels, and the Weekly Wages of the Labourer in 1850-1,
in the Midland and Western Counties, being the mixed corn and grass
districts; and in the East and South-Coast Counties, being the chief corn-
producing districts of England.

Midland and Western Counties.	Per Acre.		Labourers' Wages.	East and South-Coast Counties.	Per Acre.		Labourers' Wages.
	Rent.	Produce.			Rent.	Produce.	
	s. d.	bush.	s. d.		s. d.	bush.	s. d.
Cumberland	25 0	27	13 0	Northumber-	20 0	30	11 0
Lancashire	42 0	28	13 6	land.			
West Riding	40 0	30	14 0	Durham......	17 0	16	11 0
Cheshire ...	30 0	28	12 0	North Riding	29 0	20	11 0
Derby	26 0	33	11 0	East Riding	22 6	30	12 0
Nottingham	32 6	32	10 0	Lincoln	30 0	26	10 0
Leicester ...	35 0	21	9 6	Norfolk	25 6	32	8 6
Stafford	30 0	28	9 6	Suffolk	24 0	32	7 0
Warwick ...	32 6	30	8 6	Huntingdon	26 6	32	8 6
Northampton	30 0	28	9 0	Cambridge...	28 0	32	7 6
Bucks.........	26 0	25	8 6	Bedford	25 0	25	9 0
Oxford	30 0	25	9 0	Hertford......	22 6	22	9 0
Gloucester...	28 0	23	7 0	Essex	26 0	28	8 0
North Wilts	35 0	28	7 6	Surrey	18 6	22	9 6
Devon	30 0	20	8 6	Sussex	19 0	22	10 6
				Berks	30 0	30	7 6
				Hants	25 0	30	9 0
				South Wilts	17 6	24	7 0
				Dorset	20 0	21	7 6
Averages ...	31 5	27	10 0	Averages ...	23 8	26½	9 1

Average rent of cultivated land in all the counties   - 27s. 2d.
Average produce per acre of wheat   -   -   - 26⅔ bushels.
Average weekly wages of labourer   -   -   - 9s. 6d.

The great corn-growing counties of the east coast are thus
shown to yield an average rent of 23s. 8d., an acre ; the more
mixed husbandry of the midland counties, and the grazing
green crop, and dairy districts of the west, 31s. 5d. This striking
difference, being not less than 30 per cent., is explained chiefly
by the different value of their staple produce, as already shown :
corn, the staple of the east coast, selling at the same price as it
did 80 years ago, while dairy produce, meat, and wool have

* See illustrative Map at beginning of volume.

nearly doubled in value. The difference in rent does not arise from a greater fertility of soil, as may be seen by comparing the produce of wheat. The corn counties, in so far as they yield barley and feed or produce cattle and sheep, benefit by the rise in price.*

Leases are the exception throughout England ; and though we have found them more prevalent in the west, there has been no sufficient uniformity to account in any degree for the difference of rent.

But the size of farms has an undoubted influence on the rent. In the dry climate of the counties on the east coast, the operations of a corn farm can be carried on, with great precision and regularity, on an extensive scale. In the chalk districts especially, the fields are open and unencumbered with wood ; the dry nature of the land admits of sheep folding, and a large tract may be conveniently managed under the superintendence of one person. By this means the landlord's outlay in buildings and fences is much economised, and he finds it his interest to encourage a class of large farmers, men of capital and education.

* The table is so far incomplete, that our information in regard to the different " rates " payable by the tenant in addition to his rent, shows them to be so variable that no accurate average could be given. Some farms were tithe-free, and on others the landlords are now taking upon themselves the payment of tithe. In some parishes the poor-rates were trifling, in others exorbitant, and the same with highway rate and other county rates. The table at page 514 shows that the poor rates are, on the average, nearly equal in their pressure in both sections of the country, the average poor relief of the Midland and Western Counties being 1s. 9¾d., and of the East and South Counties 1s. 10d. per pound. And on the whole, though the tithe appears to be heaviest on the south and eastern counties, the rates, in the aggregate, may be held to be nearly alike in both divisions of the country, and will not affect the truth of the averages we have given. We have a strong feeling that landlords would find it a good plan to take upon themselves the payment of all rates except the poor-rate, letting their land, as in Scotland, at a certain rent, free of all other rates. The landlords, who, in effect, pay all the rates in diminished rent, would then have a direct interest in controlling and economising the county expenditure, for which they are both best qualified and have most time, and the tenants would know the exact extent of their engagements, and not be obliged, as at present, to reserve a wide margin for these uncertain liabilities.

As we proceed westward, the country becomes more wooded, and better adapted for pasturage; the enclosures are smaller, the farms less extensive, and the farmers more numerous. Still farther west, the moistness of the climate materially affects the mode of cultivation,— unfavourable to corn crops, especially before the introduction of tile drainage, and favourable to grass. The farms are of small extent, and held by a numerous class of tenants, who live frugally, and, in many cases, assist, with their families, in the labours of the farm. We have here all the elements necessary to make a difference in the rate of rent. The chief commodity of the western farmer is the produce of his dairy, his cattle, and his flock. The large eastern farmer looks principally to his wheat and barley. The landlord of the western and midland counties possesses the two great advantages of his soil being used for the production of the most valuable of our agricultural commodities, whilst his farms, from their size, are accessible to a larger body of competitors; in short, are in greater demand than the corn farms of the east. Our notes of the average extent of farms in the various counties, give 430 acres for the corn farms of the east, and 220 acres for the mixed farms of the midland and western district.

The geological nature of the country as affecting the character of the soil itself for fertility or otherwise, has a considerable influence on its intrinsic value. In all the lower-rented counties, except the three northernmost, chalk is the prevailing characteristic. In the higher-rented counties, red sandstone is the principal geological formation.

An attentive consideration of the above table will strike the careful reader in several new points of view. That the large capitalist farmer of the east coast, possessing the most cheaply cultivated soil, and conducting his agricultural operations with the most skill, should not only pay the lowest rent, but be the loudest complainer under the recent depression of prices, is to be accounted for by his greater dependence on the value of corn. The moistness of the climate of the west, on the other hand, discouraged corn cultivation, and compelled a greater reliance on stock. And, as the country becomes more prosperous, the difference in the relative value of corn and stock will gradually be increased.

The production of vegetables and fresh meat, hay for forage, and pasture for dairy cattle, which were formerly confined to the neighbourhood of towns, will necessarily extend as the towns become more numerous and more populous. The facilities of communication must increase this tendency. Our insular position, with a limited territory, and an increasingly dense manufacturing population, is yearly extending the circle within which the production of fresh food, animal, vegetable, and forage, will be needed for the daily and weekly supply of the inhabitants and their cattle; and which, both on account of its bulk, and the necessity of having it fresh, cannot be brought from distant countries. Fresh meat, milk, butter, vegetables, and hay, are articles of this description. They can be produced in no country so well as our own, both climate and soil being remarkably suited to them. Wool has likewise increased in value as much as any agricultural product; and there is a good prospect of flax becoming an article in extensive demand, and therefore worthy of the farmer's attention. The manufacture of sugar from beet-root may yet be found very profitable to the English agriculturist, and ought not to be excluded from consideration.* Now all these products require the employment of

---

* There are two important considerations with regard to the culture of flax and sugar beet. The farmer may not only receive a remunerative price for the fibre of the one and the saccharine matter of the other, but he retains on his farm the seed of the flax, and the refuse of the sugar manufacture, to feed his stock, and increase the quality and quantity of their manure. The uses of linseed as food for cattle are well known in this country, and in regard to the refuse of the beet-root manufacture, we may mention (on the information of the Comte De Gourcy, who has devoted several years to the personal investigation of continental agriculture) that very large stocks of cattle are fed on the sugar farms, and that a machine has been lately invented by a sugar manufacturer at Baden, which, like our thrashing machine, can be introduced at about the same expense on individual farms, and by

considerable labour, very minute care, skill, and attention, and a larger acreable application of capital than is requisite for the production of corn. So various are the objects thus requiring attention and economical arrangement, that a very large undertaking, such as is now carried on by some of the wealthier farmers of the eastern counties, could not, on this more elaborate system, be profitably conducted under the single superintendence of one person. This will inevitably lead to the gradual diminution of the largest farms, and to the concentration of the capital and attention of the farmer on a smaller space.

The individual experience of the agricultural class may be appealed to in support of this opinion. The consumption of bread in a farmer's family is not half so large an item, in the annual expenditure of his household, as butcher's meat; and milk and vegetables, if they were purchased in the market, would cost him more than bread. If he looks back for thirty years, he will find that this difference has been gradually increasing. With the great mass of consumers, bread still forms the chief article of consumption. But in the manufacturing districts where wages are good, the use of butcher's meat and cheese is enormously on the increase; and even in the agricultural districts the labourer does now occasionally indulge himself in a meat dinner, or season his dry bread with a morsel of cheese. In a gentleman's family consisting of himself, his wife, six children, and ten servants, the average expenditure for each individual per annum, for articles of food produced by the farmer, is 9*l*. 10*s*. for meat, butter, and milk, and 1*l*. 2. 4*d*. for bread. In a large public establishment containing an average throughout the year of 646 male persons, chiefly boys, the expenditure per head for meat, cheese, potatoes, butter, and milk is 4*l*. 10*s*. 6*d*., and for bread 2*l*. 1*s*. 6*d*. The price of each article is charged in both cases at the

which the sugar can be extracted from the beet, and prepared for commerce, at a price of 2½d. a pound, after paying the cost of manufacture, and a remunerative value to the farmer for his beet.

present average rates throughout England. The first example shows an expenditure in articles the produce of grass and green crops nearly nine times as great as in corn; and the second, which may be regarded as more of an average example, also shows an outlay $2\frac{1}{4}$ times greater on the former articles of produce than the latter. Here we see not only the kind of produce most in demand, but the direction in which household expenditure increases when the means permit. It is reasonable to conclude that the great mass of the consumers, as their circumstances improve, will follow the same rule. And in further illustration of this argument it may be mentioned that the only species of corn which has risen materially in price since 1770 is barley, and that is accounted for by the increasing use of beer, which is more a luxury than a necessary of life.

Every intelligent farmer ought to keep this steadily in view. Let him produce as much as he can of the articles which have shown a gradual tendency to increase in value. The farms which eighty years ago yielded 100l. in meat and wool, or in butter, would now produce 200l., although neither the breed of stock nor the capabilities of the land had been improved. Those which yielded 100l. in wheat then, would yield no more now, even if the productive power of the land had undergone no diminution by a long course of exhaustion. The clays of Durham and Cleveland, and the wealds of Surrey, Sussex, and Kent are in this state of reduced fertility. The wheat they produce brings the same price per bushel as it did eighty years ago, but the quantity each acre yields is diminished. The tenants of these and similar districts are the poorest of their class in England, and the rent of the landlords has scarcely increased. In Cheshire and Lancashire there are clays as stiff and infertile; but even if they produced no more than they did eighty years ago, their owners and tenants have increased in wealth, inasmuch as that produce of cheese and butter, the staple of their district, which then sold for 100l., is now worth 200l. But the acreable produce itself has likewise increased; and this is a most important feature in the case, for a large stock of well-fed animals every year adds fertility to the land on which they are kept; while a constant succession of corn crops, not yielding a corresponding return of manure, gradually diminishes that natural fertility. The consequence of this, and likewise an illustration of our argument, is that at present corn land in the wealds of Surrey or Sussex may be hired at 15s. or bought for 21l. an acre, while grass land of much the same quality in Cheshire lets at 30s. and sells at 45l. an acre. Nay, even in the same county, the contrast is more striking; for in Surrey a meadow lets at 3l. an acre, while tillage land, originally of the same quality, on the opposite side of the fence, shall scarcely fetch 15s.

While we thus attempt to indicate the direction in which experience seems to have shown that agricultural enterprise will for the future be most remunerative, it is proper to advert to the possible effects of such a change, on the supply of food, and the demand for labour. If more land should thereby be gradually laid to grass, or a greater extent be devoted to the production of meat and vegetables, we should expect, as the result of better cultivation, that there would be little or no diminution in the annual produce of corn, inasmuch as the smaller extent would yield a larger acreable return. But although that increased return should be found insufficient to compensate in quantity for the diminished breadth of corn crops, no anxiety need thence be felt for the bread of the people. Rest from corn-cropping is the best preparation for the future growth of corn. And if an emergency should ever arise by which, in consequence of war, we should be driven back on our own resources, we would find that we had been laying up in our rich grass fields, and well-manured green crop lands, a store of fertility which might be called into action in a single season, and which would yield ample crops of corn for consecutive years, with little labour or expense.

Experience also shows that this change of husbandry would

not prove injurious to the labourer. Green crops require more manual labour than corn; and even an increase of grass combined with green crops would probably not diminish the demand for labour. It is in the strictly corn districts of the south and east that the labourer's condition is most depressed. The dairy lands of North Wilts, the vale of Gloucester, and the vale of Aylesbury afford better wages to the labourer than the corn districts of the same counties, Salisbury Plain, the Cotswolds, and the corn farms on the Chiltern hills.

LETTER LIII.

THE LANDLORD.

Dec. 1851.

IN our last letter we showed the advantage, both to the land-lord who receives, and to the tenant who pays rent, of cultivating as much as possible that description of produce which has a tendency to increase in value in this country. We need be under no apprehension of thereby unduly diminishing the growth of corn, for the more stock an arable farm maintains, the more productive will be its yield of corn. And if corn should rise in price so considerably as to affect its comparative value, it is an easy process to extend our growth of it. With the present prices, and the knowledge of the fact that the rich corn provinces of the continent are open to us, and are daily becoming more accessible by the extension of railways and steam navigation, there seems good reason to anticipate the permanence of a low range of prices. The safe course for the English agriculturist is to endeavour, by increasing his live stock, to render himself less dependent on corn, while he at the same time enriches his farm by their manure, and is thus enabled to grow heavier crops at less comparative cost.

Before this can be done with full advantage, wet land must be thoroughly drained, unnecessary obstructions to economical tillage should be removed, convenient farm-roads be provided for economising labour, and well arranged buildings be constructed for accommodating the cattle, accumulating the manure, and manufacturing the crop. Let any man compare, as we have done, two farms in the same neighbourhood, the one of which is neither better nor worse provided in these respects than the average of the country, and the other with all these improvements effected and turned to account. The tenant of the first uses nearly double the quantity of seed which the second finds sufficient; for much seed perishes in undrained land, and much is carried off by the birds which harbour in straggling fences. Every operation of tillage is more difficult, and of course more expensive; the crop is not so good in quality nor in quantity. He has tried to grow turnips to feed stock, but the land was insufficiently prepared, and the crop a comparative failure — such as it is, it cannot be consumed where it grows, owing to the wetness of the soil; and, while the land itself is cut up, the horses are distressed in drawing it home through the miry fields; — there being no proper accommodation for the cattle, the turnips are wastefully consumed, the animals do not thrive, and the manure is imperfectly made and much of it allowed to run to waste. The other farm has been drained and trenched — the land turns up mellow and dry — it is easily and thoroughly wrought at the proper season, and the turnip seed sown — the air and rain permeate the soil and dissolve the well made manure — the tender rootlets of the plant find food, and flourish — the crop is heavy, and does not disappoint the farmer. The land is so firm and dry that a portion of the crop is left on it to be consumed by sheep — the sheep thrive, and enrich the ground. The rest of the crop is carried home easily on convenient roads, and given to stock, housed in well arranged covered boxes, where warmth and shelter economise the food, and the facilities of intercommunication cheapen the cost of attendance. The stock of all kinds thrive under this generous treatment; it is worth the farmer's while to study what is best for them; their food is given at regular intervals, — roots, corn, and cake each in due proportion; they fatten rapidly — and pay. The rich manure, all of which has been safely preserved, is laid on the ground without stint. The fields are fruitful, and the farmer prosperous.

Are the men who occupy these two farms competing with each other on equal terms? Can the man who sows three bushels of wheat and reaps twenty-four, sell it with a profit at the same price as he who sows two and reaps forty? The man who starves his scanty stock in winter, can he profit equally with the other, whose well-fed and comfortably-lodged animals leave plenty to enrich the land, and fill their owner's pocket? As well might the hand-loom compete with the power-loom, the windmill with the steam-engine, or the stage-coach with the railway.

What, then, is the actual state of England in regard to these important improvements? Drainage in the counties where it is needed has made considerable progress, the removal of useless hedgerows is slowly extending, but farm-buildings everywhere are generally defective. The inconvenient, ill-arranged hovels, the rickety wood and thatch barns and sheds, devoid of every known improvement for economising labour, food, and manure, which are to be met with in every county of England, and from which anything else is exceptional in the southern counties, are a reproach to the landlords in the eyes of all skilful agriculturists who see them. One can hardly believe that such a state of matters is permitted to exist in an old and wealthy country. Buildings of such a character that every gale of wind brings something down which the farmer must repair, and of so combustible a nature that among ill-disposed people he lives in continual dread of midnight conflagration. With accommodation adapted to the requirements of a past century, the farmer is urged to do his best to meet the necessities of the present.   The economies of arrangement and power which are absolutely necessary to ensure profit amid the active competition of manufacturers, are totally lost sight of here.   And even the waste of raw material, which would be ruinous in a cotton-mill, is continued as a necessary evil, by the farmer, whose landlord provides him neither sufficient lodging for his stock, nor in that lodging, such as it is, the power of economising food by warmth and shelter.

We do not advocate expensive buildings, or urge upon landlords a heavy expenditure without a proportionate result.   In many parts of the country we have seen money squandered on expensive and ill-contrived buildings, from which the tenant reaped little advantage.   But if the farmers of England are to be exposed to universal competition, the landlords must give them a fair chance.   If they refuse to part with the control of their property for the endurance of a lease, they must themselves make such permanent improvements as a tenant at will is not justified in undertaking.   The farmers of that part of the continent nearest our shores have far better accommodation for their stock, than the majority of English tenants.   The sub stantial and capacious farmeries of Belgium, Holland, the north of France and the Rhenish provinces, contrast most favourably with the farm buildings common in most English counties.

But how can landlords afford such an expenditure as would be required to improve their estates, and maintain and increase their value?   They must make themselves acquainted first with what is absolutely requisite, and then with the best and most economical mode of carrying that into effect.   The funds necessary must also be forthcoming, or the rent will fall to a far greater extent than the annual amount of a sinking fund to repay this outlay.

A work so necessary could not have been so long neglected, if the great body of English landlords had been practically acquainted with the management of land.   In the beginning of this century and for twenty years afterwards, young men of family very naturally and properly made the army their profession, and committed the management of their estates to agents who, in those days of high prices, had little else to do than to receive rents. But that time has passed, and we trust the young men of the present generation will not be obliged to make war their profession. War prices also are gone, and the rude practice which flourished with wheat at 80s. a quarter, is altogether ruinous with wheat at 40s. The tenant may be the first to feel this change, but the landlord is the man on whom it eventually falls. Let him learn his profession — that of a landowner. He will soon discover the benefits of improvement and therefore its necessity, the advantage of drainage, the evils of numerous hedgerows, the destructiveness of game preserves, and the economy to the farmer, and, by consequence, to himself, of good roads and well arranged buildings. He will appreciate the difference between an improving tenant and a sluggard, and will encourage the one, and get rid of the other. He will see the advantage of promoting the investment of capital in cultivation, and the necessity therefore of giving his tenant the security of a lease. He will perceive the hardship of stringent covenants to a good tenant, and their inefficacy in preventing deterioration by a bad one. And if his estate is so extensive that his personal attention is required for public as well as private objects, his knowledge will enable him to select an agent properly qualified, whose advice he will himself be capable of estimating and controlling.

The present age is eminently practical. Every business in the country requires previous application, in those who practise it, to render it profitable. The labourer must perfect himself by years of patient application in the peculiar department of work in which he hopes to excel. The tradesman must serve his apprenticeship,— the professional man must study and work hard to obtain a knowledge of his business. The success or failure of these men affects themselves only. But the landlord's influence for good or evil extends to his tenants and labourers, and in its general results regulates, in no unimportant degree, the productiveness and welfare of the country.   Yet of all classes in the community he is the only one who receives no special training.   Our great universities offer him no peculiar instruction to fit him for the important functions of his station. He comes to it frequently without knowledge of its duties, and, with a consciousness of his own inabilty to perform them, he resigns all into the hands of his agent.

The selection of a properly qualified agent or steward is, on every large estate, a matter of the utmost importance. Honesty and uprightness are indispensable, capacity and personal activity, with an inquiring and unprejudiced mind, sound judgment, and decision of character, are all necessary.   An agent should be capable of choosing a class of tenantry who would aid him in the improvement of his employer's estates; he should be able to consult with and advise them in the management of their farms, pointing out resources which they may have overlooked; he should study the proper subdivision of farms and fences; the best arrangement of farm buildings and the most economical mode of constructing them ; he should be competent to decide on the fields that require drainage, so that, while necessary improvements are not neglected, the money of his employer may not be needlessly expended.   The presence of such an agent is visible at once in the general air of comfort, activity, and progress which animates all classes connected with the estate which he superintends.   Some landlords, whilst they admire and envy the improvements effected by such a man, yet fear to employ him on account of the expense which his operations entail in the first instance.   Expense is a comparative term.   If all improvement is declined, there will of course be little outlay.   But the most bigoted are conscious that if rents are to be maintained, their farms must afford the same facilities to the farmer as those of their neighbours, and that progress must be made.   An experienced sensible agent, with the aid of a willing tenantry, will effect as much with 100*l.* as an inexperienced or incompetent man can with 200*l.* And it is not only in this way that he economises, but by timely and wise encouragement he carries the tenantry forward in a course of improvement, which enables them better to withstand the pressure of low prices of corn, by adapting their management to altered circumstances.

The loss unconsciously sustained by some of the large proprietors of land in this country, from the incompetence of agents, is quite inconceivable. We recollect having met with a weakly old gentleman, a retired officer, who had the management of a very extensive and valuable estate, in the vicinity of a great town, which, within the last twenty years, has increased immensely in population and wealth. Dairy produce and vegetables are in great demand, and there is a cheap and abundant supply of manure. The land, considering its situation and advantages, and its quality, was moderately let, and an active tenantry should have prospered amid circumstances in every way so favourable. It would have been reasonable to expect that the rental would have been gradually creeping up, where the increasing wants of a well-employed population ought to have made itself so strongly felt. But that was not so. The tenants were hereditary, and so was the system. A certain quantity of wheat paid the rent in former times, and it ought to do so still. The supplies which were now most renumerative were furnished by others, and the men who had a market, at their door, for commodities which they and others within the circuit of a few miles could alone supply, continued to look almost exclusively to an article in the production of which the whole country could compete with them. The result is what under such management might be expected. The agent felt himself bound to recommend a reduction of rent over the whole estate some years ago, and was, at the time we saw him, again prepared to advise a second reduction ; both amounting together to nearly 25 per cent. We have not a doubt that the arrangements for the gradual improvement of the agricultural

management of the estate which a competent and able agent
would have adopted, might in this case not only have prevented
any reduction in the annual value of the property, but have
laid the foundation for its gradual increase, in the prosperity of
an active and intelligent tenantry.

But there is one great barrier to improvement, which
the present state of agriculture must force on the attention
of the legislature, — the great extent to which landed property is incumbered. In every county where we found an
estate more than usually neglected, the reason assigned was the
inability of the proprietor to make improvements, on account
of his incumbrances. We have not data by which to
estimate with accuracy the proportion of land in each county in
this position, but our information satisfies us that it is much
greater than is generally supposed. Even where estates are not
hopelessly embarrassed, landlords are often pinched by debt,
which they could clear off if they were enabled to sell a portion,
or if that portion could be sold without the difficulties and
expense which must now be submitted to. If it were possible
to render the transfer of land nearly as cheap and easy as that
of stock in the funds, the value of English property would be
greatly increased. It would simplify every transaction, both
with landlord and tenant. Those only who could afford to
perform the duties of landlords would then find it prudent to
hold that position. Capitalists would be induced to purchase
unimproved properties, for the purpose of improving them and
selling at a profit. A neglected estate would thus become
a matter of choice to men of capital, and the progress of
improvement would be rapid beyond precedent. A measure
which would not only permit the sale of encumbered estates,
but facilitate and simplify the transfer of land, would be more
beneficial to the owners and occupiers of land, and to the
labourers, in this country, than any question connected with
agriculture that has yet engaged the attention of the legislature.

Nor is this matter of opinion or conjecture. We have the

experience of neighbouring countries to show that their system of registering real property, and the comparatively cheap cost of transferring it, makes it the most eligible security, either for purchase or loan. The security on which a banker in Frankfort or Hamburg, as we learn from Mr. Stewart, advances money with the greatest facility, and at the lowest rate of interest, is real property, whether houses or land. In this country, on the contrary, Consols are the most available security. The owner of real property, who is in need of temporary assistance, finds himself embarrassed at every step by technical difficulties of title, and legal doubts, which compel him either to pay a high interest, or get, if he can, collateral security, or abandon the attempt altogether. These expensive and sometimes inextricable doubts and difficulties, are the cause of the market price of land in this country being lower than on the continent, where from thirty to thirty-three years' purchase is the common rate. Now, of all countries in the world, this, with its immense and *increasing* manufacturing and commercial population, its *limited* territory, and its superabundance of capital, is the place where land should rise in relative value above all others. We have seen that the average rent of cultivated land in England has doubled within eighty years. The prospect for another period is better, from the extraordinary supplies of gold, and the increasing comfort and wealth of the mass of the people. We see no reason to doubt that if the transfer of land were simplified, its value might be increased by five years' purchase; for persons seeking investment would look not merely to an immediate return, but to the certainty of a prospective increase of value which land in this country affords. A rise in value to this extent would free many an embarrassed landlord from his difficulties; and would, at all events, enable him to borrow money for the improvement of his estate on more reasonable terms. And if many should be compelled to sell, the permanence of our national institutions, and respect for the rights of property, would be better ensured by admitting to the class of landowners sagacious and prudent men, the architects of their own fortune, than by artificially maintaining families in a position the duties of which they cannot perform.

In the transfer of land, it is necessary that the parcels be clearly identified, and any special privileges or duties attached to them intelligibly described. This, and some other public objects, might be accomplished by opening a record in which any proprietor should be entitled to have his land inserted under due precautions. An authenticated extract from this record might become the foundation of many successive transfers by short endorsements, the title being completed by the entry of these in the principal record. A system resembling this has been described to us as having long been established with success in an eastern country. And although the question may be embarrassed with the technical refinements in which it has been enveloped, it would seem to common sense that these are not necessarily inherent in the subject. Apart from these, the chief practical difficulty would be to devise a scheme of record which would avoid repetition, and yet admit of easy reference. It might be open to those proprietors who should choose to avail themselves of its advantages, without being compulsory upon others.

LETTER LIV.

THE FARMER.

EXTRAORDINARY DIFFERENCE IN THE ART OF AGRICULTURE AS PRACTISED
IN DIFFERENT COUNTIES, AND OCCASIONALLY ON NEIGHBOURING FARMS—
THE MOST PRIMITIVE, AS COMPARED BY COST AND PRODUCE, ALSO THE
MOST EXPENSIVE SYSTEM — SOMETHING TO BE LEARNED EVERYWHERE —
THE ROTATION OF CROPS EXPANDING WITH REQUIREMENTS OF DENSER
POPULATION, AND INCREASING FACILITIES OF TRANSPORT, AND NEW
SOURCES OF MANURE — THE GREAT PRINCIPLE OF ROTATIONS — NECES-
SITY FOR THE FARMER TO KEEP THIS BEFORE HIM AND FOR THE LAND-
LORD TO DO NOTHING WHICH OBSTRUCTS HIS PROGRESS — SECURITY FOR
THE FARMER'S CAPITAL — MAY BE OBTAINED EITHER BY LEASE OR
"TENANT-RIGHT" — RESULTS OF TENANT-RIGHT IN THE COUNTIES WHERE
IT IS RECOGNISED — PRODUCES FRAUD — PERPETUATES OBSOLETE PRAC-
TICES — IS INDEFINITE IN EXTENT — UNJUST TO THE LANDLORD — DISLIKED
IN SOME COUNTIES BY THE BEST TENANTS — AND DOES NOT ENSURE GOOD
FARMING — LEASES WITH LIBERAL COVENANTS THEREFORE CONSIDERED
GREATLY PREFERABLE.

Dec. 1851.

THE position of the tenant-farmers of England next demands
our attention. To show the progress which has been made
in the art of agriculture in this country, it is not necessary to go back to any authority of the last century for a description of the processes then adopted. Every county presents
contrasts abundantly instructive, the most antiquated and the
most modern systems being found side by side. The successful
practices of one farm, or one county, are unknown or unheeded
in the next. On one side of a hedge a plough with five horses
and two men, and on the other side of the same hedge, a plough
with two horses and one man, are doing precisely the same
amount of work. In adjoining fields may be seen a foul turnip
crop under ten tons an acre, and a luxuriant one above thirty.
On neighbouring farms of similar soil the wheat crop may vary
from twenty to forty bushels an acre, and most probably the

man who grows twenty pays not less than 9s. for thrashing
that quantity by hand, while the other thrashes his forty bushels
by steam for 3s. 6d.

In the preceding Letters the details of good farming are given
much more at length than instances of the reverse, as it was
from the first only that instruction could be drawn.   This was
from no want of examples of antiquated farming; for if we
spent one day in examining Sir John Conroy's farm at Arborfield, Mr. Hudson's at Castleacre, Mr. Beasley's at Overstone,
or Lord Hatherton's at Teddesley, we were almost sure to be
wandering on the next through the mazes of frequent hedgerows, gazing at five horses elaborately doing the work of two,
manure suffered to go to waste, cattle insufficiently housed and
fed, land undrained and unproductive, and farmers complaining,
not without reason, of their want of success.   One day we
learned the processes by which Mr. Huxtable economises labour,
manure, and food; and the next we saw in operation an antiquated fanning machine, precisely the same as Arthur Young
described it eighty years ago, and worthy of the days before the
Conquest; manure treated as a troublesome nuisance; and cattle
wasting their substance and their food by being kept starving
in the open fields in winter.   The same day on which we saw
the steam engine of Mr. Thomas of Lidlington in Bedfordshire, with which he is enabled to thrash his wheat crop for
1d. a bushel, we found other farmers paying four or five times
as much for the same operation, not so well done by hand.   On
one farm in Suffolk we have seen light land prepared for turnips
by skim ploughing, scarifying, and one deep furrow, at a cost
not exceeding 25s. an acre; and on another, of precisely the
same kind of soil, the farmer was compelled by covenant to give
his land four or five furrows, with repeated harrowing and
rolling, to effect the same object, at more than double the cost.

Nor are these small economies to be despised.   On the two
corn crops of a four-course rotation, the different expense of
thrashing by hand and by steam will amount to 8s. an acre, which

being saved on the half of the ploughed land of the farm, is equivalent to 4s. an acre on the whole of it; and that is equal, in many cases, to a reduction of 20 per cent. in the rent.   The saving of seed which Sir John Conroy has effected by having his land in such a high state of cultivation, is, as compared with the average quantity sown in similar districts of England, quite equal to a saving of 10 per cent. on the common rent of corn lands.   But this is a saving of more limited application, inasmuch as a very thin sown crop is later in ripening, and more subject to mildew, and, unless accompanied by the most careful and continued hoeing, more favourable to weeds, besides being more easily affected by casualties of season, — all which are serious objections in a moist or northern climate.

It would be only repeating what has been much better done by Mr. Pusey, in the 26th number of the Royal Agricultural Society's Journal, if we were to draw into one view the savings which the modern farmer can effect, by the use of improved machines, cheaper feeding stuffs and manures, and more economical and rational processes of husbandry.   There is scarcely a single county in which the agricultural reader of these Letters will not find some practice better managed than his own, some process by which he may increase his crops, or fatten his stock, at less expense than it has hitherto cost him.   Some counties are much more advanced than others, and accordingly present more numerous examples for instruction ; but the careful student will find, in the description of every county, local practices which long experience has brought to a high state of perfection.   By combining with his own what he learns of the best, and rejecting the practices of the worst, he may establish for himself a system of agriculture, suited to his particular soil and climate, founded on the experience of successful practical men.   He will find that the best farmers have not attained success by blind adherence to a given rotation, but by a constant adaptation of their plans to the growing wants of the country, taking advantage of railway or steam-boat communications to cheapen the cost of transit to the best markets, and of portable manures or cattle food to replace the exhaustion caused by the increasing abstraction of corn and stock from the farm.

The question, what is the best rotation of crops, is so variously answered in these Letters that the reader may have some difficulty in arriving at a satisfactory conclusion. The Norfolk or four-course rotation is undoubtedly the one most generally approved, but it is to its principle of alternate corn and cattle crops, rather than to a strict adherence to its original detail, that this approval is accorded. In many cases we have inspected farms managed, on a strict four-course, to the highest pitch which the land under that system would yield. Do what he could, the farmer was unable to calculate with certainty on the success of each crop in the course. The clover failed, or the turnips were diseased. The barley was too heavy and did not fill, or the wheat lost root and proved thin Farm as high as he could, his unvarying routine of crops had exhausted something from his light soil which the aids at his command did not exactly replace. He drops the half of the clover from the course and substitutes winter beans. This succeeds, and he is tempted to try again. Mangold is taken instead of a portion of his turnips, and white or yellow turnips are grown where swedes were before. In the next round the position of these crops is reversed. His green crops now flourish, and he turns his attention to the corn. He finds that, by enriching his land, he improves the wheat crop, but endangers the barley. He cannot grow heavy crops of roots without manure, and he knows that to feed his sheep with profit he must hasten them forward by the aid of corn and cake. The land must therefore be enriched, and as with such high condition the barley might be lost, he sows the ground with wheat. An excellent crop of wheat reduces this condition sufficiently to admit of a safe and productive barley crop, which costs him nothing for manure, and very little for labour. But in this process of improvement the four-course has disappeared, and been replaced by a five, so arranged that red clover, white clover and trefoil, winter beans, and mangold, swedes, and turnips, are respectively repeated on the same ground at no shorter intervals than 15 years. The course then stands thus :—

1 $\frac{1}{3}$ Clover, $\frac{1}{3}$ White Clover and Trefoil, $\frac{1}{3}$ Winter Beans.
2 Wheat.
3 $\frac{1}{3}$ Mangold, $\frac{1}{3}$ Swedes, $\frac{1}{3}$ Turnips.
4 Wheat.
5 Barley, in some cases sown after mustard, ploughed in green.

And that in the course of time will, without doubt, in its turn, give place to another, under the guidance of further experience. Near a large population, where there is a demand for vegetables, and a supply of street manure, the farmer may find himself better paid by green crops than corn. Accordingly we have found the most intelligent farmers in such situations employ two-thirds of their land in growing green crops, and one-third in corn. In the western counties the climate exercises a powerful influence, and the successful farmers of Lancashire take two corn crops and two green crops alternately. In short, the detail is every where varied by the judicious agriculturist, to suit the necessities and advantages of the particular locality, when he is permitted by his agreement, and has sufficient skill, to pursue a rational system.

The reader will see that no one system or course of husbandry is applicable to every situation. It was not because the four-course was an alternation of corn and cattle crops, that it succeeded, though that was itself a great improvement, nor because it produced regularity of system, though that is also of much importance. Nor was it owing to the mere treading of the land by the feet of the sheep, though to that much of the success of the system used to be attributed. It was because it was a step in the right direction, one of those gropings in the dark, by which the man of mere practice occasionally finds the best path. Pursuing it without the guide of science, it soon began to fail, and lead him astray. There

was no virtue in the constant round of crops or regularity
of practice to compensate the increased exhaustion occasioned
by the sale of larger produce without an equivalent return
of manure. It was because it so far fulfilled the principle
of keeping the land DRY, CLEAN, and RICH, that it was in any
degree successful.

On a full recognition of that principle rests our future agricultural progress. The landlord and the farmer must both
recognise it in their dealings with each other, and with the
land. Crops which do not pay the farmer, do not suit his
purpose, and to restrict him to the growth of such is both
impolitic and absurd. His business is to grow the heaviest
crops of the most remunerative kind his soil can be made to
carry, and, within certain limits of climate which experience
has now defined, the better he farms, the more capable his
land becomes of growing the higher qualities of grain, of
supporting the most valuable breeds of stock, and of being
readily adapted to the growth of any kind of agricultural
produce, which railway facilities or increasing population may
render most remunerative. In this country the agricultural
improver cannot stand still. If he tries to do so, he will soon
fall into the list of obsolete men, being passed by eager competitors, willing to seize the current of events and turn them
to their advantage. The four-course, or any other course
when it has served its time, must expand itself to meet the
increasing requirements of the day, by appropriating to itself
the simultaneously enlarging resources of modern science and
enterprise.

This naturally brings us to the statement of a question which
we have considered and discussed with intelligent practical
farmers in all parts of England,— security for the capital of the
farmer, whether under the designation of " Compensation for
unexhausted Improvements," or, more briefly, " Tenant-right."
The investment of a tenant's capital in land seldom contemplates an immediate return. He does not anticipate that a

large expenditure in cleaning and enriching worn-out land will be all repaid to him in the first crop. He lays the foundation for a series of good crops, which in the aggregate he expects to repay him with interest. If he drains, makes fences, or other improvements of a more permanent character, a still longer period is requisite to compensate him. But he must either be secured in the possession of his farm for a certain period, sufficiently long to enable him to receive the benefits of his investment, or have some precise agreement under which he is to be repaid, in fixed proportions, for his outlay, if his landlord should see fit to resume possession of the farm. Without either the one or the other, an improving tenant has no legal security for the capital he invests in the cultivation of another person's land.

Yet the great proportion of English farms are held on yearly tenure, which may be terminated at any time by a six months' notice on either side. It is a system preferred by the landlord, as enabling him to retain a greater control over the land, and acquiesced in by the tenants, in consideration of easy rents. During a period of high prices, moderate rents could be paid without the investment of much capital by the tenant. But low prices and universal competition compel agricultural improvement. We must either farm as well as our neighbours, or be undersold by them. The investment of tenants' capital, whether in money, skill, or industry, is now therefore more than ever necessary to success. It may be said, with perfect truth, that great agricultural improvements have been made, and the most entire confidence subsists between landlord and tenant under this uncertain tenure. That tenants do, in many instances, invest their capital largely, with no other security than their landlord's character, we most willingly testify; and the confidence which subsists between the two classes in England generally, is in the highest degree honourable to both. In no country, perhaps, in the world, does the character of any class of men for fair and generous dealing, stand higher than that of the great body of English landlords. Yet there are exceptions,

and these are unfortunately becoming more numerous. The
son does not always inherit the virtues of his father. Necessity
or education may make his views different. Family provisions
and allowances may leave him less to spend from the same
rental. The tenant too, mixing more with the world than he
used to do, or being educated at a more advanced period of
its progress, begins to dislike the dependence implied in this
relation. He knows that he must invest his capital more
freely than heretofore in the cultivation of his farm, and in
these days of change he feels that he is entitled to ask some
effective security for its repayment. That security he may
obtain, either by being guaranteed by lease in the possession of
his farm for such a number of years as will give time for his
invested capital to have full effect and be returned to him, or,
if the landlord declines to give a lease, by an agreement on a
certain basis for compensation for unexhausted improvements,
when either party wishes to terminate the connection. One
or other of these alternatives the improving farmer is fairly
entitled to expect, and for the reasons now to be given we most
strongly recommend the general adoption of leases, in preference
to tenant-right.

The only counties in which the custom of tenant-right is
fully recognised, are Surrey, Sussex, the Weald of Kent, Lincoln, North Notts, and part of the West Riding. In these
counties the custom has been so long in operation, as to have
become binding in law, and they afford us an opportunity of
judging whether the system has worked so well in practice as
to justify its extension to all the other counties of England.
In each of those counties, except the Weald of Kent, which
we apprehend to be much the same as the contiguous tract in
Surrey and Sussex, we minutely examined the state of agriculture, and the relations subsisting between landlord and tenant,
as affected by this legalised custom, and our impression of each
in its place without reference to the other, was narrated in our
former Letters. In the Wealds of Surrey and Sussex, where the

custom is most stringent, we found the state of agriculture extremely backward, the produce much below the average of
England, the tenants deeply embarrassed, and the landlords
receiving their low rents irregularly; in fact, no men connected
with the land thriving, except the appraisers, who were in
constant requisition to settle the disputed claims of outgoing
and entering tenants. We found both farmers and landlords
complaining that the system led to much fraud and chicanery,
and that an entering tenant was compelled by it to pay as
much for bad as for good farming; that intelligent farmers
were most desirous that their landlords should buy up the
tenant-right, and thus put an end to it, and landlords in many
cases were doing so.   In Lincolnshire and North Notts we
found the great improvement of agriculture of late years attributed to the system of compensation to outgoing tenants; yet,
on examining the state of agriculture itself, it seemed to us, if
not inferior, certainly in no respect superior to the proficiency
of the same class of farmers in West Norfolk, whose capital is
not protected by any compensation agreements, but by a twenty-
one years' lease.   The indefiniteness of the " custom " was also
much complained of, and its constant liability to increase.
Frauds were beginning to creep into the system, and landlords,
for their own protection, were obliged to limit and define the
custom by special agreement.   In the southern portion of the
West Riding, where tenant-right is very stringent, it is found
to lead to great fraud and abuse, there being instances of
" smart " men, who make it their business to take a farm, hold
it for a year or two, and by " working up to a quitting," as
it is termed in Surrey, make a considerable profit by the difference which their ingenuity and that of their appraiser enables
them to demand when they leave, as compared with what they
paid at their entry.   Obsolete practices are valued under this
system at their original cost, so that the plan of giving five
furrows to a light soil, in preparation for turnips, is perpetuated
and must be paid for, though, under the modern system, two

furrows on such land at the proper season are known to be not merely more economical, but really more beneficial.

The amount of these valuations varies between 3*l.* and 5*l.* an acre. A tenant entering to a farm is thus obliged to pay over a large sum, to his predecessor, for operations in the direction and execution of which he has had no voice. There can be no doubt whatever that any man would prefer to spend his money in making improvements according to his own judgment; but the advance of so much capital over and above the ordinary stock of the farm, either requires tenants of more than the means of ordinary farmers, or throws the land into the hands of men who, having expended the larger portion of their ready money in paying for their entry, are so hampered during their tenancy as to be unable to do justice to their farms. It is also obvious—for we are bound to look at the question as it affects both parties — that such a system offers great facility for combination by the tenants against their landlord. The owner of say 4000 acres in such a district, might find it very difficult to refuse the demands of his tenants, however unjust, if, during a period of agricultural depression, they offered him the alternative of getting his farms thrown on his hands, with a tenant-right to be paid down, amounting to four or five years' rental.

Without going further into the question, it must be plain that it is not the interest of the landlords, if the decision is left with them, to adopt this system. To legalise it by act of parliament, so as to render its operation general over the kingdom, it would be necessary to prove that it would promote the public welfare. We have seen in the counties where it exists that *the agriculture is on the whole inferior to that of other districts,* and in no case, even under the most favourable circumstances, superior to other well-conditioned counties which do not possess this tenant-right. In every county it has led to fraud in a greater or less degree. It perpetuates bad husbandry, by stereotyping costly practices which modern improvements have rendered obsolete. It absorbs the capital of the entering tenant, thus limiting his means for

future improvement.   It unfairly depresses the letting value of
land.—Perhaps it may be urged that we dwell on the abuses
rather than on the fair and legitimate uses of the system.   But
it is not easy to see where the line of demarcation is to be
drawn.   The difficulty has already occurred in Lincolnshire,
where landlords find it necessary to limit by special agreement
the otherwise indefinite and constantly widening objects which
this custom may be understood to embrace.   With the best and
purest intention, a farmer may lay out 1000*l.* in drainage or
manures, but, if his investment turns out disadvantageous, is it
consistent with common sense that he is to be at liberty to
relieve himself from the consequences of his own miscalculations
or imprudence, by giving up his farm, and demanding reimburse-
ment of the " unexhausted improvement " from his landlord ?
The same principle too which is applicable to the farmer in his
buildings and his farm might be equally claimed by the labourer
in his cottage, his garden, and his allotment.

The practical working of tenant-right has thus led us to the
conviction, (contrary, we admit, to our preconceived opinions,)
that it is not desirable to extend it, either legally or conventionally,
to other parts of the kingdom.   However well it may look in
theory, we should find the honest and intelligent farmers of
other counties becoming disgusted with its frauds, and, as the
same class are now doing in Surrey, North Notts, and the
West Riding, demanding its restriction, and recommending their
landlords to buy it up and get rid of it.

The wish for leases will increase, when the tenant at will discovers that security for his capital by tenant-right is neither possible nor desirable.   There is a very prevalent dislike to leases on
the part of the tenantry of England.   To a considerable extent
this was occasioned by the uncertainty of the maintenance of protection previous to the free-trade measures, but chiefly from the
fact that there was really less change of tenancy and a lower scale
of rent under a system of yearly tenure than under lease.   If a
man improved his farm during a lease, he was obliged to

pay an increased rent for it, in consequence of that improvement, when he renewed it for a second term. If he held from year to year, he either made no improvement, or, speaking generally, so little, that the difference of produce from year to year was so gradual and imperceptible that the farmer kept nearly the whole advantage to himself. In the one case there was a gradual progress caused by a greater exertion on the part of the tenant, and a larger outlay by the landlord, in the advantage of which all parties participated ; in the other an encouragement to maintain things as they are, that there might be no inducement on the part of the landlord to raise the rent.

But there are instances in many counties, and particularly the north and west, of the tenant's unassisted improvements during a lease having been taken very unfair advantage of at its conclusion. The landlord's right of preference under the law of distraint has been repeatedly urged to us as affording an embarrassed or inconsiderate landlord great facility in thus oppressing an improving tenant. In Northumberland we have given examples which prove in the strongest manner that the injury sustained by the tenants, by being induced, through unfair competition, to offer exorbitant rents, never fails to reach the landlord in the prostration of that class whose means have been thus crippled, and who are compelled to resign their farms, which are then relet at greatly diminished rents. But a landlord with ordinary foresight must see that his interest is bound up in the permanent improvement of his estate, not in a temporary and therefore uncertain rise of rent. It is both his duty and his interest to encourage sure and steady progress; and we have no hesitation in saying that, in that respect, the system of yearly tenure has proved itself in practice, as it is in theory, inferior to that of leases with liberal covenants, when fairly and judiciously tried, as in the examples we have given at Holkham and at Woburn.*

* See Note on Leases at end of volume.

LETTER LV.

THE LABOURER.

Dec. 1851.

THE last class of the agricultural body whose interests we have to consider, is the Labourer. The disparity of wages paid for the same nominal amount of work in the various counties of England, is so great as to show that there must be something in the present state of the law affecting the labourer, which prevents the wages of agricultural labour finding a more natural level throughout the country. Taking the highest rate we have met with — 15s. a week in parts of Lancashire, and comparing it with the lowest — 6s. a week in South Wilts, and considering the facilities of communication in the present day, it is surprising that so great a difference should continue. To use the words of Adam Smith, "Such a difference of prices which, it seems, is not always sufficient to transport a man from one parish to another, would necessarily occasion so great a transportation of the most bulky commodities, not only from one parish to another, but from one end of the kingdom, almost from one end of the world, to the other, as would soon reduce them more nearly to a level. After all that has been said of the levity and inconstancy of human nature, it appears evidently from experience that man is, of all sorts of luggage, the most difficult to be transported."

The table on p. 512 shows the average weekly wages, in the counties we visited, of the agricultural labourer in 1850–1. We again divide the country into the two divisions of the corn counties of the East and South coast; and the mixed corn and grass of the Midland and Western counties. The table is so constructed as also to show the wages of the Northern counties separately from those of the Southern. The black line, dotted on the map, indicates the limit southwards of the Coal formation, within which the great branches of mining and manufacturing enterprise, with the exception of Wales, Somerset, and Cornwall, may be said to be confined.

An examination of this table shows very clearly that the higher wages of the Northern counties is altogether due to the proximity of manufacturing and mining enterprise. The difference between the rates in the corn counties of the East, and the mixed husbandry of the Midland and Western counties, is not so uniform as to warrant any deduction such as showed itself so distinctly in the average rent of those districts.

The influence of manufacturing enterprise is thus seen to add 37 per cent. to the wages of the agricultural labourers of the Northern counties, as compared with those of the South. The line is distinctly drawn at the point where coal ceases to be found, to the south of which there is only one of the counties we visited in which the wages reach 10s. a week, Sussex. The local circumstances of that county explain the cause of labour

TABLE showing the Rate of AGRICULTURAL WAGES in 1850–51.*

NORTHERN COUNTIES.			
Midland and Western Counties.	Weekly Wages.	East and South Coast Counties.	Weekly Wages.
	s. d.		s. d.
Cumberland - -	13 0	Northumberland -	11 0
Lancashire - -	13 6	Durham - -	11 0
West Riding -	14 0	North Riding -	11 0
Cheshire - -	12 0	East Riding -	12 0
Derby - -	11 0	Lincoln - -	10 0
Nottingham - -	10 0		
Stafford - -	9 6		
SOUTHERN COUNTIES.			
Warwick - -	8 6	Norfolk - -	8 6
Northampton -	9 0	Suffolk - -	7 0
Bucks - -	8 6	Huntingdon -	8 6
Oxford - -	9 0	Cambridge - -	7 6
Gloucester - -	7 0	Bedford - -	9 0
North Wilts -	7 6	Hertford - -	9 0
Devon - -	8 6	Essex - -	8 0
		Berks - - -	7 6
		Surrey - -	9 6
		Sussex - -	10 6
		Hants - -	9 0
		South Wilts -	7 0
		Dorset - -	7 6
Average of West -	10 0	Average of East -	9 1

					s. d.
Average of all NORTHERN COUNTIES	-	-	-	11 6
Average of all SOUTHERN COUNTIES	-	-	-	8 5
Average over the whole	-	-	-	-	9 6

being there better remunerated; the wealthy population of
Brighton, and other places on the Sussex coast, affording an increased market for labour beyond the demands of agriculture.

A comparison with the price of labour in the same counties
in 1770 will show this influence clearly. In Cumberland, at

---

* See illustrative Map at beginning of volume.

that time, the wages of the agricultural labourer were 6*s.* 6*d.*, in the West Riding 6*s.*, in Lancashire 6*s.* 6*d.* ; in each of which counties they have since increased fully 100 per cent. In all the Northern counties the increase is about 66 per cent. The increase in the eighteen Southern counties mentioned by Young is under 14 per cent. In some of them there is no increase whatever, the wages of the agricultural labourer in part of Berkshire and Wilts being precisely the same as they were 80 years ago, and in Suffolk absolutely less. The average wages in 1770 in the Northern counties visited by Young were 6*s.* 9*d.* ; and of the Southern counties 7*s.* 6*d.*

Nothing could show more unequivocally the advantage of manufacturing enterprise to the prosperity and advancement of the farm-labourer. We constantly hear expressions of regret, on the part of those who do not look beneath the surface, that the agricultural labourer, hitherto accustomed to the peace and plenty of his Arcadian lot, is year after year being withdrawn from it by the increasing demands and more tempting wages of the manufacturer. But, when we look to the facts, we find that in the manufacturing districts agricultural rents and wages have kept pace with each other; while in the purely agricultural counties the landlords' rent has increased 100 per cent., and the labourers' wages not quite 14. In the Northern counties the labourers are enabled to feed and clothe themselves with respectability and comfort, while in some of the Southern counties their wages are insufficient for their healthy sustenance.

But the agricultural labourer in the Southern counties, while he derives from his labour the means of a very scanty existence, is almost every where felt as a burden instead of a benefit to his employer. To ascertain how far this feeling is well-founded, we have compiled the following table ; in which the counties are arranged for comparison in the same order as in our table of wages, p. 512. : —

TABLE

TABLE showing the rates of the amount expended for relief of the poor, 1st. per pound on property, 2nd. per head of population, and 3rd. the per centage of paupers to the population, vagrants excluded, in thirty-one counties, arranged for comparison, of the Eastern with the Western counties, and the Northern with the Southern. Compiled from Parliamentary Returns, 1848., No. 735.

Midland and Western counties.	Poor relief. Per £ on property		Poor relief. Per head of population.		Ratio per cent. of paupers to the population.	East and South Coast counties.	Poor relief. Per £ on property		Poor relief. Per head of population.		Ratio per cent. of paupers to the population.
NORTHERN COUNTIES.											
	s.	*d.*	*s.*	*d.*			*s.*	*d.*	*s.*	*d.*	
Cumberland	1	1	4	3	6.2	Northumberland	1	2½	5	7½	6.7
Lancashire -	1	0¾	3	7¼	7.2	Durham - -	1	3½	3	7½	5.2
West Riding	1	5½	4	1¼	6.0	North Riding -	1	1	5	10¾	6.5
Cheshire -	1	0¼	3	8	5.5	East Riding -	1	2¾	5	10¾	7.9
Derby -	1	0½	3	8½	4.2	Lincoln - -	1	2¼	6	7¾	7.5
Nottingham -	1	5	5	3	7.9						
Stafford	1	1	3	8¼	4.3						
SOUTHERN COUNTIES.											
Warwick -	1	3½	5	1½	5	Norfolk -	2	2	9	8½	12.8
Northampton	1	11½	9	2¼	11.3	Suffolk - -	2	2	9	3½	13.6
Bucks - -	2	4½	10	2¾	14.6	Huntingdon -	2	0	9	6½	11.6
Oxford -	2	5	10	4½	15.1	Cambridge -	1	9¼	9	1¼	10.7
Gloucester -	1	8¼	6	10½	9.9	Bedford - -	2	0½	7	7¾	11.6
North Wilts	2	3¼	10	5	16.1	Hertford - -	1	8½	7	9¼	11.3
Devon -	1	11¼	7	0½	10.6	Essex - -	2	1¾	9	9¼	14.2
						Berks - -	2	2¾	9	11¼	12.8
						Surrey - -	1	11	6	9	8.8
						Sussex - -	2	1¾	9	1½	12.7
						Hants - -	2	2¾	8	2¼	11.9
						South Wilts -	2	3¼	10	5	16.1
						Dorset - -	2	2¾	9	7½	15.7
Average of all Midland and Western counties - -	1	9¾	6	3	8.9	Average of East and South Coast counties	1	10	7	10½	10.9
Average of all Northern counties -	1	2	4	7¾	6.2	Average of all Southern counties, reckoning North and South Wilts once only - -	2	0½	8	8½	12.1

Here will be remarked the same broad line of demarcation which was formerly exhibited in our Table of Wages, between the Northern, or manufacturing, and the Southern, or agricultural, districts. But there is this striking difference, which is almost invariable, that the counties which stand high in the scale of poor rates, stand low in the scale of wages.

The evil effects remain of an interference by law, in 1782, to fix the rate of wages. In 1795, owing to a rise in the price of corn from 54s. to 75s., the magistrates of several of the Southern counties issued tables showing the wages which, in their opinion, every labouring man should receive, proportioned to the number of his family, and the price of bread ; and the parish officers were instructed to make up the difference between this rate and that paid by his employer! A system akin to this continued to be acted upon, as is well known, down to the passing of the Poor Law Amendment Act, and destroyed, as might have been expected, every feeling of independence on the part of the labourer. But the same system is, in effect, still in existence ; for there is little difference in principle between it and that which we have so frequently mentioned as being adopted by the rate-payers of a parish, agreeing to divide amongst them the surplus labour, not according to their respective requirements, but in proportion to the size of their farms. In such a parish, the superior skill of a farmer in economising one of the chief costs of production is arbitrarily set aside, and he is reduced to the same level with his unskilful neighbours. But it has been also proved to operate in the same disadvantageous manner upon the skilled labourer, whose capacity would enable him to do more work and earn a higher rate of wages, but which he is discouraged from doing, as the effect would be to diminish the employment of others for whom work must be found. The bad labourer is thus paid the same rate of wages as the good, emulation is discouraged, and the standard of skill and efficiency kept down. If a labourer knows that he must be employed at a certain uniform rate of wages, whatever be the quality of his

work, he has no motive to improve. But should he, notwithstanding these artificial trammels, feel within himself both the
power and the will to do better, the law of settlement tells
him that he is not at liberty to carry his skill to a better market
except on conditions which are felt to be prohibitory.

There is another evil with regard to the labourer, which
is not confined to the Southern counties, — the system of
"close" and "open" parishes, by which the large proprietors
are enabled to drive the labourer out of the parish in which he
works, to a distant village, where, property being more divided,
there is not the same combination against poverty. It is the
commonest thing possible to find agricultural labourers lodged
at such a distance from their regular place of employment, that
they have to walk an hour out in the morning, and an hour
home in the evening, — from forty to fifty miles a week. In
one county the farmers actually provide donkeys, on which
their labourers ride out and home, to prevent them tiring themselves with walking, that so they may be more vigorous at their
work. Two hours a day is a sixth part of a man's daily labour,
and this enormous tax he is compelled to pay in labour, which
is his only capital. Nor is this the sole evil of the practice,
for the labourers are crowded into villages where the exorbitant
cottage rents frequently oblige them to herd together in a
manner destructive of morality and injurious to health.

The average wages of the NORTHERN COUNTIES have been shown to be - -	11 / 6			
Their average poor relief per £ on property -	...	1/ 2		
— — per head of population - -	...	...	4 / 7¾	
Their rate per cent. of paupers to the population	...	...	...	6.2
Contrasted with which are : —				
The average wages of the SOUTHERN COUNTIES -	8 / 5			
Their average poor relief per £ on property -	...	2 / 0½		
— — per head on population - -	..	...	8 / 8½	
And their per-centage of paupers to the population	...	...	...	12.1

Here it appears that, while the average wages of the North
are 37 per cent. above those of the South, the expenditure in
the North for poor relief is about 70 per cent. lower on pro-

perty, and about 87 per cent. lower when estimated according
to population; and the difference in the per-centage of paupers
is nearly as one to two.

The redundance of labour which oppresses property and
depreciates wages in the South will not only relieve itself as
soon as freedom is restored to the labourer *to settle where he
will,* but the change, by equalising the market of labour, will
cheapen the cost, and stimulate the progress, of production in the
North.   It is obvious, however, that a change to mere union
settlements would not accomplish these desirable results; since
it is found that the marked inequality in wages and poverty
is not limited to unions or counties, but bisects the kingdom by
unmistakeable lines into two great geographical divisions.   We
may draw a line across the map of England : all to the south
of that line we shall have high poor rates and low wages, and
all to the north of it high wages and low poor rates ; — on one side
an enforced excess of labour, impoverishing and bearing down
the working man, and, by consequence, rates pressing on property with undue severity; on the other a comparative deficiency
of labour, raising its price to an unequal average, and operating
unfairly on the cost of production ; — in these two divisions
the same people, the same language, habits, and institutions, with
cheap and rapid communication between them, and no obstacle
except a law which, aggravating the natural indisposition to
move, hinders the working man from carrying his labour to
the best market.

It is not our province to discuss the Poor Laws, or to
attempt to lay down a remedy for a state of things which is
confessedly injurious both to employer and employed.   We
desire only, as strongly as we can, to direct attention to a system
fraught with so many evils, — a law of settlement which binds the
labourer to a parish in which his labour is not required, and
prevents another, where labour is deficient, from obtaining that
supply which would be to all parties so beneficial.   The importance of the subject, and the inquiry and discussion it has

recently undergone, lead us to hope that some remedial measure will be early introduced by the Legislature to enable and encourage the free circulation of labour throughout England. The over supply is, as far as we have seen, apt to be exaggerated. As labourers begin to withdraw, employers will soon discover, under the pressure of higher wages, that the surplus was not so great as they led themselves to believe. The lowest rate of wages we met with in England, 6s. a week, was in an agricultural parish in South Wilts, where one large farmer employed the whole labour of the parish, and fixed as he chose the scale of wages; and yet, in this very parish, the resident labourers were insufficient for the regular summer work of the farms, strangers from a different part of the county being introduced for a season to perform the operation of turnip hoeing, and to assist in the hay and corn harvests.

The change in the price of provisions has added greatly to the comfort of the labourer. Within the last ten years the decrease in price of the principal articles of his consumption is upwards of 30 per cent. In 1840 a stone of flour cost him 2s. 6d., which he can now purchase for 1s. 8d.; good Congou tea in 1840 was, exclusive of duty, 2s. 6d. per lb., and is now only 1s.; and the same quality of sugar which then cost him 6d. per lb., can be had now for 3½d.

Such a reduction in the price of provisions is a great boon to the labourer, because it gives him the command of additional comforts, and thereby elevates his condition. This is totally different from the effect of resorting to a lower quality of food, such as potatoes, contentment with which lowers the standard of comfort, and debases the condition of the labourer. The sure consequence of such a depression of standard is, that labourers being more cheaply produced, increase in more rapid proportion than the capital for their employment, and, the labour-market being overstocked, wages are lowered. There could no greater evil befal the English agricultural labourer than that any circumstances should compel him to depress his standard of comfort so far as to be content for his principal subsistence with the lowest species of human food in this country, the potato.

In the counties visited by Arthur Young, the rent of the labourers' cottages has increased since his time from 8*d.* to 1*s.* 5*d.* a week, being upwards of 100 per cent.; while agricultural wages in the same counties, on the average, have risen in the same period only from 7*s.* 3*d.* to 9*s.* 7*d.*, or about 34 per cent.

The great difference in the rate of wages between the Southern and Northern counties, is a sufficient proof that *the wages of the agricultural labourer are not dependent on the prices of agricultural produce.* A bushel of wheat, a pound of butter, or a stone of meat, is not more valuable in Cumberland or the North Riding, than in Suffolk or Berkshire; yet the wages of the labourer in the two former are from 60 to 70 per cent. higher than in the two latter counties. The price of bread is not higher in July and August than in May and June; yet, in every agricultural county, the wages of labour during the period of harvest are increased. Nor are better wages directly the effect of capital; for the poor farmer of the cold clays of Durham or Northumberland pays 11*s.* a week, while the large capitalist who cultivates half a parish in South Wilts or Dorset pays only 7*s.* to his labourer. The higher rate is unmistakeably due to the increased demand for labour. This has been greatest in the manufacturing and mining districts of the North, and near the commercial towns and great seaports, whose prodigious increase of business has attracted and been followed by a similar increase of wealth and population. The increase of population shown by the census of 1851, during the last ten years, in the twelve counties in the foregoing table where wages are highest, exceeds by 6 per cent. the increase of the Southern counties during the same period. It thus appears that the welfare of the agricultural labourer is, more than that of any class in the community, dependent on the continued progress of our manufacturing and mercantile industry.

LETTER LVI.

CONCLUSION.

Dec. 1851.

HAVING thus concluded the consideration of the three great interests connected with agriculture — the landlord, the tenant, and the labourer — we have felt most strongly the want of any agricultural statistics, on which full reliance can be placed, for estimating the yield of food for the people. There are statistical returns on almost every other subject connected with the business or welfare of the country, but that which may be well regarded as the most important of all, — the annual supply of food, — is still left to conjecture. The only information we possess as to the extent of cultivated land in England, is the result of a survey made by Mr. Couling, civil engineer and surveyor, in 1827. Since that period, much of the pasture land has been broken up for the cultivation of corn, and a great inroad has been made en the portion set down as uncultivated. Applying our figures to the calculation, and testing them by the return to parliament (March, 1845) of the annual rent of land in every parish of England assessed to the property tax in 1843, we divide the area of England as follows:—

27,000,000 acres of cultivated land, including meadow
        and arable pasture grounds, at 27s. 2d.
            per acre of rent -   -   - £36,675,000
2,000,000 acres uncultivated, at 5s. per acre -   -   500,000
3,160,000 acres moor and mountain, at 1s. 6d.   -   237,000

32,160,000 acres   -   -   -   -   - £37,412,000

From this estimate the acreage and rent of Middlesex is
excluded, the circumstances of the metropolitan county being
exceptional. This tallies very nearly with the actual rental *,
and, so far as we have seen, is as accurate a basis on which
to found an estimate of the cultivated and uncultivated land of
England as the present sources of information afford. These
twenty-seven million acres of cultivated land we have subdivided according to the various crops, in the following table.—
(See next page.)

This table shows the average annual produce of wheat in
England, after deducting seed, to be 10,250,000 quarters, which
gives, from our own soil, a fraction less than five bushels to each
individual of the present population. But it is based upon the
assumption that, of the twenty-seven million acres of land in
England capable of cultivation, one-third in the Eastern division,
and two-thirds in the Midland and Western division, of the
country, are in permanent grass and meadow, making upwards
of thirteen million acres altogether, not used for the growth of
corn. This may or may not be near the truth, as we know of
no accurate data by which to test it, and have therefore been
obliged to guide ourselves to a general result in accordance with
personal observation and information. The average produce of
wheat in England we have already shown to be 26⅔ bushels an
acre, which is considerably below the estimate given in the leading
statistical works of the day, in which four quarters or 32 bushels

---

   * The land rental of England as determined by the assessment under the
Property and Income Tax Acts in 1842–3 was   -   - £37,795,905
   Of which the rental of Middlesex was   -   -   -   387,861

Land rental of England, excluding Middlesex   -   - £37,408,044

Assuming the Midland and Western counties to be two-thirds in grass, and one-third in tillage, we have of 13,000,000 acres of cultivated land - - - -
And assuming the Eastern division to be one-third in grass and two-thirds tillage, we have of 14,000,000 acres of cultivated land - - - -

	Acres in Grass.	Acres in Tillage.
	8,666,000	4,333,000
	4,666,000	9,334,000
	13,332,000	13,667,000

The 13,667,000 acres in tillage is divided thus:—

	Acres.	Produce per Acre. Bush.	Total Produce. Qrs.	Deduct Seed per Acre. Bush.	Seed. Qrs.	Produce under deduction of Seed. Qrs.
One fourth in wheat - - - -	3,416,750	× 27 =	11,531,531	3	1,281,281 =	10,250,250
One fourth in barley, oats, and rye, viz.:—						
Barley - - - 1,416,750	-	× 38 =	6,729,562	4	708,375 =	6,021,187
Oats and rye - - 2,000,000	-	× 44 =	11,000,000	5	1,250,000 =	9,750,000
	3,416,750					
One-fourth in clover, "seeds," beans and pease, viz.:—						
One-sixth of whole area in clover and "seeds" - - 2,277,750						
One-twelfth beans and pease - 1,139,000	3,416,750	× 30 =	4,271,250	4	569,500 =	3,701,750
One-fourth in turnips, mangold, potatoes, rape and fallow, viz.:—						
In turnips, mangold, and potatoes 2,116,750						
In rape and fallow - - 1,300,000	3,416,750					
	13,667,000		33,532,343		3,809,156	29,723,187

are assumed as the average produce. Our figures we believe to be more correct, both because they are the average of figures given to us by competent judges in different parts of all the counties we visited, and also from the knowledge that four quarters as an average produce is to be met with only on farms where both soil and management are superior to the present average of England.

The necessity for accurate information on this question cannot be better shown than by this, — that notwithstanding all our progress in agriculture, our command of manures, and our improved processes, the total produce of corn of all kinds in England is, according to the estimates of the most eminent writers, less now by two millions of quarters than it was stated to be in 1770 by Arthur Young! We cannot doubt that Young greatly over-rated the produce of his time, but we have no certain knowledge that we are right ourselves. The difference of seasons, too, influences the average materially. The same land, with the same management in every way, will sometimes vary in its yield more than ten bushels an acre, according as the season proves favourable or otherwise. If we estimate the total produce of corn of all kinds in England, in a good year, at thirty million quarters, it is not too much to say that it may vary to the extent of one-fourth, or $7\frac{1}{2}$ million quarters, by the effects of a cold ungenial summer. For it is not merely a variation of five bushels of wheat above or below the average. A diminution of produce to the extent of five bushels an acre, caused by a cold season, implies also a falling off in quality from 62 lb. to 58 lb. per bushel, (the corresponding figures, when converted into flour, being 53 lb. and 47 lb.), making a difference of 48 lb. of flour per quarter of wheat, or equivalent to a farther decrease, in the average produce, of three bushels per acre, or eight bushels altogether. The only government agricultural statistics in the United Kingdom — the returns made by the constabulary in Ireland — show a falling off of ten bushels an acre in the produce of wheat in 1848, as compared with 1847. The climate of Ireland is certainly more uncertain for wheat than that of England, yet the experience of most practical men in this country will attest that a fluctuation of five bushels an acre above or below the average is not uncommon. This difference is equivalent to the sustenance of a sixth part of the population, and yet it has not been thought of sufficient importance to warrant the government in incurring the moderate expense necessary to obtain such information.

It is therefore on grounds of public policy, not less than for the special benefit of agriculture, that we venture to insist on the advantage of obtaining correct agricultural statistics. Independent, however, of their general value, we believe the information obtained might be so classified and arranged as to throw much light on the relative progress of agricultural improvement in different counties, and the causes which encourage or retard that progress. It is sometimes objected that the farmers might be unwilling to give the minute information desired ; but we must bear testimony, from personal experience, that in the thirty-two counties which we visited, during a period of strong political excitement, and coming in contact day after day with men of all parties, we can recall only two instances in which information was refused. On all other occasions we were received with the utmost frankness, and every facility was afforded, both by personal inspection and inquiry, for obtaining the amplest information. The returns might embrace particulars illustrative of the state of rural economy, by which the productiveness of one district could be contrasted with another, and a calm and indisputable statement of facts be made to carry conviction to the mind of the most prejudiced or unthinking, whilst it would be a guide of the most useful kind to the earnest agricultural improver.

When it was finally resolved that protective duties were to be taken away from agricultural produce, it was thought by some that Parliament should have interfered to relieve tenants from contracts made under an artificial system, on the permanence of which they had been encouraged to rely. A measure so obviously just was probably deemed unnecessary, from the almost uniform system of yearly tenure in England, which put it in the power of a farmer to terminate his contract when he chose. But this is not a fair alternative to give to an improving tenant who had sunk money on the faith of protection; and the duty becomes the more imperative on landlords to share the difficulties of the crisis in a fair and honourable spirit with their tenants, since the legislature has left the question to individual arrangement. On the purely corn farms it is obvious that a reduction of price, in the only commodity the farmer has to sell, must for a time incapacitate him from paying his former rent,— the costs of production, as in his case, remaining the same. And on the more kindly soils, where it is admitted that capital may be employed with advantage in increasing their productiveness, and on which a variety of remunerative crops may be introduced, this change cannot be effected in a season. A considerate landlord will therefore take care that his tenants, during the period of transition, do not bear an undue share of a burden, the heaviest part of which ought assuredly to fall upon him who has the largest and most permanent interest in the land.

On the same principle it would have been just that the tithe, which was calculated as a certain proportion of the produce of "cultivated corn land," should have been reconverted to the same proportion of the diminished gross produce which such land yields when laid to grass. For it is clear that if low prices of corn shall compel the owners and occupiers of the stiff wheat soils of low quality to lay them to grass—though such a course may be really more remunerative to both—yet the gross produce of each acre will, under this system, be considerably diminished. If the tenth part of the average produce was converted to a rent-charge of 6s. an acre when the land was under corn, it would obviously be unfair to throw on the owner or occupier the same charge when measures for the public welfare compelled a change of cultivation which reduced that propor-

tion of the produce to 4s. an acre. This change is not compensated by the fluctuation of the rent-charge according to the
average prices of corn, for that applies a lower price to a fixed
quantity of produce, whereas in reality it should be a lower price
to a diminished quantity of produce. The basis of conversion
has become erroneous, and to that extent presses unfairly on
the owners and occupiers of such lands.

Our observations must now draw to a close. To education
in its widest sense we look as the most powerful aid in the
farther progress of British agriculture. Knowledge, — of his
business, and true interest to the landlord and the tenant, —
and of the best mode of promoting his own welfare to the
labourer, — is the first requisite to obtain an improvement of
their condition. A wise pursuit of individual interest will, we are
persuaded, be most conducive to their own and the general
welfare. It is by individual energy that this is to be developed.
And while no exclusive protection is granted by the legislature,
the agricultural interest has a right to demand that all trammels
on their enterprise and industry should now be withdrawn.

The measures of a public character which, in addition to those
within the power of individual landlords or farmers, we have
indicated, are these : —

1. The cheapening and facilitating the transfer of land.
2. The sale of overburdened estates.
3. The encouragement of leases, with liberal covenants.
4. An alteration of the law of settlement.
5. The collection of agricultural statistics.

We see no reasons to despond, but many to encourage hope
in the future prospects of British agriculture. All the evidence
we have collected tends to show, that in the districts where the
increase of manufacturing and commercial enterprise and wealth
has been greatest, there the rent of the landlord, the profits of
the tenant, and the wages of the labourer have most increased.
" It is manufacture that creates the town, that peoples the moor,
that sinks the shaft, that fills the port, that swells the metro-

polis, that rewards husbandry, and that draws the redundant population of the hamlet to the less picturesque labyrinth of the town." The recent gold discoveries of California and Australia will give increased development to the manufacturing interests, and at the same time will enhance the value both of land and its produce. The resources of our limited territory will every year be called upon for an increased supply of the most remunerative crops. Science points out, and commerce renders available, new manures to enrich the soil; engineering skill is applied to the construction of machinery for economising the cost of production, and lightening the toil of the labourer. The reaping-machine has been introduced with every prospect of its successful application. The steam-engine has been yearly extending its benefits in cheapening the processes of agriculture; and there is every reason to believe that not many years will elapse before it will be successfully applied to the direct cultivation of the soil. New products are recommended as objects more remunerative than corn; and the culture of flax for spinning, and possibly of beet for sugar, may ere long transfer to the Southern counties a portion of that manufacturing enterprise which the expense of transporting coal has hitherto confined to the North. Look in what direction we may, we have good prospects of success for the agriculturist, — increasing markets, without which his produce would be valueless; increasing wealth to enhance its price; resources of manure at home and abroad to maintain and add fertility to his farm; — he wants nothing but a reasonable adjustment of his rent, intelligent co-operation on the part of his landlord, and the security of a lease, to render well-directed enterprise successful.

But, besides the evidences given in most of the foregoing Letters, observing men of all parties concur in attesting that at no former period has the general progress of agricultural improvement been greater than at present. On every side increased exertions are being made by both landlord and tenant. How otherwise is it possible to account for the entire

absorption of that mass of labourers who were three years
ago employed in the construction of railways, and the prospect
of whose dismissal was looked forward to with dismay? Then
if we consider the great body of highly intelligent and wealthy
farmers scattered over the country, and whose applied science is
concentrated in the "Journal of the Royal Agricultural Society," and at the great annual shows of live stock and implements; or to the eminent men both at home and abroad,—
Liebig, Buckland, Daubeny, and Playfair, Professors Henslow,
Lindley, Johnston, and Way, Mr. Lawes, and Dr. Anderson, of
Edinburgh, who are bringing pure science to the aid of agriculture; or to our most influential public writers on farming, Mr.
Pusey and Mr. Morton, whose practical experience enables them
to test or illustrate the subjects they discuss,— we feel that we
may speak with confidence and hope of the future.

In concluding this extensive inquiry, we beg to express
our heartfelt thanks to the landlords, agents, and farmers
throughout the country, who so cordially aided our humble
endeavours to make it trustworthy and complete. It has been
our aim to give a faithful description of English agriculture as
it is, dwelling in detail on such examples as were most instructive, offering suggestions where we felt ourselves competent
to do so, and not hesitating, where necessary, to criticise
careless indifference on the part of the landlord, or indolent
waste of opportunity by the tenant. This was a delicate and
difficult duty, but we have endeavoured to discharge it im-
partially and without offence. We rise from our task conscious of its imperfect execution, but with a firm persuasion
that, though there are many exceptions, the great body of
the landlords and tenants of England have, by mutual co-
operation, energy and capacity sufficient to meet, and by
degrees to adapt themselves to a change, which, in its extraordinary effect on the welfare of all other classes of the community, will sooner or later bear good fruits also to them.

NITRATE OF SODA AND SALT AS A TOP-DRESSING FOR WHEAT.

Referred to at Page 167.

HAVING read in the " Times " Commissioner's Report, that some very good farmers in Norfolk made a practice of top-dressing their wheat in spring with nitrate of soda, I determined once more to try this salt, which, as the older members of our Society will remember, was once a very fashionable manure, but the use of which was discontinued by its advocates in consequence of its tendency to lay the corn and to produce mildew. These two serious faults, it now appears, may be corrected by mixing with the nitrate a moderate quantity of common sea salt, which, when used in heavy doses, destroys the life of grass, and may therefore be readily supposed to counteract the dangerous suddenness of vegetation that nitrate produces. Thus common salt may prevent mildew, and is known certainly, on some soils, to strengthen the straw. The nitrate was sown, as directed, at the rate of 1 cwt. per acre, mixed with 1 cwt. of common salt ; but this quantity was not given at once, being divided, as enjoined, into two doses, applied at a fortnight's interval and in showery weather. It was so applied to a ten-acre piece of white wheat, a portion thereof being, however, passed over. The whole produce has been thrashed out already, in order to test the effect. A portion was top-dressed, not with nitrate, but guano. The result is as follows : —

	Bushels per acre.	Increase in bushels.	Cost of dressing.		Value of increased produce.	
			s.	d.	s.	d.
Undressed	21	—	—		—	
Guano, 2 cwt. . . .	24	3	20	0	15	0
Nitrate 1 cwt. and salt 1 cwt.	25½	4½	17	0	22	6

The other trial was made on an eight-acre piece of red wheat, following barley. The wheat had begun to appear very blue and spindling, notwithstanding a good coat of dung given it in the autumn to make up for cross cropping. The improvement was immediate, and has stood the test of thrashing ; for the account is as

M M

follows. Two acres were thrashed, one on each side adjoining the half-acre in the middle, on which no nitrate of soda was sown.

	Bushels per acre.	Increase in bushels.	Cost of dressing.		Value per acre of increased produce.		Profit per acre.	
			s.	d.	s.	d.	s.	d.
Undressed . . .	19⅜	—	—		—		—	
Nitrated. . . .	27¾	8⅜	17	0	42	0	25	0

The profit on this piece is certainly more than the value of the rent of the land, which is a poor blowing sand. The theory of this action is now clearly established by Mr. Lawes's experiments ; for nitrogen, whether as ammonia in guano or whether in a nitrate, is proved to be the food generally wanted by wheat. — *Mr. Pusey, in Royal Agricultural Society's Journal.*

NOTE ON LEASES, p. 509.

WE have not thought it necessary to enter into a discussion, in this work, of the best form of leases, or the period of their duration, though to both points due reference has been made in describing the best practices we have met with. A lease should be long enough to permit the tenant to reap the full benefit of the capital he invests in improvement, and at the same time not so long as to admit of any delay in commencing his improvements with vigour. Twenty-one years may be reckoned, and is found in practice, a term which sufficiently secures the interests of the tenant.

The system of leasing land, though far too little practised in England, is of very ancient date. We are indebted to the kindness of J. R. M'Culloch, Esq., for making us acquainted with an Attic lease, dated in the 108th Olympiad, 345 years before the Christian era, which is so curious and instructive that we are glad to have an opportunity of laying it before the agricultural public. It is a lease by the Æxonians, the townspeople or *demos* of Æxone, of a piece of land called the Philais, near Mount Hymettus, to a father and his son for 40 years, for 152 drachmas a year.

" The demos of Æxone let on lease the Philais, to Autocles, the son of Anteas, and to Anteas, the son of Autocles, for forty years, for 152 drachmas a year; the said land to be farmed by them, or planted with trees, as they please; the rent to be paid in the month of Hecatombæon. If they do not pay it they forfeit their security, and as much of the produce as they stand in arrear. The Æxonians not to sell nor to let the said land to any one else, until the forty years have elapsed. In case of a loss on the part of the tenants by hostile invasion, no rent to be paid, but the produce of the land to be divided between the Æxonians and the tenants. The tenants are to deliver up half the land fallow, and all the trees upon the land; for the last five years the Æxonians may appoint a vine dresser. The lease to begin with respect to the corn land with Eubulus the Archon entering into office; but with respect to the wood, not before Eubulus goes out of office. The lease to be cut upon stone, to be set up by the magistrates, one copy in the temple of Hebe, the other in the Lesche; and boundary stones to be set up upon the land, not less than two tripods on each side. And if a tax should be paid for the land to government, the said tax to be paid by the Æxonians; or, if paid by the tenants, to be deducted from the rent. No soil to be carried away by digging of the ground, except from one part of the land to another. If any person makes a motion in contravention of this contract, or puts it to the vote, he shall be answerable to the tenants for the damage."

This inscription was brought, with several others, from Greece, many years ago, and is now in the University of Leyden, and believed to be of unquestionable authenticity. It shows that the mode of letting land for a fixed number of years, with a money rent and certain conditions of management, must have been well understood in Attica, at that remote period. The terms of the lease indicate a very considerable knowledge of agriculture. For five years previous to its expiration, the landlord was to appoint a vine dresser, and at the termination the tenants were bound to have half the land in fallow. The landlord, on the other hand, was not to sell or let the land during the period of the lease, — he was to share in the tenants' loss in case of hostile invasion, — and he was to pay, or allow the tenant to deduct from his rent, the taxes required by government. In this short lease there is much of that plain common sense which fairly recognises the interest and duties of both parties, and which is so often lost sight of in the perplexing labyrinth of modern legal phraseology.

INDEX.

THE END.

LONDON:

SPOTTISWOODES and SHAW,
New-street-Square.